Communications in Computer and Information Science 1750

Dmitry Balandin · Konstantin Barkalov ·
Iosif Meyerov (Eds.)

Mathematical Modeling and Supercomputer Technologies

22nd International Conference, MMST 2022
Nizhny Novgorod, Russia, November 14–17, 2022
Revised Selected Papers

Springer

Editors
Dmitry Balandin 🆔
Lobachevsky State University of Nizhny
Novgorod
Nizhny Novgorod, Russia

Konstantin Barkalov 🆔
Lobachevsky State University of Nizhny
Novgorod
Nizhny Novgorod, Russia

Iosif Meyerov 🆔
Lobachevsky State University of Nizhny
Novgorod
Nizhny Novgorod, Russia

ISSN 1865-0929   ISSN 1865-0937 (electronic)
Communications in Computer and Information Science
ISBN 978-3-031-24144-4   ISBN 978-3-031-24145-1 (eBook)
https://doi.org/10.1007/978-3-031-24145-1

This Springer imprint is published by the registered company Springer Nature Switzerland AG
The registered company address is: Gewerbestrasse 11, 6330 Cham, Switzerland

Preface

The 22nd International Conference and School for Young Scientists "Mathematical Modeling and Supercomputer Technologies" (MMST 2022) was held during November 14–17, 2022, in Nizhni Novgorod, Russia. The conference and school were organized by the Mathematical Center "Mathematics of Future Technologies" and the Research and Educational Center for Supercomputer Technologies of the Lobachevsky State University of Nizhni Novgorod. MMST 2022 was organized in partnership with the International Congress "Russian Supercomputing Days".

The topics of the conference and school cover a wide range of problems of mathematical modeling of complex processes and numerical methods of research, as well as new methods of supercomputing aimed to solve state-of-the-art problems in various fields of science, industry, business, and education.

This edition of the MMST conference was dedicated to Professor Victor Gergel, who passed away in 2021. Victor Gergel chaired the Program Committee of the conference from 2001 to 2020 and was a brilliant scholar and innovator. Since his student years he had developed mathematical models, methods, and software systems to solve global and multi-criteria optimization, pattern recognition, and classification problems. Starting from the early 2000s, Professor Gergel became involved in the world's rapidly developing area of parallel computing, where he achieved a great deal of success and international recognition. He led one of Russia's first research and education centers for supercomputing technologies at the Lobachevsky State University of Nizhni Novgorod, which became part of the national supercomputing research and training system. Professor Gergel was a member of Program Committees, gave plenary speeches at major conferences on supercomputing in Russia and worldwide, and contributed to many expert groups on the subject. His pioneering ideas in the field of programming education have played a significant role in the training of software professionals at the Lobachevsky State University of Nizhni Novgorod. The educational packages and textbooks on parallel programming developed by Professor Gergel and his co-authors are used throughout Russia.

The scientific program of the conference featured the following plenary lectures given by outstanding researchers and practitioners:

- Yuri Boldyrev (Peter the Great St. Petersburg Polytechnic University)—On the fundamentals of the digital economy (a case study of material production).
- Alexander Leytman (TIAA) and Mikhail Soloveitchik (Independent Analyst)—Mathematical modeling of organizations' risks in the global financial industry using supercomputer technologies.
- Alexander Naumov (Huawei Technologies Co.), The use of neural networks for solving first-order partial differential equations.
- Mikhail Yakobovskiy (Keldysh Institute of Applied Mathematics of the Russian Academy of Sciences)—Supercomputer Simulation—Processor Load Balancing

- Sergey Yakushkin (Syntacore Company)—Open and free RISC-V processor architecture—from microcontrollers to supercomputers

These proceedings contain 20 full papers and five short papers carefully selected to be included in this volume from the main track and special sessions of MMST 2022. The papers accepted for publication were reviewed by three referees from the members of the MMST 2022 Program Committee and independent reviewers in a single blind process.

The proceedings editors would like to thank all members of the conference committees, especially the Organizing and Program Committee members as well as external reviewers for their contributions. We also thank Springer for producing these high-quality proceedings of MMST 2022.

November 2022

Dmitry Balandin
Konstantin Barkalov
Iosif Meyerov
Publication Chairs

Organization

Program Committee Chair

Balandin, D. V.   Lobachevsky State University of Nizhni
                                   Novgorod, Russia

Program Committee

Barkalov, K. A.   Lobachevsky State University of Nizhni
                                   Novgorod, Russia
Belykh, I. V.   Georgia State University, USA
Boccaletti, S.   Institute of Complex Systems, Italy
Dana, S. K.   Indian Institute of Chemical Biology, India
Denisov, S. V.   Oslo Metropolitan University, Norway
Feygin, A. M.   Institute of Applied Physics, RAS, Russia
Ghosh, D.   Indian Statistical Institute, India
Gonchenko, S. V.   Lobachevsky State University of Nizhni
                                   Novgorod, Russia
Hramov, A. E.   Innopolis University, Russia
Ivanchenko, M. V.   Lobachevsky State University of Nizhni
                                   Novgorod, Russia
Kazantsev, V. B.   Lobachevsky State University of Nizhni
                                   Novgorod, Russia
Koronovskii, A. A.   Saratov State University, Russia
Kurths, J.   Potsdam Institute for Climate Impact Research,
                                   Germany
Malyshkin V. E.   Institute of Computational Mathematics and
                                   Mathematical Geophysics, SB RAS, Russia
Mareev, E. A.   Institute of Applied Physics, RAS, Russia
Moshkov, M. Y.   King Abdullah University of Science and
                                   Technology, Saudi Arabia
Meyerov, I. B.   Lobachevsky State University of Nizhni
                                   Novgorod, Russia
Nekorkin, V. I.   Institute of Applied Physics, RAS, Russia
Osipov, G. V.   Lobachevsky State University of Nizhni
                                   Novgorod, Russia
Pikovsky, A. S.   University of Potsdam, Germany
Sergeyev, Ya. D.   University of Calabria, Italy; Lobachevsky State
                                   University of Nizhni Novgorod, Russia

Turaev, D. V.	Imperial College London, UK
Wyrzykowski, R.	Czestochowa University of Technology, Poland
Yakobovskiy, M. V.	Institute of Applied Mathematics, RAS, Russia
Zaikin, A. A.	University College London, UK

Organizing Committee

Zolotykh, N. Yu. (Chair)	Lobachevsky State University of Nizhni Novgorod, Russia
Balandin, D. V.	Lobachevsky State University of Nizhni Novgorod, Russia
Barkalov, K. A.	Lobachevsky State University of Nizhni Novgorod, Russia
Kozinov, E. A.	Lobachevsky State University of Nizhni Novgorod, Russia
Lebedev, I. G.	Lobachevsky State University of Nizhni Novgorod, Russia
Meyerov, I. B.	Lobachevsky State University of Nizhni Novgorod, Russia
Oleneva, I. V.	Lobachevsky State University of Nizhni Novgorod, Russia
Sysoyev, A. V.	Lobachevsky State University of Nizhni Novgorod, Russia

Contents

Computational Methods for Mathematical Models Analysis

Computation in Optimization and Optimal Control

Supercomputer Simulation

Computational Methods
for Mathematical Models Analysis

Diffusion in the Phase Space of the Autooscillatory System, Demonstrating the Stochastic Web in the Conservative Limit: Numerical Investigation

Alexander V. Golokolenov[(✉)] [iD] and Dmitry V. Savin[(✉)] [iD]

Saratov State University, Astrakhanskaya 83, 410012 Saratov, Russia
golokolenovav@gmail.com, savin.dmitry.v@gmail.com

Abstract. The paper investigates diffusion in the phase space of the weakly dissipative version of the pulse-driven Van der Pol system. Amplitude of external pulses depends on the dynamical variable in the same way as in the Zaslavsky generator of the stochastic web, and the system under investigation transforms into the stochastic web generator in the conservative limit. Whilst the conservative system demonstrates the unbounded diffusion in the phase space through the stochastic layer, trajectories of the autooscillatory system converge to several attractors, and diffusion can be obtained only in some limited time interval. The trajectories demonstrating diffusion properties were detected using the finite-time Lyapunov exponents, and for an ensemble of such trajectories dependence of average energy on time was analyzed. Whilst in the conservative system average energy grows linearly versus time, in the autooscillatory system this dependence appears to be rather complex. In the time interval associated with existence of diffusion it can be, however, approximated with the power law. The dependence of it's exponent on the dissipation parameter value and on the initial energy of the ensemble was investigated. The exponent increases with the decrease of dissipation and decreases up to 0 with the increase of the initial ensemble energy. Dependence on the initial energy have the same shape in wide interval of dissipation parameter values.

Keywords: Stochastic web · Weak dissipation · Diffusion · Autooscillations

1 Introduction

The stochastic web is a special type of organization of phase space of conservative systems, which are degenerate in sense of the KAM theorem [9]. Originally found by Zaslavsky in a model, derived from a plasma physics problem [12], it was widely studied from both theoretical and applied points of view (see, e.g., [11,13]). The main feature of systems, demonstrating existence of a stochastic web in phase space, is a possibility of an unbounded diffusion through the

© The Author(s), under exclusive license to Springer Nature Switzerland AG 2022
D. Balandin et al. (Eds.): MMST 2022, CCIS 1750, pp. 3–14, 2022.
https://doi.org/10.1007/978-3-031-24145-1_1

stochastic layer. Parameters and features of this diffusion process were also thoroughly studied by different authors (see, e.g., [1,8]). The simplest model, demonstrating the uniform stochastic web, is the pulse-driven linear oscillator with amplitude of pulses depending on the dynamical variable [13]. Such oscillator-based models are widely used in theoretical nonlinear dynamics and can demonstrate various dynamical phenomena. Particularly, addition into such systems small dissipative terms, which can be done very naturally, allows to use them as rather convenient model for analyzing the weakly dissipative, or almost conservative dynamics, which is characterized by extreme multistability [4,5] and extremely long transients [7]. Usually dynamics of such systems is studied for models which are nondegenerate in the sense of the KAM theorem [9]. In recent years several works appeared where the dynamics of weakly dissipative systems demonstrating the stochastic web in the conservative limit was analyzed, but the main focus was held on the long-time behavior and structure of their coexisting attractors [2,3,10]. On the other hand, since one of the main features of stochastic web is a diffusion through the phase space, analysis of transient processes in dissipative versions of such systems and their connection with the properties of conservative trajectories seems to be an interesting task. In the present paper we analyze diffusion properties of trajectories of the system, generating the uniform stochastic web in the conservative case, with an addition of small dissipation of autooscillating type.

2   Methods of the Investigation

We consider the pulse-driven Van der Pol oscillator

$$\ddot{x} - (\gamma - \mu x^2)\dot{x} + x = \sum_{n=-\infty}^{\infty} F(x)\delta(t - nT). \tag{1}$$

where $F(x)$ is following Zaslavsky [13] chosen to be $\lambda \cos x$, and hence in the conservative case ($\gamma = \mu = 0$) system (1) demonstrates the uniform stochastic web in the phase space (Fig. 1a). We will further consider the frequency of an external pulses $2\pi/T$ four times greater than the natural frequency of the autonomous system, in this case the phase space of the conservative system has a crystal-type symmetry with the rotation angle $\pi/2$ [13]. For the purpose of numerical investigation we will use the stroboscopic section map, integrating the Eq. (1) between the external pulses with the Runge-Kutta method of 4th order with integration step $0.01T$—one iteration of such a map corresponds to one period of the external force.[1] In order to investigate diffusion properties of the

---

[1] We also have tested the symplectic Forest-Stremer-Verlet (FSV) method [6] for integration of (1) in the conservative case and in the case of small dissipation—in the latter case the fourth-order FSV method was modified for systems with small non-Hamiltonian perturbation, and values of the dissipation parameters were typical for further investigation—but did not found any differences visually comparing the structure of the phase portraits and attractors obtained via both methods.

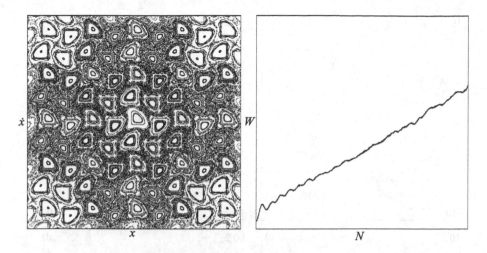

Fig. 1. Illustrations of the dynamics of the system (1) in the conservative case $\gamma = \mu = 0$: a) phase portrait of the stochastic web in the stroboscopic section; b) dependence of the average ensemble energy W on the number of the stroboscopic map iterations N.

trajectories of the dissipative system we will analyze trajectories of an ensemble of initial conditions, following the approach used previously for the conservative system [1]. In the conservative case an average energy of an ensemble of points chosen inside the stochastic layer grows linearly with time [1] (Fig. 1b), and hence we suppose that it will be a valuable characteristic of properties of trajectories for the dissipative system also. For (1) the energy is obviously defined as

$$W = (x^2 + \dot{x}^2)/2 \qquad (2)$$

and is in fact determined by a radial coordinate of the point on the phase plane $(x, \dot{x})$.

In the dissipative system all trajectories converge to periodic attractors after transient process, but some of them demonstrate diffusion properties. In this case a point travel through the regions of the phase space, which correspond to the stochastic layer in the conservative case, before transition to regular convergence to an attractor. For further investigation is therefore necessary to select such trajectories, which we will hereafter designate as demonstrating the residual diffusion. Obviously the existence of diffusion is closely connected with the presence of chaos, and it seems useful—in order to discover such trajectories—to analyze their stability in certain time interval. For this purpose we employ the finite-time Lyapunov exponent (FTLE), which we will determine here as

$$L = \frac{\sum\limits_{i=1}^{N} \ln(||\tilde{\mathbf{x}}_i||/||\tilde{\mathbf{x}}_0||)}{N} \qquad (3)$$

where $\tilde{\mathbf{x}}$ is a perturbation vector and N is a number of iterations of the stroboscopic map counted from the starting point of the trajectory. Figure 2 shows the

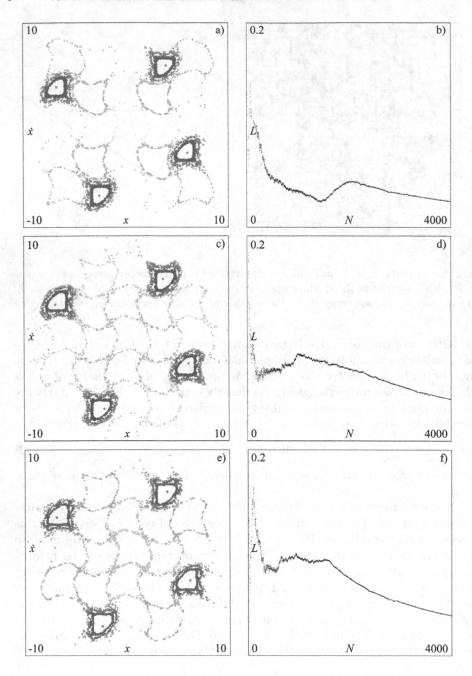

Fig. 2. Three trajectories of the system (1) demonstrating residual diffusion: the phase portraits in the stroboscopic section (left column, a, c, e), the color intensity is proportional to the number of the stroboscopic map iterations N, the red dots correspond to the attractors; the dependences of the FTLE values on time $L(N)$ for corresponding trajectories (right column, b, d, f). Values of parameters: $\lambda = 1.2$, $\gamma = 0.00001$, $\mu = 0.00001$, the length of the trajectories $N_{max} = 4000$. (Color figure online)

phase portraits and the time dependences of the FTLE value (3) for the typical trajectories demonstrating residual diffusion. The phase portraits demonstrate that the complex behavior of the trajectory after some transient switches to regular convergence, and the positive value of the FTLE begins to decrease.

Time dependences of FTLE similar to shown in right column in the Fig. 2 were obtained for a variety of other trajectories demonstrating the residual diffusion: after short initial part with very complex behavior of L a plateau exist at certain positive level until regular convergence starts. The length and level of this plateau can be used to determine a criterion of presence of the residual diffusion. Based on obtained data, we found that at the 500th iteration the transition to regular convergence has not yet occurred for the majority of trajectories, and hence it is possible to use the FTLE value (3) at this point as such a criterion. In order to determine the boundary value we use distribution of L for an ensemble of systems. Since dissipation level in the autooscillating system as well as diffusion properties of the trajectories in the conservative case both depend on the energy of the system, one can expect that properties of trajectories and dependence of average ensemble energy on time can also depend on the interval, where the initial energy—or, because of (2), the radial coordinate on the phase plane—of the ensemble elements is distributed. Several ensembles of systems were created by choosing initial conditions with the radial coordinates inside certain intervals, different for each ensemble. Examples of such ensembles with different initial values of the radial coordinate are shown in the Fig. 3, and distributions of the FTLE values (3) for three of such ensembles are shown in the Fig. 4. Distribution in all cases looks like a narrow "bell" located inside the interval $(-0.01; 0.02)$, with the maximum of distribution around 0.01, and

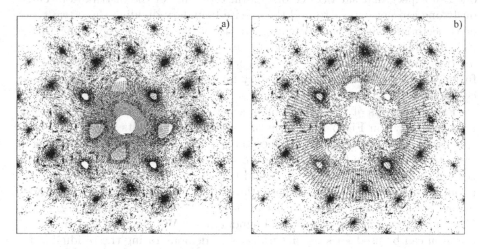

Fig. 3. Distribution of initial conditions on the phase plane (red dots) and corresponding phase portraits (black dots) for different ensembles. Initial values of the radial coordinate are distributed inside the interval (1–4) (a); (4–7) (b). (Color figure online)

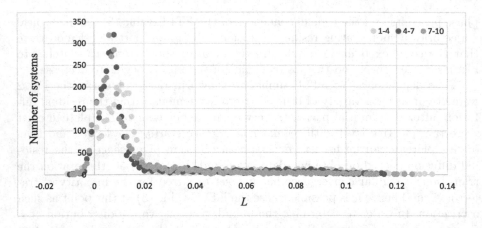

Fig. 4. The distributions of the FTLE values (3) at the 500th iteration for three different ensembles of systems. Initial values of the radial coordinate are distributed inside the intervals (1–4), (4–7), (7–10), color correspondence is specified in the figure. (Color figure online)

long and thin "tail" of distribution in the area of essentially positive values of L exists also in all cases. Since the behavior of trajectories corresponding to regular convergence to an attractor should be qualitatively the same, one can expect that these trajectories form the "bell", while trajectories corresponding to the residual diffusion can demonstrate more diversified behavior due to the presence of transient chaos and fluctuations in the stochastic layer, and, consequently, one can expect that such trajectories form the "tail" of the distribution. Obviously one can not expect to find the precise boundary dividing these two types of trajectories, but for an estimation we use value of L corresponding to the right boundary of the "bell" of the distribution—0.02—as the lower boundary for trajectories demonstrating the presence of transient chaos and residual diffusion, and 0.015 as the upper boundary for trajectories demonstrating regular convergence.

Phase portraits of trajectories of these two types selected via described criteria are shown in Fig. 5. The form of these phase portraits is in good correspondence with the structure of the stochastic web in the conservative case: trajectories corresponding to the "bell" of the distribution are located inside the cells of the stochastic web (Fig. 5a), and trajectories from the "tail" of the distribution correspond to the stochastic layer (Fig. 5b). Examples of trajectories of the latter type were shown earlier in the Fig. 2, where the values of L at the 500th iteration are far above chosen boundary 0.02. We conclude that suggested criterion can be used for selecting trajectories demonstrating the residual diffusion, and in further investigation we will choose the trajectories with value of L more that 0.02.

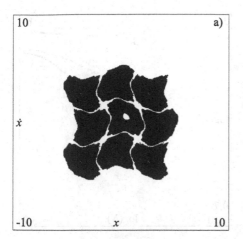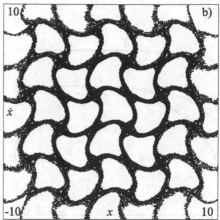

Fig. 5. The phase portraits for ensembles of initial conditions with different FTLE values (3) at 500th iteration of the stroboscopic map: a) less than 0.015; b) more than 0.02. Values of parameters: $\lambda = 1.2$, $\gamma = 0.00001$, $\mu = 0.00001$.

3   Dependence of Average Ensemble Energy on Time

Existence of the criterion for selection of the trajectories demonstrating the residual diffusion in the dissipative system allows to investigate the behavior of ensembles of such trajectories. Figure 6 shows the time dependences of average energy for ensembles of systems with different initial energies, chosen as described above. The graphs plotted in the linear scale (Fig. 6a) show nonlinear growth of energy during the transient process and saturation at large time, where trajectories converge to attractors. In order to reveal the form of dependence during the transient time the graphs were plotted in the double logarithmic scale (Fig. 6b). Except for a short interval in the very beginning, where fluctuations are rather strong, the graph in the rest part of the transient time can be approximated with a linear function, which means that average ensemble energy depends on time by the power law $W \sim N^{\alpha}$, where α is the slope of the linear dependence. The result of the approximation via least square method is shown in the Fig. 7.

It can be seen that the dependences are approximated by the linear function rather well, and the values of the standard deviation of the approximating linear function from the original dependence Δ are rather low. Since the right point of the linear part of the dependence corresponds to the beginning of the regular convergence to attractors for the most of trajectories, length of the linear part can be used as an estimation of the diffusion duration. If the original dataset is well described by a linear dependence, the Δ should be small, therefore the right boundary of the interval corresponding to linear dependence should be chosen providing minimal values of the standard deviation. We constructed a set of approximation functions using different intervals where dependence is supposed to be linear, and choose those corresponding to the minimal values of Δ. Figure 8

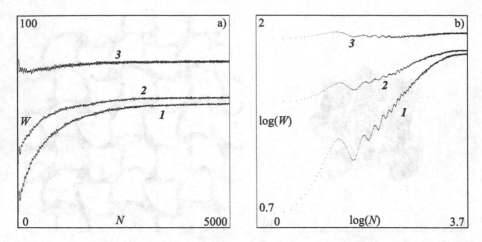

Fig. 6. Energy growth vs. number of iterations of the stroboscopic map in linear (a) and double logarithmic (b) scales for different ensembles with radial coordinates of initial conditions distributed inside the intervals (1–4) (1), (4–7) (2), (7–10) (3). Values of parameters: $\lambda = 1.2$, $\gamma = 0.00001$, $\mu = 0.00001$.

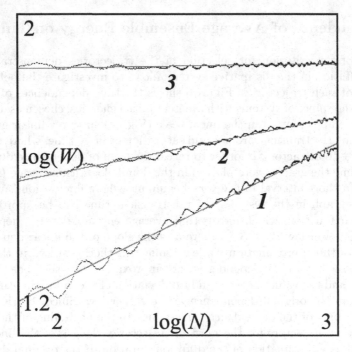

Fig. 7. Energy growth vs. number of iterations of the stroboscopic map in double logarithmic scale for different ensembles with radial coordinates of initial conditions distributed inside the intervals (1–4) (1), (4–7) (2), (7–10) (3): the linear approximations. Values of slope α and standard deviation Δ: 1) $\alpha = 0.3921$; $\Delta = 9.4541 \cdot 10^{-5}$, 2) $\alpha = 0.1589$; $\Delta = 2.1173 \cdot 10^{-5}$, 3) $\alpha = 0.0112$; $\Delta = 1.3295 \cdot 10^{-5}$. Values of parameters: $\lambda = 1.2$, $\gamma = 0.00001$, $\mu = 0.00001$.

shows the dependence of the standard deviation Δ on time with the minimum points marked. We found that as the initial energy of the ensemble increases, the diffusion duration also increases, and the absolute values of Δ decreases.

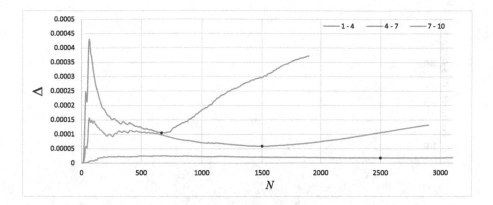

Fig. 8. The dependence of the standard deviation of linear dependence Δ on time for ensembles of systems with initial values of the radial coordinate distributed inside the intervals (1–4), (4–7), (7–10). Color correspondence is specified in the figure, the minimum points are marked with black dots. (Color figure online)

Distributions of energy for ensembles under consideration are shown in the Fig. 9. While the average ensemble energy demonstrates growth with the increase of time, peaks of energy distributions inside each ensemble become wider and lower because of diffusion. It is also worth mentioning here that even after rather long transients energy distributions remain sufficiently different for different ensembles (Fig. 9b).

As we mentioned earlier, the behavior of the system depends on the interval Δr, where initial values of the radial coordinate of the ensemble elements are distributed. In order to investigate this dependence in more detail it seems productive to use greater number of intervals, making the intervals themselves smaller. When approaching the conservative case, the maximal value of the power exponent α increases. At the same time an increase of the initial energy causes the decrease of α, which approaches 0 at certain values of initial energy, as shown in the Table 1. Let us designate an average radial coordinate of initial points in such an ensemble as the saturation radius r_{sat}. The saturation radius defined this way also increases as the dissipation decreases. For the convenience of analyzing data for different values of μ we normalize radial coordinates to r_{sat}, which is different for different values of μ. Figure 10 shows the dependences of the power exponent α on the normalized radial coordinate $\rho = r/r_{sat}$ for different values of the nonlinear dissipation parameter μ. Note that they have the same form in wide range of μ values.

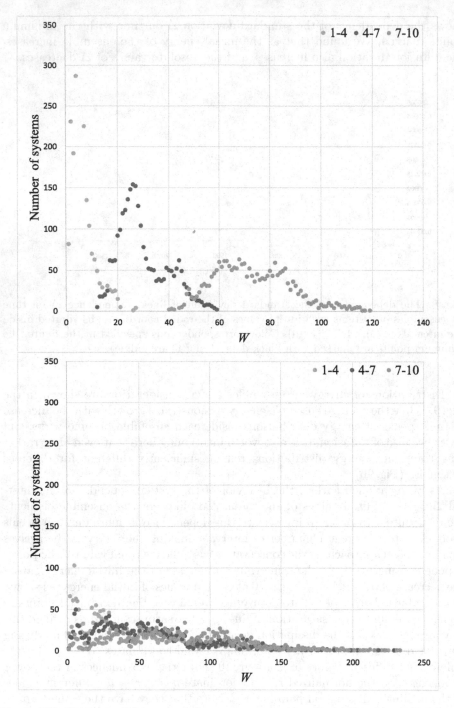

Fig. 9. Distributions of the energy of systems at starting point (a) and at 1000th iteration of the stroboscopic map (b) for several ensembles with initial values of the radial coordinate distributed inside the intervals (1–4), (4–7), (7–10). Color correspondence is specified in the figure. (Color figure online)

Table 1. Dependence of the power exponent α on the interval where the radial coordinate of the initial conditions is distributed for different values of the nonlinear dissipation parameter μ; $\gamma = 0.00001$ in all cases.

μ	1e$-$06	5e$-$05	1e$-$05
Δr	α		
1-2	0.724	0.5913	0.525
2-3	0.652	0.5019	0.43
3-4	0.566	0.4356	0.311
4-5	0.525	0.3179	0.216
5-6	0.4	0.1956	0.216
6-7	0.359	0.1481	0.082
7-8	0.317	0.1132	0.038
8-9	0.234	0.0706	0.017
9-10	0.188	0.0365	
10-11	0.169	0.0243	
11-12	0.112		
12-13	0.094		
13-14	0.074		
14-15	0.048		

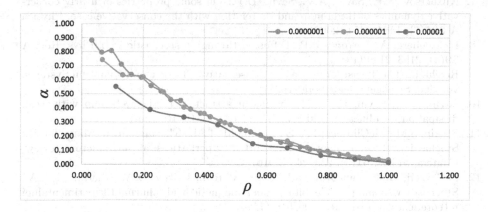

Fig. 10. Normalized graphs of the dependence of the power exponent α on the interval, where initial values of the radial variable of the ensemble elements are distributed, at different values of μ: 0.0000001, 0.000001, 0.00001 (color correspondence is specified in the figure); $\gamma = 0.00001$ in all cases. (Color figure online)

4 Conclusion

The presence of trajectories demonstrating the diffusion through the stochastic web in the phase space of the weakly dissipative system was shown. The criterion for the presence of such residual diffusion, based on the analysis of the finite-time Lyapunov exponent values, was introduced, and properties of dependence of the average ensemble energy on time were analyzed for ensembles of such trajectories. This energy depends on time by the power law. A power exponent decreases down to 0 with the increase of the initial ensemble energy, and it's dependence on the initial ensemble energy has the similar shape in wide interval of the parameter of nonlinear dissipation.

References

1. Daly, M.V., Heffernan, D.M.: Chaos in a resonantly kicked oscillator. J. Phys. A: Math. Gen. **28**(9), 2515–2528 (1995)
2. Felk, E.V.: The effect of weak nonlinear dissipation on structures of the ≪stochastic web≫ type. Izvestiya VUZ. Appl. Nonlinear Dyn. **21**(3), 72–79 (2013). (in Russian)
3. Felk, E.V., Kuznetsov, A.P., Savin, A.V.: Multistability and transition to chaos in the degenerate Hamiltonian system with weak nonlinear dissipative perturbation. Physica A **410**, 561–572 (2014)
4. Feudel, U.: Complex dynamics in multistable systems. Int. J. Bifurcation Chaos **18**(06), 1607–1626 (2008)
5. Feudel, U., Grebogi, C., Hunt, B.R., Yorke, J.A.: Map with more than 100 coexisting low-period periodic attractors. Phys. Rev. E **54**(1), 71–81 (1996)
6. Forest, E., Ruth, R.D.: Fourth-order symplectic integration. Physica D **43**(1), 105–117 (1990)
7. Kuznetsov, A.P., Savin, A.V., Savin, D.V.: On some properties of nearly conservative dynamics of Ikeda map and its relation with the conservative case. Physica A **387**(7), 1464–1474 (2008)
8. Lichtenberg, A.J., Wood, B.P.: Diffusion through a stochastic web. Phys. Rev. A **39**(4), 2153–2159 (1989)
9. Reichl, L.: The Transition to Chaos: Conservative Classical and Quantum Systems, vol. 200. Springer, Heidelberg (2021)
10. Savin, A.V., Savin, D.V.: The basins of attractors in the web map with weak dissipation. Nonlinear World **8**(2), 70–71 (2010). (in Russian)
11. Soskin, S.M., McClintock, P.V.E., Fromhold, T.M., Khovanov, I.A., Mannella, R.: Stochastic webs and quantum transport in superlattices: an introductory review. Contemp. Phys. **51**(3), 233–248 (2010)
12. Zaslavskii, G.M., Zakharov, M.I., Sagdeev, R.Z., Usikov, D.A., Chernikov, A.A.: Stochastic web and particle diffusion in a magnetic field. Zhurnal Eksperimentalnoi i Teoreticheskoi Fiziki **91**, 500–516 (1986)
13. Zaslavsky, G.M.: The Physics of Chaos in Hamiltonian Systems. World Scientific (2007)

Aberrator Shape Identification from 3D Ultrasound Data Using Convolutional Neural Networks and Direct Numerical Modeling

Alexey Vasyukov(✉)🆔, Andrey Stankevich🆔, Katerina Beklemysheva🆔,
and Igor Petrov🆔

Moscow Institute of Physics and Technology, Dolgoprudny, Russia
a.vasyukov@phystech.edu

Abstract. The paper considers the problem of silicon aberrator shape identification from 3D medical ultrasound data using convolutional neural networks.

This work demonstrates that it is possible to obtain high quality numerical 3D ultrasound images using direct numerical modeling methods. Current study models reflections from long smooth boundaries and individual large reflectors, as well as background noise from point reflectors. The synthetic computational data obtained in this way can be used to develop convolutional neural networks for 3D ultrasound data.

This work shows that 3D convolutional neural network can identify position and shape of the silicone aberrator boundary from an ultrasound data. The papers covers the cases of strong noise and significant signal distortions. It is demonstrated that 3D network can handle the distortions and correctly distinguish the boundary of materials from the responses of individual large reflectors. This possibility of the network is due to its three-dimensional architecture, which uses all spatial information from all directions.

Keywords: Ultrasound · Matrix probe · Numerical modeling · Inverse problem · Convolutional neural networks

1 Introduction

Ultrasound is one of the widely used methods of medical studies. Recent advances of ultrasound equipment significantly enhanced both resolution and contrast of traditional· ultrasonic images. Another important area of research is a development of matrix probes that allow to obtain 3D volumetric ultrasonic scans that provide significant new capabilities compares with classical 2D ultrasound images obtained with linear probes.

Traditional 2D ultrasound images are widespread. The methods for automated analysis of these images are well developed, including the techniques based

Supported by RSF project 22-11-00142.

on machine learning methods and convolutional neural networks. This area of research is covered in modern scientific literature very well. Several papers to mention demonstrate the convolutional network's feasibility for different biomedical imaging problems. The paper [1] studied an application of deep learning to artifacts correction on single-shot ultrasound images obtained with sparse linear arrays. The initial results demonstrated an image quality comparable or better to that obtained from conventional beamforming. The work [2] used neural networks to identify the shape of the aberration prism that distorts the ultrasonic signal. The work [3] applied convolutional networks to elasticity imaging to distinguish benign tumors from their malignant counterparts based on measured displacement fields on the boundary of the domain.

It should be noted that the vast majority of modern works use 2D problem statements and 2D ultrasound images. This fact naturally raises the question of generalizing similar machine learning approaches to a fully three-dimensional case. This should allow to fully unlock the potential of modern ultrasound equipment. The papers devoted to this area show that this approach is really promising. The work [4] covered 3-class classification problem for 3D images of human thyroid. The paper [5] used 3D neural networks for ovary and follicle detection from ultrasound volumes. The authors of [6] studied the application of 3D convolutional neural networks for super-resolution of microvascularity visualization. The study [7] compared directly the results of 2D and tracked 3D ultrasound with an automatic segmentation based on a deep neural network regarding inter- and intraobserver variability, time, and accuracy.

Processing of ultrasound images with convolutional neural networks is typically based on U-Net architecture [8] for 2D cases. 3D cases mostly use the same approach implemented in 3D [9]. Different modifications of these approaches exist. For example, the work [10] used three U-Nets that segmented the 3D ultrasound images in the axial, lateral and frontal orientations, and these three segmentation maps were consolidated by a separate segmentation average network.

However, full 3D neural networks for medical ultrasound are still not well covered. One of the reasons limiting the research in this area of research is a limited amount of open 3D ultrasound data available for development and testing of machine learning algorithms. The limited amount of data is especially important because it can lead to overfitting of the network to the data from the existing samples [11]. The current paper demonstrates an approach for the development of convolutional neural networks for 3D ultrasound image processing using direct numerical modeling methods.

This paper is organized as follows. Section 2 is devoted to the direct problem of 3D ultrasound images modeling - problem statement, mathematical model and numerical method are described. This works covers for the direct problem the following artifacts: the reflection from the boundary between acoustically contrasting layers, the noise caused by small bright reflectors in the media, the distortions caused by large reflectors. Section 3 presents the approach for the inverse problem of determining the position of the boundary between layers using convolutional neural networks. The architecture of the network, the dataset creation procedures, the training of the network are covered. The results of numerical experiments and their discussion are given in Sect. 4. The concluding remarks are given in Sect. 5.

2 Direct Problem

2.1 Problem Statement

This paper considers the problem of forming a 3D ultrasound image in the area containing the boundary between acoustically contrasting layers. The computational domain is a parallelepiped. The upper face of the parallelepiped corresponds to the outer border of the body, on which the 3D matrix ultrasound probe is located. This face is considered to be a free surface outside the area of the contact with the sensor. The external pressure condition is set under the sensor.

The boundary between two layers is smooth, the shape of the boundary is taken arbitrary, the boundary is at an arbitrary depth ranging from 10% to 90% of the total depth of the computational domain. The upper layer contains many reflective objects that distort the final ultrasound image. Small point reflectors and large pores are considered. Small reflectors are used to simulate imperfections in the top layer material. Each individual small reflector generates a weak echo response. However, a large number of such reflectors (from hundreds to tens of thousands in different calculations) leads to a significant noise in the ultrasound image. Large pores describe macroscopic inclusions. Echo responses from these inclusions can be comparable to the reflection from the boundary between the layers.

2.2 Mathematical Model and Numerical Method

This paper uses an acoustics model to describe the medium. According to the acoustic model [12], the ultrasound signal propagation is described by the following equations:

$$\rho(\mathbf{x})\frac{\partial \mathbf{v}(\mathbf{x},t)}{\partial t} + \nabla p(\mathbf{x},t) = 0 \qquad\qquad \text{in } \Omega, \qquad (1)$$

$$\frac{\partial p(\mathbf{x},t)}{\partial t} + \rho(\mathbf{x})c^2(\mathbf{x})\nabla \cdot \mathbf{v}(\mathbf{x},t) = -\alpha(\mathbf{x})c(\mathbf{x})p(\mathbf{x},t) \text{ in } \Omega, \qquad (2)$$

where Ω is the computational domain, $\rho(\mathbf{x})$ is the material density, $\mathbf{v}(\mathbf{x},t)$ is the velocity vector, $p(\mathbf{x},t)$ is the acoustic pressure, $c(\mathbf{x})$ is the speed of sound, $\alpha(\mathbf{x})$ is the Maxwell's attenuation coefficient [13].

The acoustic model takes into account longitudinal (pressure) waves in soft tissues and does not account transverse (shear) waves. This model is conventional in diagnostic ultrasound simulations since the attenuation coefficient for shear waves in soft tissues is four orders of magnitude greater than that for pressure waves at MHz frequencies [14].

The wavefront construction method is used for the numerical solution. The implementation uses the modifications described in [15]. This numerical method is focused exclusively on acoustic equations, which is an acceptable limitation for the present work. The paper [15] demonstrated that calculations using this method allow to obtain numerical ultrasound images that match the experimental data qualitatively and quantitatively. The method allows one to describe the reflection from long boundaries and from point reflectors. The boundary between the layers and the boundaries of large pores are described using long boundaries approach. The small reflectors are considered as point ones. Signal processing and B-scan image generation follow the algorithms described in [16].

2.3 Numerical Results for the Direct Problem

The statement of the direct problem extends the results presented in [15]. The previous work studied the scan of the medical phantom through the silicone aberrator prism. That study was performed for the 2D setup with the linear probe and showed the applicability of the wavefront construction method acoustic model to this problem. Current research considers a similar problem statement in full 3D using a matrix probe.

The matrix sensor considered in this work is square and consists of an arraty of 24×24 elements. The transmitted signal has the frequency $\omega = 3\,\mathrm{MHz}$. The signal is digitized using $45\,\mathrm{MHz}$ frequency. The size of the obtained volumetric 3D image is $24 \times 24 \times 1024$.

The speed of sound in both layers is fixed. The outer layer is more rigid, the speed of sound is $3.0\,\mathrm{km/s}$. The second layer is softer, the speed of sound is $1.5\,\mathrm{km/s}$, this value is typical for the soft tissues of the human body. The number of small reflectors in the volume of the outer layer under the probe varied from 100 to 2500. The number of large pores ranged from 5 to 50.

Figure 1 shows an example image for the problem statement without reflectors that distort the signal. Reflection from the boundary between two layers is clearly visible for all slices, the location of the boundary is obvious. Figure 2 demonstrates an example of an image with a large number of pores and small reflectors in the upper layer. In this particular image, the boundary between the layers is easily distinguishable, but this is due solely to the fact that all the reflectors were artificially localized near the surface, and the signal from them was weakly superimposed on the signal from the boundary. The general view of this image allows to estimate the level of noise and distortion compared with the echo from the boundary. When noise and reflection from pores are superimposed on the echo signal from the boundary, the interpretation of the response will become extremely difficult.

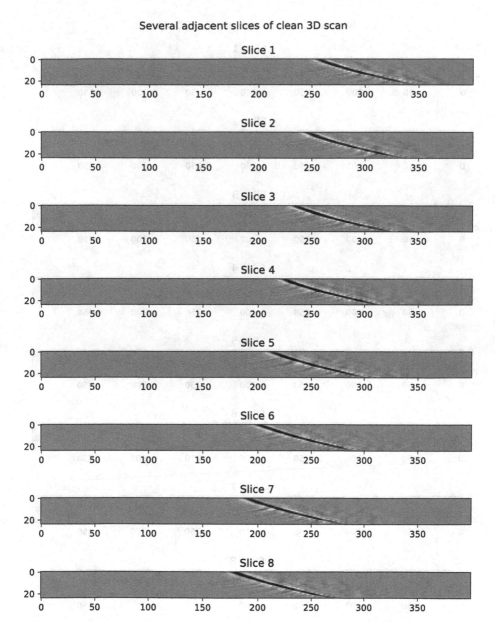

Fig. 1. A sample of 3D scan without distortions. The first eight slices are presented. The image is zoomed to the area around the border for better visibility.

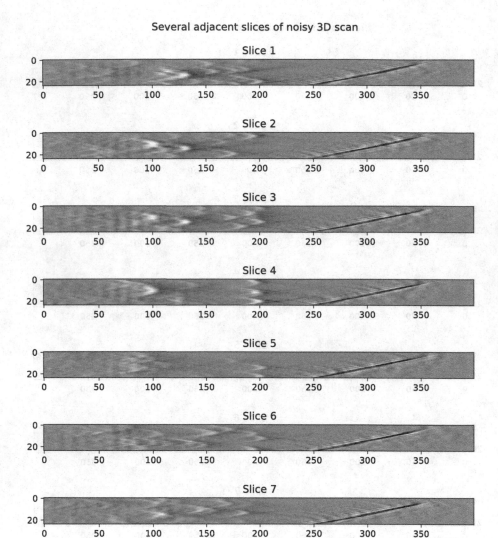

Fig. 2. A sample of 3D scan with distortions. The first eight slices are presented. The image is zoomed to the area around the border for better visibility.

3   Inverse Problem

3.1   Problem Statement

The inverse problem in this work is to determine the position of the boundary between two layers based on 3D ultrasound data. The input data of the inverse problem is a three-dimensional array of dimensions $24 \times 24 \times 1024$. The first two dimensions correspond to the numbers of elements in the matrix probe along two axes, and the third dimension corresponds to the depth in time domain. The output of the inverse problem is the position of the interface between two media in the volume presented as the input.

As shown above, the quality of the input images can vary a lot. In the absence of signal distortions, the inverse problem becomes trivial, the position of the boundary is easily determined both visually and by the simple signal processing algorithms. The samples with strong distortions can be extremely difficult to interpret both visually and automatically. This work uses convolutional neural networks to solve the inverse problem.

3.2   Neural Network Architecture

There are three possible approaches to building a convolutional neural network for volumetric 3D ultrasound data.

1. A 3D image can be represented as a set of 2D slices. This allow to use a traditional 2D convolutional network for each of the slices. The network has one input channel that takes a slice data as a grayscale image. This approach is easy to implement, but it has obvious drawbacks. When working with a 2D slice, the network fundamentally cannot use the full spatial information that was originally contained in the full 3D image.
2. A 2D network with multiple input channels can be used to overcome the obvious disadvantage of the previous option. The implementation of this approach can be based on a large number of ready to use technical solutions, since multichannel 2D convolutional neural networks are routinely used for color image processing. Classical pipeline uses several channels to input separate RGB color components of the image. However, there is an experience showing that similar approach is applicable to a significantly different nature of data in different channels. For example, the work [17] considered seismic inversion and used multichannel 2D network to encode the data acquired for the same object using different positions of the source of the signal. In the case of 3D ultrasound, separate input channels can be used to represent adjacent 2D slices of the full 3D image. This will allow the network to use inter-slice information to some extent. For example, the work [18] used this approach to classify 3D ultrasound data.
3. It is possible to implement a fully 3D convolutional network using 3D convolution kernels. This approach is much more difficult to implement technically, since there are very few ready to use technical components for such networks.

However, there are no fundamental obstacles to the implementation of such a network. In this case, the input data of the network is a 3D volume representing the original data as a 3D grayscale image. The use of 3D convolution seems to be the most promising option. This approach does not require an artificial separation of spatial variables and makes it possible to fully reflect the physical nature of the problem being considered.

This paper uses a 3D convolutional neural network following the UNet architecture generalized to the 3D case. The input data is elongated along the depth axis, this fact is caused by the nature of the data. The quantitative value of the input tensor depth depends on (a) a target scanning depth represented in time domain, (b) a sampling frequency of the sensor. This work uses the depth of 1024 samples. Two other axes can have different dimensions depending on the sensor structure. This work uses patch-based approach that is commonly used for biomedical image segmentation [19,20]. The size of the patch is 16×16 in the sensor plane and 512 along the depth of the image. The patch stride is 8 for both in-plane dimensions and 256 for depth, the patches overlap to ensure smooth results for the complete prediction. This patch-based approach reduces the memory requirements for the network and allows to process significantly larger images if necessary. The main network hyperparameters are presented in the Table 1.

Table 1. Network hyperparameters.

Parameter	Value
Activation	ReLU
Normalization	Batch
Convolution kernel size	3
Convolution stride	1
Convolution mode	Same (middle)
Number of up/down blocks	4
Input dimensions (patch)	$16 \times 16 \times 512$
Output dimensions (patch)	$16 \times 16 \times 512$
Input channels	1
Output channels	1

3.3   Dataset and Training

The dataset was created by computing numerical samples as described in the previous section. The shape and position of the interface between two media was randomly generated for each sample. All boundary points for all samples were located between the depth levels of 100 and 900 in the time domain. The number of small reflectors varied from 100 to 2500, the number of large pores varied from 5 to 50.

The input data contained noisy and distorted synthetic ultrasound images. Ground truth images contained true position of the border between two layers. A separate ground truth data manual preparation was not required since an exact border position is knows for synthetic data.

A total of 10 000 numerical samples were prepared. They were divided into training and validation datasets using the ratio of 75:25. Separate samples were used for final testing, they were generated additionally and not used in any way during network training.

The Adam optimization algorithm was used for training with a constant learning rate of 1e−4 and the BCEWithLogits loss function. The training was performed from scratch, pretrained networks were not used. Training was performed for 100 epochs with the batch size of 12 samples. The loss function was observed to become stable around the 30th–40th epoch, the value of the loss only slightly oscillated for the next epochs around the achieved values. We used the weights from the epoch at which the best value of the loss function was achieved on the validation set. Typical training time was about 12 h.

4   Numerical Experiments and Discussion

Figure 3 shows an example of an image with strong distortions. The response from the boundary between the layers is heavily noisy with pores and point reflectors. It is difficult to determine the location of the boundary from this image both visually and using simple signal processing algorithms. The real location of the boundary for this case is shown in Fig. 4. The results of the predictions of the convolutional neural network are shown in Fig. 5. Note that this data sample belongs to the set prepared for the final testing and was not used in any way during network training.

This example shows that the resulting neural network can restore the position of the boundary between layers with high accuracy, including the cases of significant noise and distortion in the input signal. It should be noted that the network correctly ignores relatively large bright reflectors at depths of 50–150. Separately, it should be noted that the network does not react to the large pore visible on slices 7 and 8 at depths of 220–250, although this reflection coincides in intensity with the response from the target media boundary and is close to it in location. However, the network successfully determines the true position of the boundary near this pore. This result is due to the chosen 3D network architecture, which allows full use of the information of all spatial directions, which makes it possible to distinguish a separate large reflector from a large smooth interface between the media.

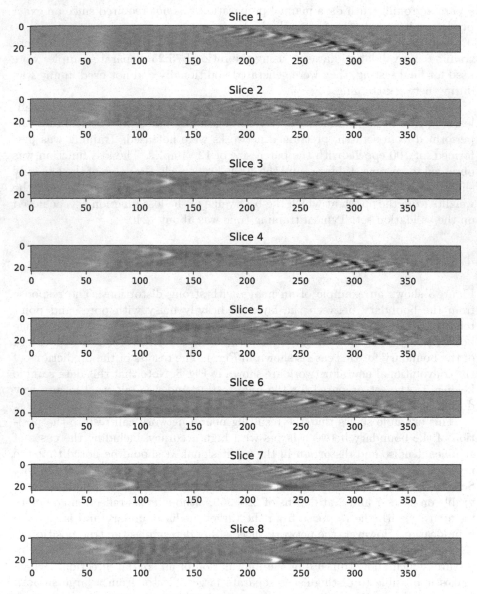

Fig. 3. A sample of 3D scan with high level of distortions. The image is zoomed to the area around the border for better visibility.

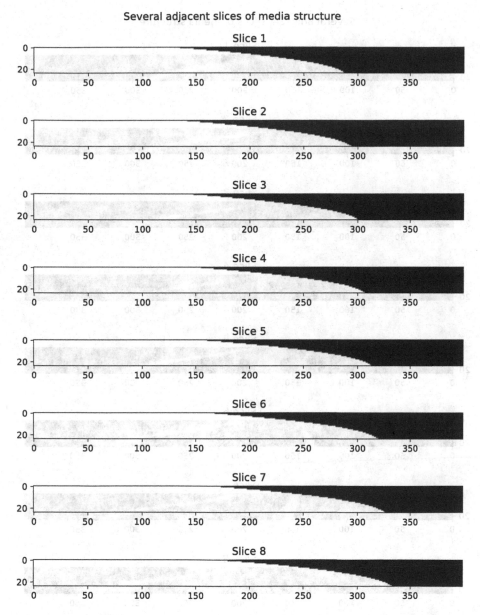

Fig. 4. Ground truth position of the border between two layers for the sample with high level of distortions. The image is zoomed to the area around the border for better visibility.

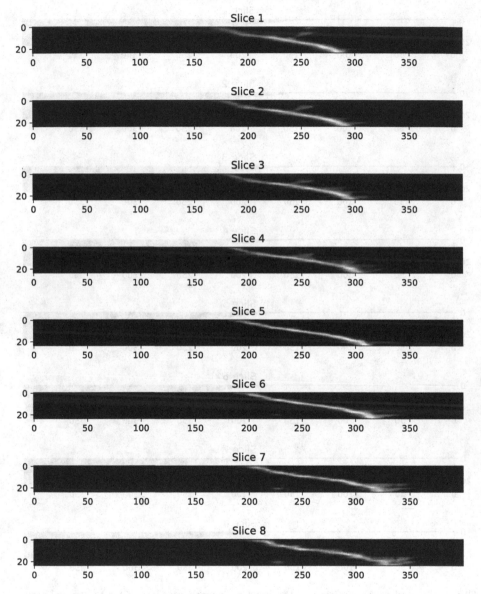

Fig. 5. Prediction of the border between two layers for the sample with high level of distortions. The image is zoomed to the area around the border for better visibility.

5   Conclusions and Future Work

This work shows that it is possible to obtain numerical 3D ultrasound images of sufficiently high quality using direct numerical modeling methods. This work considers reflections from smooth boundaries and individual large reflectors, as well as background noise from point reflectors. The synthetic computational data obtained in this way can be used to explore the possibilities of 3D ultrasound, as well as to develop convolutional neural networks for 3D ultrasound for the cases that lack real experimental data available.

This work shows that 3D convolutional neural network can identify position and shape of the silicone aberrator boundary based on the ultrasound data, including the cases of strong noise and significant distortions. It is shown that the network can correctly distinguish the boundary of materials from the responses of individual large reflectors comparable in brightness. This possibility of the network is due to its three-dimensional architecture, which uses all spatial information from all directions.

Further steps should include an implementation of an elastic material model and a contact between elastic (bone) and acoustic (soft tissues) media. This should allow to simulate the statements that cannot be described in terms of a purely acoustic approximation. It is possible that after the transition to the full elasticity model, it will be necessary to revise the parameters of the convolutional network, since a significant complication of the wave pattern should be expected.

References

1. Dimitris, P., Manuel, V., Florian, M., Marcel, A, Jean-Philippe, T.: Single-shot CNN-based ultrasound imaging with sparse linear arrays. In: 2020 IEEE International Ultrasonics Symposium (IUS), pp. 1–4 (2020)
2. Stankevich, A.S., Petrov, I.B., Vasyukov, A.V.: Numerical solution of inverse problems of wave dynamics in heterogeneous media with convolutional neural networks. In: Favorskaya, M.N., Favorskaya, A.V., Petrov, I.B., Jain, L.C. (eds.) Smart Modelling for Engineering Systems. SIST, vol. 215, pp. 235–246. Springer, Singapore (2021). https://doi.org/10.1007/978-981-33-4619-2_18
3. Patel, D., Tibrewala, R., Vega, A., Dong, L., Hugenberg, N., Oberai, A.: Circumventing the solution of inverse problems in mechanics through deep learning: application to elasticity imaging. Comput. Methods Appl. Mech. Eng. **353**, 448–466 (2019)
4. Lu, H., Wang, H., Zhang, Q., Yoon, S., Won, D.: A 3D convolutional neural network for volumetric image semantic segmentation. Procedia Manuf. **39**, 422–428 (2019)
5. Potočnik, B., Šavc, M.: Deeply-supervised 3D convolutional neural networks for automated ovary and follicle detection from ultrasound volumes. Appl. Sci. **12**(1246) (2022)
6. Brown, K., Dormer, J., Fei, B., Hoy, K.: Deep 3D convolutional neural networks for fast super-resolution ultrasound imaging. In: Proceedings of the SPIE 10955, Medical Imaging 2019: Ultrasonic Imaging and Tomography, p. 1095502 (2019)
7. Krönke, M., et al.: Tracked 3D ultrasound and deep neural network-based thyroid segmentation reduce interobserver variability in thyroid volumetry. PLoS ONE **17**(7), Article e0268550 (2022)

8. Ronneberger, O., Fischer, P., Brox, T.: U-Net: convolutional networks for biomedical image segmentation. arXiv:1505.04597 (2015)
9. Çiçek, Ö., Abdulkadir, A., Lienkamp, S.S., Brox, T., Ronneberger, O.: 3D U-Net: learning dense volumetric segmentation from sparse annotation. arXiv:1606.06650 (2016)
10. Jiang, M., Spence, J.D., Chiu, B.: Segmentation of 3D ultrasound carotid vessel wall using U-Net and segmentation average network. In: 42nd Annual International Conference of the IEEE Engineering in Medicine and Biology Society (EMBC), pp. 2043–2046. IEEE (2020)
11. Zheng, Y., Liu, D., Georgescu, B., Nguyen, H., Comaniciu, D.: 3D deep learning for efficient and robust landmark detection in volumetric data. In: Proceedings of 2015 IEEE Medical Image Computing and Computer-Assisted Intervention, pp. 565–572. IEEE (2015)
12. Mast, T.D., Hinkelman, L.M., Metlay, L.A., Orr, M.J., Waag, R.C.: Simulation of ultrasonic pulse propagation, distortion, and attenuation in the human chest wall. J. Acoust. Soc. Am. 6, 3665–3677 (1999)
13. Beklemysheva, K., et al.: Transcranial ultrasound of cerebral vessels in silico: proof of concept. Russ. J. Numer. Anal. Math. Model. 31(5), 317–328 (2016)
14. Madsen, E.L., Sathoff, H.J., Zagzebski, J.A.: Ultrasonic shear wave properties of soft tissues and tissuelike materials. J. Acoust. Soc. Am. 74(5), 1346–1355 (1983)
15. Beklemysheva, K., Grigoriev, G., Kulberg, N., Petrov, I., Vasyukov, A., Vassilevski, Y.: Numerical simulation of aberrated medical ultrasound signals. Russ. J. Numer. Anal. Math. Model. 33, 277–288 (2018)
16. Vassilevski, Y., Beklemysheva, K., Grigoriev, G., Kulberg, N., Petrov, I., Vasyukov, A.: Numerical modelling of medical ultrasound: phantom-based verification. Russ. J. Numer. Anal. Math. Model. 32(5), 339–346 (2017)
17. Stankevich, A., Nechepurenko, I., Shevchenko, A., Gremyachikh, L., Ustyuzhanin, A., Vasyukov, A.: Learning velocity model for complex media with deep convolutional neural networks. arXiv:2110.08626 (2021)
18. Paserin, O., Mulpuri, K., Cooper, A., Abugharbieh, R., Hodgson, A.: Improving 3D ultrasound scan adequacy classification using a three-slice convolutional neural network architecture. In: Zhan, W., Baena, F. (eds.) CAOS 2018 (EPiC Series in Health Sciences), vol. 2, pp. 152–156 (2018)
19. Coupeau, P., Fasquel, J.-B., Mazerand, E., Menei, P., Montero-Menei, C.N., Dinomais, M.: Patch-based 3D U-Net and transfer learning for longitudinal piglet brain segmentation on MRI. Comput. Methods Programs Biomed. 214, 106563 (2022)
20. Ghimire, K., Chen, Q., Feng, X.: Patch-based 3D UNet for head and neck tumor segmentation with an ensemble of conventional and dilated convolutions. In: Andrearczyk, V., Oreiller, V., Depeursinge, A. (eds.) HECKTOR 2020. LNCS, vol. 12603, pp. 78–84. Springer, Cham (2021). https://doi.org/10.1007/978-3-030-67194-5_9

Complex Dynamics of the Implicit Maps Derived from Iteration of Newton and Euler Methods

Andrei A. Elistratov[1,2]([✉]) [iD], Dmitry V. Savin[2]([✉]) [iD],
and Olga B. Isaeva[1,2]([✉]) [iD]

[1] Kotel'nikov Institute of Radio-Engineering and Electronics of RAS,
Saratov Branch, Zelenaya 38, Saratov 410019, Russia
zhykreg@gmail.com, isaevao@rambler.ru
[2] Saratov State University, Astrakhanskaya 83, Saratov 410012, Russia
savin.dmitry.v@gmail.com

Abstract. Special exotic class of dynamical systems—the implicit maps—is considered. Such maps, particularly, can appear as a result of using of implicit and semi-implicit iterative numerical methods. In the present work we propose the generalization of the well-known Newton-Cayley problem. Newtonian Julia set is a fractal boundary on the complex plane, which divides areas of convergence to different roots of cubic nonlinear complex equation when it is solved with explicit Newton method. We consider similar problem for the relaxed, or damped, Newton method, and obtain the implicit map, which is non-invertible both time-forward and time-backward. It is also possible to obtain the same map in the process of solving of certain nonlinear differential equation via semi-implicit Euler method. The nontrivial phenomena, appearing in such implicit maps, can be considered, however, not only as numerical artifacts, but also independently. From the point of view of theoretical nonlinear dynamics they seem to be very interesting object for investigation. Earlier it was shown that implicit maps can combine properties of dissipative non-invertible and Hamiltonian systems. In the present paper strange invariant sets and mixed dynamics of the obtained implicit map are analyzed.

Keywords: Julia set · Hamiltonian system · Implicit map

1 Introduction

One of the wide-known examples of fractal sets—Newtonian Julia set (Fig. 1)—arises as a boundary of areas of convergence to different roots of the cubic polynomial equation on the complex plane

$$\phi(z) = z^3 + c = 0, \tag{1}$$

_____
Supported by Russian Science Foundation, Grant No. 21-12-00121.

when it is solved with the Newton method [18]. This problem, first suggested by Cayley [5], allows generalization, if relaxed, or damped Newton method

$$z_{n+1} = z_n - h\frac{\phi(z_n)}{\phi'(z_n)} = z_n - h\frac{z_n^3 + c}{2z_n^2} \tag{2}$$

is used [13,14]. At the same time the iteration process (2) can be considered as a trivial discretization of the ordinary differential equation

$$\dot{z} = f(z) \tag{3}$$

with the Euler method

$$z_{n+1} = z_n + hf(z_n), \tag{4}$$

where

$$f(z) = -\frac{\phi(z)}{\phi'(z)} = -\frac{z^3 + c}{3z^2}. \tag{5}$$

Roots of the polynomial Eq. (1) are the stable nodes of the ODE (3). At the same time these roots are the attractors of the Newton and Euler iteration process (2), at least, when the discretization step h is small enough. The boundary between basins of attraction of the iteration process (2)—the fractal Newtonian Julia set—corresponds to a separatrix of (3), which should be smooth when $f(z)$ is defined by (5) (see Fig. 1). Fractalization of the separatrix occures due to the Euler discretization. This is a numerical artifact, which can be considered as neglectable for practical applications at $h \to 0$.

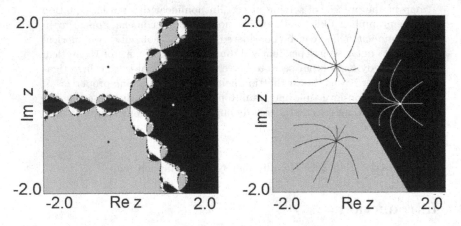

Fig. 1. Newtonian Julia set—fractal boundary between white, black and gray regions, which are the basins of attraction of three roots of the Eq. (1) with $c = -1$ (left panel) and Newtonian pie—phase plane of the flow dynamical system (3) (right panel): three roots of (1) are the stable nodes of the ODE (3), its basins of attraction are divided by the smooth separatrix. Euler discretization (4) of (3) results in emergence of fractal basin boundary, like one on the left panel, which degenerates to the true smooth linear separatrix at $h \to 0$.

While in general discretization of flow dynamical systems caused by numerical time-integration can lead to emergence of solutions which do not represent the dynamics of the original system and manifest themselves in changes of the phase space structure, as in example discussed above, or changes in bifurcation diagrams etc. (see, e.g., [12,16,17] and references there), this problem can be considered from another angle: such discretization of well-known flow systems can be regarded as a fruitful approach, which is widely used in modern nonlinear theory and allows to generate new model maps (see, e.g., [1,2,11,22] and references there). Due to emergence of numerical artifacts mentioned above dynamical systems generated this way can demonstrate various nontrivial phenomena. The discretization step h is usually defined in such models in a wide range—moreover, it can be complex [18]. This approach can also establish a background for introducing and considering of a new class of systems, one example from which is proposed in present work.

Simple construction (2), being considered as an abstract dynamical system, allows a wide spectrum of generalizations and parametrizations. Types of dynamical behavior and obtained phenomena can also be rather diverse. Particularly, the implicit dynamics, when every point in the phase space of the system has both several images and several preimages [3,4,15], is also possible. Such implicit correspondences have wide spectrum of applications besides implicit numerical schemes of equations solving. Implicit functions can occur in problems of reconstruction of a multidimensional object (or system) from its projection [6], in the theory of generalized synchronization [19], in economics [7,10], computer graphics [20], chaos control techniques [8], topology [21].

In the present paper we try to give an example of generalization of the iteration process (2) and to investigate obtained system from the point of view of theoretical nonlinear dynamics. In the Sect. 2 we present the procedure of deriving an implicit map using the modified Euler method. In the Sect. 3 we analyze the fractalization of both unstable and stable invariant sets of such exotic system.

2   Basic Model

Among the numerical recipes of the ODE integration the semi-implicit Euler method is listed:

$$z_{n+1} = z_n + h(f(z_n) + f(z_{n+1}))/2. \tag{6}$$

Let us generalize this scheme by parameterization:

$$z_{n+1} = z_n + h((1 - \alpha)f(z_n) + \alpha f(z_{n+1})). \tag{7}$$

In case when $f(z)$ chosen in form (5) this equation can be rewrited as following:

$$(\alpha h + 3)z_{n+1}^3 z_n^2 + ((1 - \alpha)h - 3)z_{n+1}^2 z_n^3 + (1 - \alpha)hcz_{n+1}^2 + \alpha hcz_n^2 = 0. \tag{8}$$

This is also some iteration process, but, in contrast to (4), not traditional for the nonlinear dynamical system theory. This is an implicit map with the evolution operator looking like

$$\Psi(z_{n+1}, z_n) = 0. \tag{9}$$

Both forward and backward iterations of this map are defined by multi-valued functions,

$$z_{n+1} = \Psi_+(z_n) \tag{10}$$

and

$$z_n = \Psi_-(z_{n+1}) \tag{11}$$

respectively.

Two examples of implicit maps are described in [4,9]. Below we are trying to study the implicit dynamics on the new example of such map (8).

3   Numerical Simulation

3.1   Repellers

We will start investigation from studying backward dynamics of the map (8) at $\alpha = 0$. In this case the map is single-valued time-forward and multi-valued time-backward. We will study structure of its repellers, which form boundaries of basins of attraction. It is worth mentioning here that since the map (8) is defined by the cubic polynomial, solutions of the Eq. (11) can be found analytically. The repellers of the map (8) at different values of $|h| \leq 1$ are shown in the Fig. 2. At $h = 1$ we obtain the classical Newtonian Julia set, other positive values of h correspond to transformations of this fractal (see Fig. 2a, b). Repellers in this case still define boundaries of areas of convergence of the Newton method (2) to different roots of (1)—or of the Euler method to different nodes of the ODE (3). At negative values of h the process of search of repellers of the map (8) corresponds to solving the ODE (3) with the time reversed with the Euler method. The result of this process is a fractal set similar to the Sierpińsky gasket (see Fig. 2d, e). $h = 0$ is a degenerate case corresponding transition between these two situations (see Fig. 2c, f).

The repellers for $h > 1$ are shown in the Fig. 3, here $h = 3$ (Fig. 3a) is also a degenerate case—which follows from the structure of function (8), where one of the terms becomes in this case equal to 0.

A useful tool for quantitative analysis of the phase space structure transformations is a fractal dimension of the basin boundaries. Figure 4 demonstrates the dependence of the box-counting dimension on the parameter h value. In the vicinity of $h = 0$ an abrupt change of the value of dimension occurs, which indicates a phase transition of the Julia set. In the vicinity of the degenerate case $h = 3$ the fractal dimension value tends to 1—it corresponds to degeneration of the Julia set, which looks in this case like a smooth circle (Fig. 3a). In the region near $h = 4$ the value of dimension grows almost up to 2, and Julia set almost becomes a fat fractal (Fig. 3b). When $h \geq 7$, the value of dimension decreases down to values below 1. The attracting invariant set undergoes here crisis, and basin boundaries degenerate to the fractal dust.

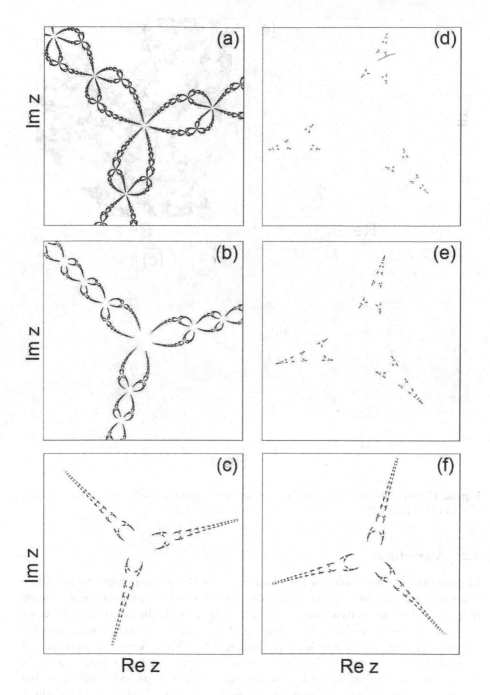

Fig. 2. The repellers in the phase space of the map (8) with $h = 1.0$ (a), $h = 0.5$ (b), $h \to +0$ (c) and $h = -1.0$ (d), $h = -0.5$ (e), $h \to -0$ (f). Parameter α is equal to zero. Parameter c is not essential, here and further it is fixed being equal to $(1 - i)/2$.

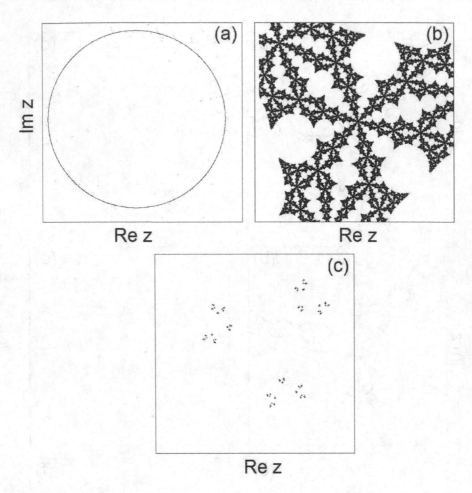

Fig. 3. The repellers in the phase space of the map (8) with $\alpha = 0$ and different values of h: 3 (a), 5 (b), 7 (c).

3.2   Attractors

In general case every point in the phase space of the implicit map (8) has three images—roots of qubic polynomial equation. Time-forward dynamics of such map is, as backward dynamics also, not single-valued. To study time-forward dynamics it seems productive to apply methods which are usually employed for an analysis of repellers. Particularly, we apply the "chaos game" algorithm [18] in order to find attracting chaotic trajectories. To find periodic trajectories we choose the roots of (8) at each iteration according to a periodic sequence. We construct symbolic codes, consisting of characters «1», «2» and «3»—which correspond to the choice of the first, the second or the third root respectively. The period of dynamical orbit should be in this case equal or larger than a minimal period of such a sequence.

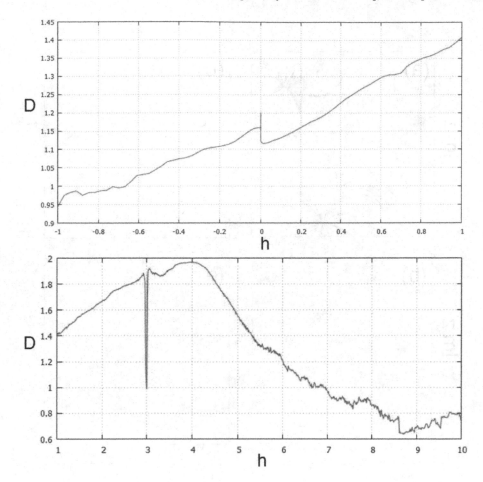

Fig. 4. Box-counting dimension of the repellers of the implicit map (8) with $\alpha = 0.0$ for $|h| \leq 1$ (upper panel) and $1 \leq h \leq 10$ (bottom panel).

Let us start from the illustration of the evolution of attractors in special case $\alpha = 0$, which is shown in the Fig. 5. This situation corresponds to the use of the explicit Euler method, and the forward-time dynamics is in this case single-valued. Figure 5a represents three attracting nodes at $h \to 0$, while an increase of parameter h causes several period doubling bifurcations (Fig. 5c–e), and the transition to chaos occurs (Fig. 5f).

The bifurcation diagram shown in the Fig. 6 gives more complete picture of the forward-time behavior of the system (8). Here the real part of the complex variable z on the attracting invariant set is plotted versus parameter h. At this picture the evolution of one of three attracting fixed points is demonstrated. It undergoes several period-doubling bifurcations, transition to chaos and back to the periodic regime, and finally is destroyed via crisis. In the vicinity of the point

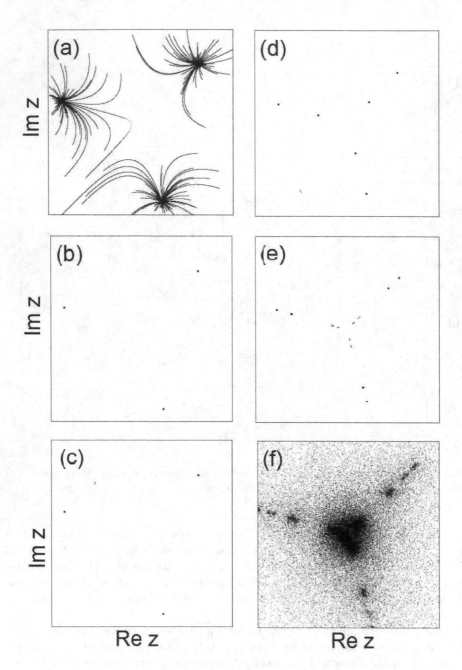

Fig. 5. Transformation of the attracting invariant sets of the implicit map (8) with $\alpha = 0.0$ and with $h \to 0$ (a), $h = 0.5$ (b), 1.0 (c), 2.15 (d), 2.75 (e), 2.792 (f).

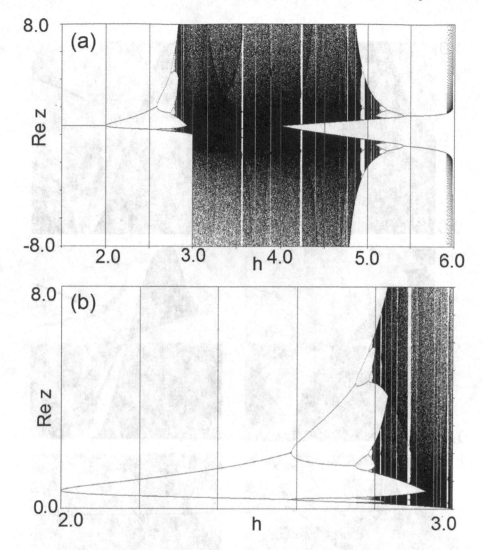

Fig. 6. The bifurcation diagram for one of attracting invariant sets of the implicit map (8) with $\alpha = 0.0$ (a) and its enlarged fragment (b).

$h = 4$, where, as we mentioned above, Julia set almost becomes a fat fractal, time-forward dynamics is chaotic.

Next Fig. 7 demonstrates the picture of dynamical regimes on the parameter plane (h, α). When $\alpha \neq 0$, the map (8) becomes an implicit one, which means that its dynamics is now muti-valued both time-forward and time-backward. Except the simplest case, when the symbolic sequence has period 1 (Fig. 7a), charts for different periodic sequences demonstrate similar features. Areas of periodic dynamics in the quadrant $(h > 0, \alpha < 0.5)$ have typical "tongue" shape

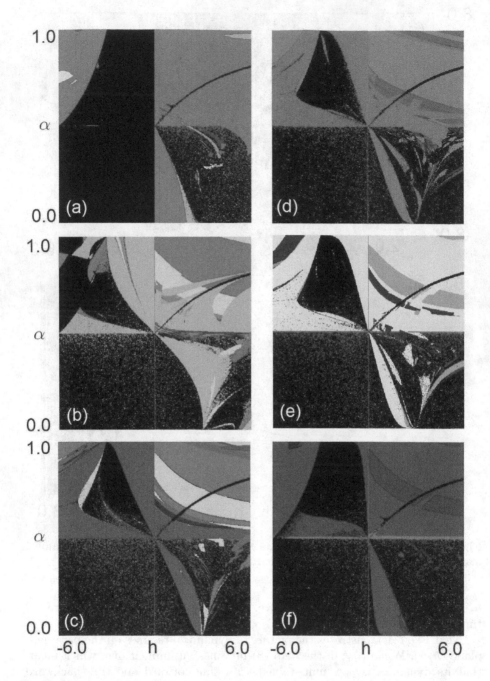

Fig. 7. The charts of dynamical regimes of the implicit map (8) for the following periodic sequences of time-forward roots choice: 1 (a), 12 (b), 112 (c), 1112 (d), 11112 (e), 1(×8)2 (f).

with spikes in the point $\alpha = 0, h = 3$, where the map (8) is degenerate. Another typical feature is partial symmetry of the parameter plane: borders of areas of aperiodic dynamics are in many cases symmetrical with respect to the point ($h = 0, \alpha = 0.5$) in quadrants ($h > 0, \alpha < 0.5$) and ($h < 0, \alpha > 0.5$), and points with periodic dynamics are symmetrical to points with aperiodic dynamic, especially for $|h| \leq 3$. It is a consequence of specific symmetry of the map (8), which is invariant with respect to the transformation $h \to -h, \alpha \to 1 - \alpha, z_n \leftrightarrow z_{n+1}$.

4 Conclusion

In this paper we present a short preliminary view on the dynamics of one example of implicit systems, namely, the map (8). We demonstrate some approaches for studying of such systems. Advanced investigation of this implicit map should clarify the structure of its invariant sets. It seems a promising and very interesting direction of research, since in implicit systems, in contrast to traditionally studied explicit ones, an infinite number of trajectories can coexist in forward time, which makes their attracting invariant sets very complicated. Moreover, complexification of the parameter h in the map (8) leads to the possibility of a situation, when $\Psi(z_{n+1}, z_n) = -\Psi^*(z_n, z_{n+1})$.[1] For the map (8) this happens at $h = \pm i$, $\alpha = 1/2$. This situation, defined in [4,9] as generalized unitarity, manifests the emergence of phenomena typical for Hamiltonian and almost Hamiltonian systems. In this context the implicit maps, being an artificial construct, can help to describe strong multistability, mixed dynamics and other complex phenomena of nonlinear dynamics.

References

1. Adilova, A.B., Kuznetsov, A.P., Savin, A.V.: Complex dynamics in the system of two coupled discrete Rossler oscillators. Izvestiya VUZ. Appl. Nonlinear Dyn. **21**(5), 108–119 (2013). (in Russian)
2. Arrowsmith, D.K., Cartwright, J.H., Lansbury, A.N., Place, C.M.: The Bogdanov map: bifurcations, mode locking, and chaos in a dissipative system. Int. J. Bifurcation Chaos **3**(04), 803–842 (1993)
3. Bullett, S.R.: Dynamics of quadratic correspondences. Nonlinearity **1**(1), 27–50 (1988)
4. Bullett, S.R., Osbaldestin, A.H., Percival, I.C.: An iterated implicit complex map. Physica D **19**(2), 290–300 (1986)
5. Cayley, A.: Desiderata and suggestions: no. 3. The Newton-Fourier imaginary problem. Am. J. Math. **2**(1), 97 (1879)
6. DiFranco, D.E., Cham, T.J., Rehg, J.M.: Reconstruction of 3D figure motion from 2D correspondences. In: Proceedings of the 2001 IEEE Computer Society Conference on Computer Vision and Pattern Recognition, CVPR 2001, vol. 1, pp. 307–314. IEEE (2001)

---

[1] Here $\Psi^*(z_{n+1}, z_n) = (\Psi(z^*_{n+1}, z^*_n))^*$.

7. Gardini, L., Hommes, C., Tramontana, F., De Vilder, R.: Forward and backward dynamics in implicitly defined overlapping generations models. J. Econ. Behav. Organ. **71**(2), 110–129 (2009)
8. Hill, D.: Control of implicit chaotic maps using nonlinear approximations. Chaos: Interdisc. J. Nonlinear Sci. **10**(3), 676–681 (2000)
9. Isaeva, O.B., Obychev, M.A., Savin, D.V.: Dynamics of a discrete system with the operator of evolution given by an implicit function: from the Mandelbrot map to a unitary map. Russ. J. Nonlinear Dyn. **13**(3), 331–348 (2017). (in Russian)
10. Kennedy, J.A., Stockman, D.R.: Chaotic equilibria in models with backward dynamics. J. Econ. Dyn. Control **32**(3), 939–955 (2008)
11. Kuznetsov, A.P., Savin, A.V., Sedova, Y.V.: Bogdanov-Takens bifurcation: from flows to discrete systems. Izvestiya VUZ. Appl. Nonlinear Dyn. **17**(6), 139–158 (2009). (in Russian)
12. Lóczi, L., Páez Chávez, J.: Preservation of bifurcations under Runge-Kutta methods. Int. J. Qual. Theory Differ. Equ. Appl. **3**, 81–98 (2008)
13. Magreñán, Á.A., Gutiérrez, J.M.: Real dynamics for damped Newton's method applied to cubic polynomials. J. Comput. Appl. Math. **275**, 527–538 (2015)
14. McLaughlin, J.: Convergence of a relaxed Newton's method for cubic equations. Comput. Chem. Eng. **17**(10), 971–983 (1993)
15. Mestel, B.D., Osbaldestin, A.H.: Renormalisation in implicit complex maps. Physica D **39**(2–3), 149–162 (1989)
16. Páez Chávez, J.: Discretizing bifurcation diagrams near codimension two singularities. Int. J. Bifurcation Chaos **20**(5), 1391–1403 (2010)
17. Páez Chávez, J.: Discretizing dynamical systems with generalized Hopf bifurcations. Numer. Math. **118**(2), 229–246 (2011)
18. Peitgen, H.O., Richter, P.H.: The Beauty of Fractals: Images of Complex Dynamical Systems. Springer, Heidelberg (1986)
19. Pikovsky, A., Rosenblum, M., Kurths, J.: Synchronization: A Universal Concept in Nonlinear Sciences. Cambridge Nonlinear Science Series. Cambridge University Press (2001)
20. Sclaroff, S., Pentland, A.: Generalized implicit functions for computer graphics. ACM SIGGRAPH Comput. Graph. **25**(4), 247–250 (1991)
21. Vlasenko, I.Y.: Internal maps: topological invariants and their applications. In: Proceedings of Institute of Mathematics of Ukrainian NAS, Institute of Mathematics of Ukrainian NAS (2014). (in Russian)
22. Zaslavsky, G.M.: The Physics of Chaos in Hamiltonian Systems. World Scientific (2007)

Recognition of Vertical Migrations for Two Age Groups of Zooplankton

O. Kuzenkov(iD) and E. Ryabova(✉)(iD)

Lobachevsky State University, Gagarin Avenue 23, 603950 Nizhny Novgorod, Russia
{oleg.kuzenkov,elena.ryabova}@itmm.unn.ru

Abstract. The purpose of the work is to predict the appearance of significant vertical movements of two age zooplankton groups as a result of adaptation to habitat conditions. The methodological basis for solving the problem is the maximization of the fitness function. Vertical migrations are considered as a strategy that ensures the achievement of the greatest value of this function for given environmental conditions. The environmental factors influencing the appearance of vertical migrations are the availability of food and predator activity, water temperature and hydrogen sulfide concentration. Experimentally recorded data on these factors have always some noise and are of a discrete selective nature. In this regard, machine learning tools are used to solve the problem. The most important stage of the work is the construction of the training sample. For this purpose, the results of an analytical and numerical search for the optimal behavior are used with linear-quadratic and hyperbolic approximations of external factors.

As a result of the study, a neural network was built that solves the problem of classifying input data sets into four classes corresponding to the presence or absence of significant vertical movements in young and adult individuals under these conditions. This makes it possible to fairly accurately recognize the presence or absence of significant vertical migrations for young and adult zooplankton individuals according to approximately specified environmental factors.

Keywords: Fitness function · Neural networks · Pattern recognition · Zooplankton diel vertical migration

1 Introduction

Currently, the importance of mathematical modeling is increasing in the studying the behavior of living systems [1–4]. In particular, modeling of zooplankton diel vertical migration (DVM) is of great interest [5]. This phenomenon was discover more than two hundred years ago [6]. It was established that zooplankton daily rises at night to the near-surface layers of water and descends into the depths during the day [7]. It is known that the daily migrations of zooplankton represent one of the most significant synchronous movements of biomass on earth and, as a result, affect the carbon exchange and climate of the planet [8–13]. The

D. Balandin et al. (Eds.): MMST 2022, CCIS 1750, pp. 41–54, 2022.
https://doi.org/10.1007/978-3-031-24145-1_4

presence of vertical migrations was tried to be explained by the influence of a predator, solar radiation, etc. [7]. However, attempts to explain this phenomenon on the basis of traditional approaches of biology have encountered significant difficulties [7]. This is due to the wide variety of zooplankton hereditary behavior modes. In particular, it is known that some species can carry out significant vertical migrations, while others cannot [7]. Moreover, different age groups of the same species may have different behaviors [14–16]. Then, it is necessary to use the methods of mathematical modeling [8, 10, 11, 17–20]. Mathematical modeling allows us to explain the quantitative characteristics of this behavior and the dependence of the implemented migration strategy on the age of the organism [21–23].

Now, the Darwinian idea of the survival of the fittest spicies is effectively used for modeling biological processes [24–26]. DVM of zooplankton are considered as an evolutionarily stable behavior, that is, the behavior that is preserved in the population as the result of the struggle for existence [27]. Evolutionarily stable behavior ensures maximum adaptability to the conditions of existence. In the mathematical modeling of such behavior, the main difficulty is to identify the fitness function, which numerically characterizes the competitive advantages of different strategies.

The problem of deriving evolutionary fitness is still far from a final solution. Different authors offer different definitions of fitness, for example, the expected individual reproductive value [7, 21], generalized entropy [28], etc. Different definitions of the fitness function sometimes lead to conflicting predictions of evolutionary results [21, 29]. One of the most general approaches to the formalization of the fitness function was proposed by A. Gorban [30–32], who defined fitness in the equations of the dynamics of measures with inheritance as the average time value of the specific reproduction rate. This approach was later developed in [33–35], where a technique for constructing a fitness function for wide classes of models was proposed [36].

Maximization of the fitness function by classical methods of calculus of variations makes it possible to construct an evolutionarily stable strategy for zooplankton migrations depending on environmental conditions [22, 29].

However, the experimentally recorded data of the external environment are always inaccurate and have a discrete selective character; only their estimates are available in a certain range. At the same time, the synchronous movement of a population always admits only a statistical description; inevitably there is a random variation in the behavior of individuals. From this point of view, it is of interest not to build an optimal movement strategy, but only to predict its main qualitative characteristics, such as the presence or absence of significant vertical movements. Classical methods for solving optimization problems are not suitable here. The problem is solved by using the technology of learning neural networks [37–39].

The purpose of this work is to predict the presence or absence of significant vertical migrations of zooplankton according to approximately given environmental factors, taking into account the age of the individual. In this work, the presence of significant migrations is predicted separately in young and adult individuals using a neural network.

2   Materials and Methods

We consider a community consisting of m different species of zooplankton

$$\{v_1, \ldots, v_m\}.$$

Each species of zooplankton has two age groups - juveniles and adult mature individuals, differing in the hereditary migration regime, which is a continuous periodic (with a period of one day) function of the vertical position x depending on the time of day τ. Here $\tau = 0$ corresponds to the noonday, $\tau = 0.5$ corresponds to the midnight, and $\tau = 1$ corresponds to the noon of the next day; x - depth of migration, measured in meters, level $x = 0$ corresponds to the water surface. The function $x_{1i}(\tau)$ corresponds to the mode of movement of young individuals, the function $x_{2i}(\tau)$ corresponds to the mode of movement of adults of the i-th zooplankton species. It is assumed that these functions are continuously differentiable on the interval $[0,1]$ and satisfy the conditions $x_{ji}(0) = x_{ji}(1)$, $i = \overline{1, m}$, $j = 1, 2$.

Let $z_{1i}(t)$ be the number of young individuals of the i-th zooplankton species, $z_{2i}(t)$ - the number of adults of the i-th zooplankton species. The following population model is used, which describes the interaction between young and adult individuals, taking into account their competition

$$
\begin{aligned}
z'_{1i} &= -p_i z_{1i} - q_i z_{1i} + r_i z_{2i} - z_{1i} \sum_{j=1}^{m} (z_{1j} + z_{2j}), \\
z'_{2i} &= p_i z_{1i} - s_i z_{2i} - z_{2i} \sum_{j=1}^{m} (z_{1j} + z_{2j}), \quad i = \overline{1, m}.
\end{aligned}
\tag{1}
$$

Here r_i is the breeding rate, q_i is the juvenile mortality rate, s_i is the adult mortality rate due to predation; p_i is the proportion of young individuals that have passed into a mature state (coefficient of maturation); it is assumed that these coefficients do not depend on time. The last term reflects the intraspecific competition. This model is consistent with the general approach to modeling population dynamics [40].

To construct the fitness function, the technique proposed in [36] are used. Its essence is as follows. Let

$$z_i = z_{1i} + z_{2i}$$

be the number of individuals of all ages of species i. It is known that $z_i \neq 0$ if $z_i(0) \neq 0$ [40,41]. Species i will be better (or fitter) than species j if the ratio of the number of j-th species to the number of i-th species tends to zero over time:

$$\lim_{t \to \infty} z_j / z_i = 0.$$

In this case, species i will displace species j from the community. Thus, an order relation (ranging) is introduced on the set of species. A numerical function $J(v_i)$ defined on this set that preserves the introduced relation (that is, $J(v_i) > J(v_j)$ when v_i is better than v_j) is called a fitness function.

As it was shown in [36], the fitness function has the form of the time-averaged specific growth rate for $z_i(t)$:

$$<z_i'(t)/z_i> = \lim_{T \to \infty} \frac{1}{T} \int_{t_0}^{T} z_i'(t)/z_i \, dt$$

or any other equivalent function.

Empirical evidence suggests that the diel vertical migrations of zooplankton are mainly determined by the following environmental factors: the distribution of food (phytoplankton) $E(x)$ by depth x, the distribution density of predators $S_x(x)$, the distribution of unfavorable habitat factors (temperature and hydrogen sulfide) $G(x)$, as well as daily predator activity $S_\tau(\tau)$ [7,15,21,36]. All these factors can be considered as continuous functions of the vertical coordinate x or the time of day τ.

To study vertical migrations, we used empirical data collected in the northeastern part of the Black Sea in summer of 2011 [15]. The data from [7,15,21,36] were used to describe environmental factors such as the depth distribution of predators and their activity during the day. The simplest linear and quadratic approximations of the functions of external factors E, S_x and G were used:

$$E = \sigma_1(x + D), \ S_x = \sigma_2(x + D), \ -D < x < 0; \ G = (x + D_0)^2, \qquad (2)$$

as well as the approximation of the function S_τ in the form of a sinusoidal dependence:

$$S_\tau = \cos(2\pi\tau) + 1, \ 0 < \tau < 1.$$

Here D is the depth of the hydrogen sulfide layer, the lower limit of zooplankton habitation, D_0 is the optimal depth of zooplankton habitation in terms of temperature, σ_1 and σ_2 are constant coefficients of decrease in the amount of food and predator with increasing depth.

We also used hyperbolic approximations of environmental factors:

$$\begin{aligned} E(x) &= \sigma_1(\tanh(\xi_1(x + D)) + 1), \\ S_x(x) &= \sigma_2(\tanh(\xi_2(x + D)) + 1), \\ G(x) &= \cosh(\xi_3(x + D_0)), \end{aligned} \qquad (3)$$

that are more precise. Here ξ_i is the growth rate of the function, the steepness of its slope at zero.

The primary analysis of the data made it possible to identify eight key macroparameters that have the greatest impact on the population growth of young ($j = 1$) and adult ($j = 2$) individuals: daily average energy gains as a result of food intake M_{1j}, energy costs for vertical movements M_{3j} (kinetic energy proportional to the square of the speed of movement), losses due to predation M_{2j}, losses due to unfavorable external conditions M_{4j}.

$$M_{1j}(v_i) = \int_0^1 E(x_{ji}(\tau))\, d\tau,$$

$$M_{2j}(v_i) = -\int_0^1 S_\tau(\tau) S_x(x_{ji}(\tau))\, d\tau,$$

$$M_{3j}(v_i) = -\int_0^1 (x'_{ji}(\tau))^2\, d\tau,$$

$$M_{4j}(v_i) = -\int_0^1 G(x_{ji}(\tau))\, d\tau.$$

Accordingly, the fitness function must be a function of eight key parameters $M = (M_{11}, \ldots, M_{41}, M_{12}, \ldots, M_{42})$: $J = J(M(v_i))$.

In model (1), the coefficients r_i, q_i, s_i, p_i are determined by these parameters. We will use the simplest linear approximations of the coefficients

$$r_i = \theta_{12}M_{12} + \theta_{32}M_{32} + \theta_{42}M_{42},$$
$$s_i = \theta_{22}M_{22},$$
$$p_i = \theta_{11}M_{11} + \theta_{31}M_{31} + \theta_{41}M_{41},$$
$$q_i = -\theta_{22}M_{22}.$$

Here, the weight coefficients θ_{kj} reflect the influence of each key factor and do not depend on the implemented strategies.

The hereditary behavioral strategies of young and adult individuals are approximated by harmonic oscillations $x_j = A_j + B_j \cos(2\pi\tau)$, where A_j is the average depth of immersion, B_j is the oscillation amplitude during the day, τ is the time of day, varying from 0 to 1, $j = 1$ corresponds to the strategy of young individuals, $j = 2$ corresponds to the strategy of adults. Notice, that

$$-D < A_j + B_j \cos(2\pi\tau) < 0, \quad -\frac{D}{2} < A_j < 0, \; 0 < B_j < \frac{D}{2}.$$

For given simplest linear and quadratic approximations of the behavior strategy and environmental factors, formulas for key parameters can be derived

$$M_{1j} = \sigma_1(A_j + D),$$
$$M_{2j} = -\sigma_2(A_j + D + \tfrac{B_j}{2}),$$
$$M_{3j} = -\pi^2 B_j^2,$$
$$M_{4j} = -((A_j + D_0)^2 + \tfrac{B_j}{2}).$$

3 Results

3.1 Fitness Identification

First of all, we obtain an explicit expression for the fitness function in the considered model (1) in terms of its coefficients. Let's make a change of variables in system (1): $\xi_i = z_{1i}/z_i$, $\eta_i = z_{2i}/z_i$. Obviously, $\eta_i = 1 - \xi_i$. Then system (1) is transformed to the form

$$\xi_i' = (r_i + q_i - s_i)\xi_i^2 - (p_i + q_i - s_i + 2r_i)\xi_i + r_i,$$

$$z_i' = z_i(-(r_i + q_i - s_i)\xi_i + r_i - s_i - \sum_{j=1}^{m} z_j), \quad i = \overline{1, m}.$$

It is clear that each of the first m equations of this system does not depend on the others and can be considered independently. Let $r_i + q_i - s_i \neq 0$, then the values ξ_i tend to

$$\xi_i^* = \frac{p_i + q_i - s_i + 2r_i - \sqrt{(p_i + q_i - s_i)^2 + 4p_i r_i}}{2(r_i + q_i - s_i)}.$$

Last m equations describe the dynamics of the total population of the species i, the reproductive factor for z_i is

$$z_i'(t)/z_i = -(r_i + q_i - s_i)\xi_i + r_i - s_i - \sum_{j=1}^{m} z_j.$$

Since the summand $\sum_{j=1}^{m} z_j$ does not depend on the index i and the mean time of ξ_i coincides with its limit ξ_i^*, then the best strategy i corresponds to the maximum of the following function

$$J = -(r_i + q_i - s_i)\xi_i^* + r_i - s_i = \frac{-p_i - q_i - s_i + \sqrt{(p_i + q_i - s_i)^2 + 4p_i r_i}}{2}.$$

This function is the fitness function in this model.

Knowing the fitness function, environmental functions, and coefficients of influence of environmental factors on model parameters, one can easily obtain optimal, i.e., evolutionarily stable behavioral strategies for young and adult individuals. To do this, we need to solve the problem of the calculus of variations, taking the fitness function as the objective functional. Then, by analyzing the obtained strategies, one can find the amplitudes of vertical displacements.

It can be seen that for different functions of external factors and for different weight coefficients, the oscillation amplitude will be different. In some cases, there will be noticeable vertical movements of young or adult zooplankton during the day, in other cases they will be practically indistinguishable against the background of constant random fluctuations. For the researcher, the most important thing is to answer the question not so much about the exact profile of the resulting movements, but about the presence or absence of significant vertical migrations of juveniles or adults.

To answer this question, we need to set the threshold value of the vertical movement, starting from which the movement is considered noticeable. Then separately compare the amplitudes of the obtained movements of individuals of different ages with a threshold value. If the amplitude exceeds the threshold value, then the movements will be significant. Then the sets of external factors can be divided into four classes, each of which corresponds to one possible case: the absence of significant movements of zooplankton, the presence of significant movements only in young individuals, the presence of significant movements only in adults, the presence of significant movements in zooplankton of any age.

Practically, it is very important to recognize the presence or absence of such movements by known external factors, even without an accurate determination of the optimal trajectory of movement. The significance of this problem is aggravated by the fact that our knowledge of external factors is always incomplete and approximate, and because of this, it is usually impossible to construct an exact solution.

3.2 Neural Network

The second part of the work consisted of the creation of a software package that makes it possible to recognize the presence or absence of significant migrations of zooplankton, using only approximate characteristics of the environment. This problem was solved using artificial neural networks and machine learning methods [42, 43].

A neural network was built using libraries Keras3 and Tensorflow2 based on Python. The input data is a discrete set of values of four external functions of the state of the environment, eight weight coefficients, as well as a threshold value of vertical displacements. The output information is a response regarding the presence/absence of pronounced vertical movements separately in juveniles and adults in given environmental conditions relative to a given threshold value. The architecture of the artificial neural network is shown in Fig. 1.

Each environmental function as a set of 104 values alternately passes through two pairs of convolution (conv) and pooling (maxpool) layers. This highlights information about the behavior of each of the functions in separate parts of the scope. Further, the received information passes through 5 fully connected layers of the neural network (dense). In parallel, a vector of weight coefficients of 8 values is introduced (4 for each age group). Each group of weight coefficients is supplemented with previous information obtained from the analysis of environmental functions, and independent processing of each group of data is carried out through two fully connected layers. Further, to study the possible mutual influence of one age group on another, all information is passed through another 3 fully connected layers. The result is 2 numbers that determine the presence/absence of vertical movements for each of the 2 age groups of zooplankton. The constructed classifier based on the neural network assigns each new set of environmental functions to one of four classes.

3.3 Construction of the Training Sample

An important point in the work is the construction of a training sample. For this, the analytical and numerical results of the approximation of environmental conditions with the help of linear-quadratic and hyperbolic functions were used.

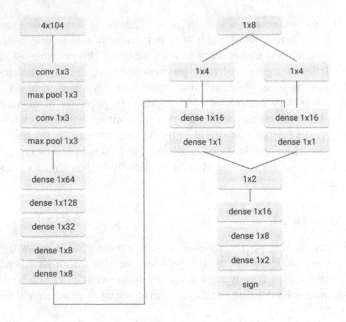

Fig. 1. Neural network architecture.

At the same time, the analytical functions were converted into tabular ones according to the discrete nature of the recorded environmental conditions.

The training sample is built as follows: a series of quadruples of initial functions are set in turn, and the oscillation amplitudes of young and adult individuals are found using optimization procedures [44]. The found values are compared with the threshold. If the amplitude value is higher than the threshold, then this precedent is considered as a case of the presence of a significant oscillation. If not, then the movement is considered not to be expressed against the background of the inevitable random fluctuations of zooplankton. Each four external factors is assigned a two-dimensional vector, each coordinate of which is a logical value "yes/no" - the presence/absence of significant migrations for the corresponding age group. Accordingly, the quadruples of external factors are divided into four non-overlapping classes. Fragments of the training sample are given in Tables 1, 2 for a threshold value of 10.

Table 1. Fragment of the training sample for linear and quadratic approximations (2).

D	D_0	σ_1	σ_2	θ_{11}	θ_{12}	θ_{31}	θ_{32}	θ_{21}	θ_{22}	θ_{41}	θ_{42}	DVM	
												Juv	Ad
140	44.3	8.61	7.53	16.8	23.7	0.16	0.04	4.94	1	0.67	0.81	Yes	Yes
140	37.8	1.54	8.99	59	29.8	0.027	0.02	3.38	0.84	0.61	0.67	Yes	No
140	29.3	1.21	5.49	31.8	72.7	0.034	0.03	2.74	0.09	0.26	0.68	No	Yes
140	37.8	78.2	1.95	3	24.5	32.8	0.04	0.03	2.95	0.26	0.38	No	No

Table 2. Fragment of the training sample for hyperbolic approximations (3).

D	D_0	σ_1	σ_2	θ_{11}	θ_{12}	θ_{31}	θ_{32}	θ_{21}	θ_{22}	θ_{41}	θ_{42}	ξ_1	ξ_2	ξ_3	DVM	
															Juv	Ad
140	66.5	2.68	4.06	100.7	61.2	0.05	0.07	4.81	4.83	0.27	0.23	1.6	1.6	1.2	Yes	Yes
140	64.7	1.62	6.24	84.2	39	0.02	0.009	4.67	3.82	0.64	0.49	1.04	8.52	4.67	Yes	No
140	44.2	2.34	6.92	8.5	27.2	0.009	0.045	0.148	0.89	0.32	0.23	1.12	7.76	6.86	No	Yes
140	114.7	1.56	5.81	63.5	93.7	0.04	0.03	1.51	1.82	0.37	0.65	1.39	9.18	6.72	No	No

We should note that hyperbolic functions approximate the real environment much more accurately then linear-quadratic functions. Then the training set should contain mainly hyperbolic approximations to train the neural network for the work with real data. Linear-quadratic approximations are used only as rare auxiliary cases. The power of the training sample for functions given by linear-quadratic approximations was 1200 generated pairs of elements; for functions defined by hyperbolic approximation – 40 000 generated pairs of elements. 70% of the received records were used to train the neural network, the remaining 30% were used to test its work. At the same time, the frequency of correct answers on the original functions (Fig. 2) was approximately 0.97.

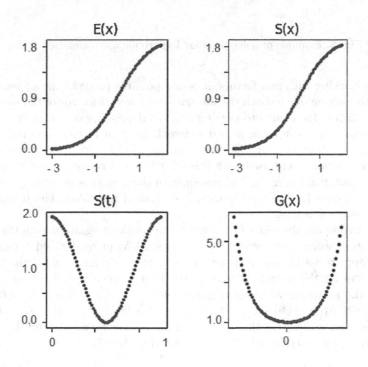

Fig. 2. Example of input data for hyperbolic approximation (3) with $\sigma_1 = \sigma_2 = 1$, $\xi_1 = \xi_2 = \xi_3 = 1$, $D = D_0 = 0$.

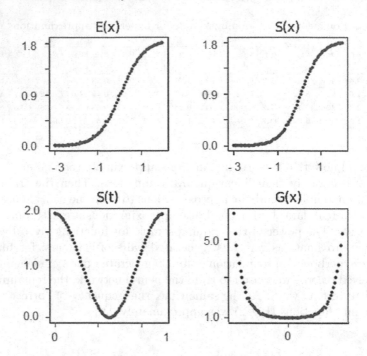

Fig. 3. Example of noisy data for hyperbolic approximation (3).

When working with real factors, it is not possible to avoid the appearance of noise. This may be due to both the human factor and the error of the tools used. Noise can distort the results to varying degrees depending on the circumstances. It is desirable to set up the neural network in such a way that noisy input data does not affect the correctness of the answer. As noisy data, the original analytical functions were used with the addition of a random noise value up to 5% of the main value (Fig. 3). The response of the neural network was compared with the response for the corresponding analytical functions. The frequency of correct answers was 0.94.

We can compare the result of the neural network recognition with the results of some other works. The work [45] presents DVM of adults and juveniles for different approximations of environmental factors. We have solved the recognition problem for the same functions on the base of the created neural network. We used the following values of fitness coefficients: $\theta_{11} = 0.16$, $\theta_{31} = 0.00007$, $\theta_{21} = 0.008$, $\theta_{41} = 0.0016$, $\theta_{12} = 0.6$, $\theta_{32} = 0.0000075$, $\theta_{22} = 0.4$, $\theta_{42} = 0.006$.

In this case, we have got the presence of migrations for adults and the absence of migrations for juveniles which are consisted with [45].

4   Summary

In this work, we develop a new method and software to explore evolutionary stable strategies of zooplankton DVM. The method combines the theoretical approach to reveal evolutionary fitness in the model and the computational tools. As a result of the study, the neural network was built. It solves the problem of classifying input data sets into four classes corresponding to the presence or absence of significant vertical movements in young and adult individuals. The important stage of the work is the construction of the training sample. For this purpose, the results of an analytical and numerical search for the optimal behavior are used with linear-quadratic and hyperbolic approximations of external factors.

The trained neural network fairly accurately recognizes the presence or absence of detectable vertical migrations of two age groups of zooplankton according to approximately specified observable information about environmental factors (distribution of food and predators in water layers, predator behavioral responses, temperature distribution of water). As future extensions, we are planning to analyse a large number of empirical cases of DVM and include more complicated theoretical models of zooplankton population growth to better train neural networks.

The results of the research were implemented into the educational process within the framework of the academic discipline "Mathematical modeling of selection processes" for 3rd year students of the studying area "Fundamental Informatics and Information Technology" [46,47].

The materials of the work formed the basis of the educational and research project of UNN students. Some results of the project are used in this work. We thank UNN students (Kolesnikov I. V., Ivina A. S., Feoktistov A. A., Yarov M. A.) for technical help of the work.

References

1. Stucchi, L., et al.: A general model of population dynamics accounting for multiple kinds of interaction. Complexity **2020**, 7961327 (2020). https://doi.org/10.1155/2020/7961327
2. Frisman, E.Y., Zhdanova, O.L., Kulakov, M.P., Neverova, G.P., Revutskaya, O.L.: Mathematical modeling of population dynamics based on recurrent equations: results and prospects. Part I. Biol. Bull. **48**(1), 1–15 (2021). https://doi.org/10.1134/S1062359021010064
3. McBride, J.M., Nimphius, S.: Biological system energy algorithm reflected in subsystem joint work distribution movement strategies: influence of strength and eccentric loading. Sci. Rep. **10**, 12052 (2020). https://doi.org/10.1038/s41598-020-68714-8
4. Demidova, A., Druzhinina, O., Jaćimović, M., Masina, O., Mijajlovic, N.: Problems of synthesis, analysis and optimization of parameters for multidimensional mathematical models of interconnected populations dynamics. In: Jaćimović, M., Khachay, M., Malkova, V., Posypkin, M. (eds.) OPTIMA 2019. CCIS, vol. 1145, pp. 56–71. Springer, Cham (2020). https://doi.org/10.1007/978-3-030-38603-0_5
5. Ringelberg, J.: Diel Vertical Migration of Zooplankton in Lakes and Oceans. Springer, Dordrecht (2010)
6. Bandara, K., et al.: Two hundred years of zooplankton vertical migration research. Biol. Rev. **96**(4), 1547–1589 (2021). https://doi.org/10.1111/brv.12715

7. Clark, C., Mangel, M.: Dynamic State Variable Models in Ecology: Methods and Applications. Oxford University Press, Oxford (2000)
8. Ducklow, H., Steinberg, D., Buesseler, K.: Upper ocean carbon export and the biological pump. Oceanography **14**(4), 50–58 (2001). https://doi.org/10.5670/oceanog.2001.06
9. Hays, G.: A review of the adaptive significance and ecosystem consequences of zooplankton diel vertical migrations. Hydrobiologia **503**(1–3), 163–170 (2003). https://doi.org/10.1023/B:HYDR.0000008476.23617.b0
10. Kaiser, M., et al.: Marine Ecology: Processes, Systems and Impacts. Oxford University Press, Oxford (2005)
11. Buesseler, K., et al.: Revisiting carbon flux through the ocean's twilight zone. Science **316**(5824), 567–570 (2007). https://doi.org/10.1126/science.1137959
12. Isla, A., Scharek, R., Latasa, M.: Zooplankton diel vertical migration and contribution to deep active carbon flux in the NW Mediterranean. J. Mar. Syst. **143**, 86–97 (2015). https://doi.org/10.1016/j.jmarsys.2014.10.017
13. Archibald, K.M., Siegel, D.A., Doney, S.C.: Modeling the impact of zooplankton diel vertical migration on the carbon export flux of the biological pump. Glob. Biogeochem. Cycles **33**, 181–199 (2019). https://doi.org/10.1029/2018GB005983
14. Ohman, M.D.: The demographic benefits of diel vertical migration by zooplankton. Ecol. Monogr. **60**(3), 257–281 (1990). https://doi.org/10.2307/1943058
15. Morozov, A., Kuzenkov, O., Arashkevich, E.: Modelling optimal behavioral strategies in structured populations using a novel theoretical framework. Sci. Rep. **9**, 15020 (2019). https://doi.org/10.1038/s41598-019-51310-w
16. Baumgartner, M.F., Tarrant, A.M.: The physiology and ecology of diapause in marine copepods. Annu. Rev. Mar. Sci. **9**, 387–411 (2017). https://doi.org/10.1146/annurev-marine-010816-060505
17. Hansen, A.N., Visser, A.W.: Carbon export by vertically migrating zooplankton: an optimal behavior model. Limnol. Oceanogr. **61**(2), 701–710 (2016). https://doi.org/10.1002/lno.10249
18. Arcifa, M.S., et al.: Microcrustaceans and predators: diel migration in a tropical lake and comparison with shallow warm lakes. Limnetica **35**(2), 281–296 (2016). https://doi.org/10.23818/limn.35.23
19. Hafker, N.S., Meyer, B., Last, K.S., Pond, D.W., Huppe, L., Teschke, M.: Circadian clock involvement in zooplankton diel vertical migration. Curr. Biol. **27**(14), 2194.e3–2201.e3 (2017). https://doi.org/10.1016/j.cub.2017.06.025
20. Guerra, D., Schroeder, K., Borghini, M., et al.: Zooplankton diel vertical migration in the Corsica channel (North-western Mediterranean sea) detected by a moored acoustic doppler current profiler. Ocean Sci. **15**(3), 631–649 (2019). https://doi.org/10.5194/os-15-631-2019
21. Fiksen, O., Giske, J.: Vertical distribution and population dynamics of copepods by dynamic optimization. ICESJ Mar. Sci. **52**, 483–503 (1995). https://doi.org/10.1016/1054-3139(95)80062-X
22. Morozov, A., Kuzenkov, O.: Towards developing a general framework for modelling vertical migration in zooplankton. J. Theor. Biol. **405**, 17–28 (2016)
23. Sandhu, S.K., Morozov, A., Kuzenkov, O.: Revealing evolutionarily optimal strategies in self-reproducing systems via a new computational approach. Bull. Math. Biol. **81**(11), 4701–4725 (2019). https://doi.org/10.1007/s11538-019-00663-4
24. Gyllenberg, M., Metz, J., Service, R.: When do optimisation arguments make evolutionary sense? In: Chalub, F., Rodrigues, J. (eds.) The Mathematics of Darwin's Legacy. MBI, pp. 233–268. Springer, Basel (2011). https://doi.org/10.1007/978-3-0348-0122-5_12

25. Birch, J.: Natural selection and the maximization of fitness. Biol. Rev. **91**(3), 712–727 (2016). https://doi.org/10.1111/brv.12190
26. Gavrilets, S.: Fitness Landscapes and the Origin of Species (MPB-41). Princeton University Press, Princeton (2004)
27. Wilfried, G., Bernhard, T.: Vertical migration of zooplankton as an evolutionarily stable strategy. Am. Nat. **132**(2), 199–216 (1988). https://doi.org/10.1086/284845
28. Klimenko, A.Y.: Entropy and equilibria in competitive systems. Entropy **16**(1), 1–22 (2014). https://doi.org/10.3390/e16010001
29. Parvinen, K., Dieckmann, U., Heino, M.: Function-valued adaptive dynamics and the calculus of variations. J. Math. Biol. **52**, 1–26 (2006). https://doi.org/10.1007/s00285-005-0329-3
30. Gorban, A.: Equilibrium Encircling. Equations of Chemical Kinetics and Their Thermodynamic Analysis. Nauka, Novosibirsk (1984). (in Russian)
31. Gorban, A.: Selection theorem for systems with inheritance. Math. Model. Nat. Phenom. **2**(4), 1–45 (2007). https://doi.org/10.1051/mmnp:2008024
32. Gorban, A.N.: Self-simplification in Darwin's systems. In: Gorban, A., Roose, D. (eds.) Coping with Complexity: Model Reduction and Data Analysis. LNCSE, vol. 75, pp. 311–344. Springer, Berlin, Heidelberg (2011). https://doi.org/10.1007/978-3-642-14941-2_17
33. Kuzenkov, O.: Investigation of a dynamical system of Radon probability measures. Differ. Equ. **31**(4), 549–554 (1995)
34. Kuzenkov, O., Ryabova, E.: Variational principle for self-replicating systems. Math. Model. Nat. Phenom. **10**(2), 115–129 (2015). https://doi.org/10.1051/mmnp/201510208
35. Kuzenkov, O., Novozhenin, A.: Optimal control of measure dynamics. Commun. Nonlinear Sci. Numer. Simul. **21**(1–3), 159–171 (2015). https://doi.org/10.1016/j.cnsns.2014.08.024
36. Kuzenkov, O., Morozov, A.: Towards the construction of a mathematically rigorous framework for the modelling of evolutionary fitness. Bull. Math. Biol. **81**(11), 4675–4700 (2019). https://doi.org/10.1007/s11538-019-00602-3
37. Abiodun, O.I., et al.: State-of-the-art in artificial neural network applications: a survey. Heliyon **4**(11), e00938 (2018). https://doi.org/10.1016/j.heliyon.2018.e00938
38. Kuzenkov, O., Morozov, A., Kuzenkova, G.: Recognition of patterns of optimal diel vertical migration of zooplankton using neural networks. In: International Joint Conference on Neural Networks, Budapest Hungary, IJCNN 2019 (2019). https://doi.org/10.1109/IJCNN.2019.8852060
39. Kuzenkov, O.: Information technologies of evolutionarily stable behavior recognition. Mod. Inf. Technol. IT Educ. **1201**, 250–257 (2020). https://doi.org/10.1007/978-3-030-46895-8-20
40. Kuzenkov, O., Ryabova, E.: Mathematical modeling of selection processes. Lobachevsky State University, Nizhnii Novgorod (2007). (in Russian)
41. Kuzenkov, O.A., Kuzenkova, G.V.: Optimal control of self-reproduction systems. J. Comput. Syst. Sci. Int. **51**(4), 500–511 (2012). https://doi.org/10.1134/S1064230712020074
42. Mohri, M., Rostamizadeh, A., Talwalkar, A.: Foundations of Machine Learning. The MIT Press, Cambridge (2012)
43. Bishop, Ch.M.: Pattern Recognition and Machine Learning. Springer, Heidelberg (2006)

44. Morozov, A.Y., Kuzenkov, O.A., Sandhu, S.K.: Global optimization in Hilbert spaces using the survival of the fittest algorithm. Commun. Nonlinear Sci. Numer. Simul. **103**, 106007 (2021). https://doi.org/10.1016/j.cnsns.2021.106007
45. Kuzenkov, O., Morozov, A., Kuzenkova, G.: Exploring evolutionary fitness in biological systems using machine learning methods. Entropy **23**(1), 35 (2021). https://doi.org/10.3390/e23010035
46. Kuzenkov, O., Kuzenkova, G., Kiseleva, T.: The use of electronic teaching tools in the modernization of the course "mathematical modeling of selection processes". Educ. Tech. Soc. **21**(1), 435–448 (2018). https://www.elibrary.ru/item.asp?id=32253185. (in Russian)
47. Kuzenkov, O., Kuzenkova, G., Kiseleva, T.: Computer support of training and research projects in the field of mathematical modeling of selection processes. Educ. Technol. Soc. **22**(1), 152–163 (2019). https://www.elibrary.ru/item.asp?id=37037790. (in Russian)

Nonintegrability of the Problem of Motion of an Ellipsoidal Body with a Fixed Point in a Flow of Particles

Alexander S. Kuleshov$^{(\boxtimes)}$ [ID] and Maxim M. Gadzhiev

Department of Mechanics and Mathematics, Lomonosov Moscow State University,
Moscow, Russia
kuleshov@mech.math.msu.su

Abstract. The problem of motion in the free molecular flow of particles of a rigid body with a fixed point, bounded by the surface of an ellipsoid of revolution is considered. This problem is similar in many aspects to the classical problem of motion of a heavy rigid body about a fixed point. In particular, this problem possesses the integrable cases, correspond to the classical Euler – Poinsot, Lagrange and Hess cases of integrability of equations of motion of a heavy rigid body with a fixed point. Equations of motion of the body in the flow of particles are presented in hamiltonian form. Using the theorem on the Liouville – type nonintegrability of Hamiltonian systems near elliptic equilibrium positions we present the necessary conditions for the existence in the considered problem of an additional analytic first integral independent of the energy integral. We proved that the obtained necessary conditions are not fulfilled for the rigid body with a mass distribution corresponding to the classical Kovalevskaya integrable case in the problem of motion of a heavy rigid body with a fixed point.

Keywords: Rigid body with a fixed point · Free molecular flow of particles · Hamiltonian system · Nonintegrability

1 Introduction. V. V. Kozlov's Theorem on the Nonexistence of Analytic First Integral Near the Equilibrium Position of Hamiltonian System

In 1976 V.V. Kozlov in his paper [1] (see also [2,3]), proved the theorem, which gives the sufficient conditions of the nonexistence for the Hamiltonian system the analytic with respect to canonical variables first integral, independent with Hamilton function H. Below we give the statement of the problem using the notations from [1] and the formulation of the corresponding theorem.

Let us consider the system of canonical equations

$$\frac{dx_i}{dt} = \frac{\partial H}{\partial y_i}, \quad \frac{dy_i}{dt} = -\frac{\partial H}{\partial x_i}, \quad i = 1, \ldots n, \quad n \geq 2 \tag{1}$$

This research was supported financially by the RFBR (grant no. 20-01-00637).

with the Hamilton function $H(y_1, \ldots, y_n, x_1, \ldots, x_n, \varepsilon)$, depending analytically on the variables $\boldsymbol{y} = (y_1, \ldots, y_n)$, $\boldsymbol{x} = (x_1, \ldots, x_n)$ and on the parameter ε, which takes values in some connected domain $D \in \mathbb{R}^r$. Suppose that for all ε the point $y_i = 0$, $x_i = 0$, $(i = 1, \ldots, n)$ be an equilibrium position of the system (1). In the vicinity of an equilibrium position $y_i = 0$, $x_i = 0$, $(i = 1, \ldots, n)$ the Hamilton function H can be represented as follows:

$$H = H^{(2)} + H^{(3)} + \cdots,$$

where $H^{(s)}$ is a homogeneous form of degree s with respect to $\boldsymbol{y} = (y_1, \ldots, y_n)$ and $\boldsymbol{x} = (x_1, \ldots, x_n)$. The coefficients of this expansion are analytic functions of the parameter ε. Let us assume that for all $\varepsilon \in D$ the frequencies of linear oscillations $\boldsymbol{\omega}(\varepsilon) = (\omega_1(\varepsilon), \ldots, \omega_n(\varepsilon))$ do not satisfy any resonant relation

$$(\boldsymbol{m} \cdot \boldsymbol{\omega}) = m_1 \omega_1 + \cdots + m_n \omega_n = 0$$

of order $|m_1| + \cdots + |m_n| \leq m - 1$. Using Birkhoff's normalization method (see, for example [4,5]), we can find a canonical transformation $(\boldsymbol{y}, \boldsymbol{x}) \to (\boldsymbol{p}, \boldsymbol{q})$, such that in the new variables

$$H^{(2)} = \frac{1}{2} \sum_{i=1}^{n} \omega_i \rho_i, \quad H^{(k)} = H^{(k)}(\rho_1, \ldots, \rho_n, \varepsilon), \quad k \leq m - 1,$$

where $\rho_i = p_i^2 + q_i^2$. The corresponding transformation is analytic in ε. Now we introduce the canonical action – angle variables $(\boldsymbol{I}, \boldsymbol{\varphi})$ by the formulas:

$$I_i = \frac{\rho_i}{2}, \quad \varphi_i = \arctan \frac{p_i}{q_i}, \quad (1 \leq i \leq n).$$

In the canonical variables $(\boldsymbol{I}, \boldsymbol{\varphi})$ we have

$$H = H^{(2)}(\boldsymbol{I}, \varepsilon) + \cdots + H^{(m-1)}(\boldsymbol{I}, \varepsilon) + H^{(m)}(\boldsymbol{I}, \boldsymbol{\varphi}, \varepsilon) + \cdots$$

We represent the trigonometric polynomial $H^{(m)}$ as a finite Fourier series

$$H^{(m)} = \sum h_{\boldsymbol{k}}^{(m)}(\boldsymbol{I}, \varepsilon) \exp(i(\boldsymbol{k} \cdot \boldsymbol{\varphi})).$$

Theorem 1 (V. V. Kozlov [1–3]). *Let $(\boldsymbol{k} \cdot \boldsymbol{\omega}(\varepsilon)) \not\equiv 0$ for all $\boldsymbol{k} \in \mathbb{Z}^n \backslash 0$. Suppose that for some $\varepsilon_0 \in D$ the resonant relation $(\boldsymbol{k}_0 \cdot \boldsymbol{\omega}(\varepsilon_0)) = 0$, $|\boldsymbol{k}_0| = m$ is satisfied and $h_{\boldsymbol{k}_0}^{(m)} \not\equiv 0$. Then the canonical Eqs. (1) with Hamilton function $H = \sum H^{(s)}$ do not have a complete set of (formal) integrals $F_j = \sum F_j^{(s)}$, whose quadratic terms $F_j^{(2)}(\boldsymbol{y}, \boldsymbol{x}, \varepsilon)$ are independent for all $\varepsilon \in D$.* $\square$

Remark 1. Note that under the assumptions of the V. V. Kozlov's Theorem 1 there may exist independent integrals with dependent (for certain values of ε)

quadratic parts of their Maclaurin expansions. Here is a simple example: the canonical equations with Hamilton function

$$H = \frac{1}{2} \left(x_1^2 + y_1^2 \right) + \frac{\alpha}{2} \left(x_2^2 + y_2^2 \right) + 2x_1 y_1 y_2 - x_2 y_1^2 + x_1^2 x_2$$

have a first integral

$$F = x_1^2 + y_1^2 + 2 \left(x_2^2 + y_2^2 \right).$$

For $\alpha = 2$, it is dependent on the quadratic form $H^{(2)}$. However, all conditions of the Theorem 1 are satisfied.   □

The advantage of the V. V. Kozlov's Theorem 1 consists in the absence of preliminary restrictive assumptions regarding the parameters of the system. This advantage substantially compensates for the fact that the additional integral must belong to the class of analytic functions, the quadratic part of which are functionally independent with the quadratic part of the Hamilton function.

V. V. Kozlov's Theorem 1 was successfully applied for proving the nonexistence of an additional first integral in the plane circular restricted three body problem [1–3]; for studying the integrability of the problem of motion about a fixed point of a dynamically symmetric rigid body with the center of mass lies in the equatorial plane of the ellipsoid of inertia [1,3,6]; for proving the nonexistence of an additional integral in the problem of motion of a plane heavy double pendulum [6–8]; for obtaining the necessary conditions for the existence of an additional first integral in the problem of motion of a dynamically symmetric ellipsoid on a smooth horizontal plane [9]; for the study of nonintegrability of the Kirchhoff equations of motion of a rigid body in a fluid [10,11].

In this paper V. V. Kozlov's Theorem 1 is used to derive necessary conditions for the existence of an additional analytic integral in the problem of motion in the flow of particles of a rigid body with a fixed point bounded by the surface of an ellipsoid of revolution.

2 Formulation of the Problem. Hamilton Function of the Problem

Equations of motion of a rigid body with a fixed point, bounded by the surface of an ellipsoid and exposed by the flow of particles, have the form [12,13]:

$$A_1 \dot{\omega}_1 + (A_3 - A_2) \omega_2 \omega_3 = \rho v_0^2 \pi a_1 a_2 a_3 \sqrt{\frac{\gamma_1^2}{a_1^2} + \frac{\gamma_2^2}{a_2^2} + \frac{\gamma_3^2}{a_3^2}} \left(h_2 \gamma_3 - h_3 \gamma_2 \right),$$

$$A_2 \dot{\omega}_2 + (A_1 - A_3) \omega_1 \omega_3 = \rho v_0^2 \pi a_1 a_2 a_3 \sqrt{\frac{\gamma_1^2}{a_1^2} + \frac{\gamma_2^2}{a_2^2} + \frac{\gamma_3^2}{a_3^2}} \left(h_3 \gamma_1 - h_1 \gamma_3 \right), \qquad (2)$$

$$A_3 \dot{\omega}_3 + (A_2 - A_1) \omega_1 \omega_2 = \rho v_0^2 \pi a_1 a_2 a_3 \sqrt{\frac{\gamma_1^2}{a_1^2} + \frac{\gamma_2^2}{a_2^2} + \frac{\gamma_3^2}{a_3^2}} \left(h_1 \gamma_2 - h_2 \gamma_1 \right);$$

$$\dot{\gamma}_1 = \omega_3 \gamma_2 - \omega_2 \gamma_3, \quad \dot{\gamma}_2 = \omega_1 \gamma_3 - \omega_3 \gamma_1, \quad \dot{\gamma}_3 = \omega_2 \gamma_1 - \omega_1 \gamma_2.$$

Here A_1, A_2, A_3 are the moments of inertia of the body about the principal axes of inertia $Ox_1x_2x_3$ with the origin at the fixed point O; $\boldsymbol{\omega} = (\omega_1, \omega_2, \omega_3)$ is the angular velocity vector of the body; $\boldsymbol{\gamma} = (\gamma_1, \gamma_2, \gamma_3)$ is the unit vector directed along the flow of particles; ρ is the constant density of the flow of particles; v_0 is the constant velocity of particles in the flow, a_1, a_2, a_3 are the lengths of the semiaxes of the ellipsoid, bounding a rigid body; $\boldsymbol{h} = (h_1, h_2, h_3)$ is the vector directed from a fixed point to the center of the ellipsoid bounding the rigid body.

For any values of parameters Eqs. (2) possess the first integrals:

$$J_1 = A_1\omega_1\gamma_1 + A_2\omega_2\gamma_2 + A_3\omega_3\gamma_3 = c_1 = \text{const}, \quad J_2 = \gamma_1^2 + \gamma_2^2 + \gamma_3^2 = 1. \quad (3)$$

Let us assume that the center of the ellipsoid lies on the first principal axis of inertia Ox_1 with the origin at the fixed point O, at a distance l from the fixed point. In other words, in the Eqs. (2) we put

$$h_1 = l, \quad h_2 = 0, \quad h_3 = 0.$$

We also assume that the ellipsoid bounding the rigid body is an ellipsoid of revolution with the axis of symmetry passing through the fixed point O. Therefore in the Eq. (2) we put

$$a_1 = b, \quad a_2 = a_3 = a.$$

In addition we assume, that the body is dynamically symmetric, and the axis of dynamical symmetry of the body does not coincide with the axis of symmetry of the ellipsoid, that bounds the body. In other words we assume, that

$$A_1 = A_2 = A, \quad A_3 = C.$$

Then the equations of motion in the flow of particles of a rigid body with a fixed point bounded by the surface of an ellipsoid of revolution will be rewritten as follows:

$$\begin{aligned}
&A\dot{\omega}_1 + (C - A)\,\omega_2\omega_3 = 0, \\
&A\dot{\omega}_2 + (A - C)\,\omega_1\omega_3 = -\rho v_0^2 \pi a^2 bl \sqrt{\frac{1 - \gamma_1^2}{a^2} + \frac{\gamma_1^2}{b^2}}\,\gamma_3, \\
&C\dot{\omega}_3 = \rho v_0^2 \pi a^2 bl \sqrt{\frac{1 - \gamma_1^2}{a^2} + \frac{\gamma_1^2}{b^2}}\,\gamma_2; \\
&\dot{\gamma}_1 = \omega_3\gamma_2 - \omega_2\gamma_3, \quad \dot{\gamma}_2 = \omega_1\gamma_3 - \omega_3\gamma_1, \quad \dot{\gamma}_3 = \omega_2\gamma_1 - \omega_1\gamma_2.
\end{aligned} \quad (4)$$

We multiply the first equation of system (4) by ω_1, the second—by ω_2, the third—by ω_3 and add them. As a result we get the following equation:

$$A\left(\omega_1\dot{\omega}_1 + \omega_2\dot{\omega}_2\right) + C\omega_3\dot{\omega}_3 = \rho v_0^2 \pi a^2 bl \sqrt{\frac{1 - \gamma_1^2}{a^2} + \frac{\gamma_1^2}{b^2}}\,(\omega_3\gamma_2 - \omega_2\gamma_3) = \rho v_0^2 \pi a^2 bl\dot{\gamma}_1 \sqrt{\frac{1 - \gamma_1^2}{a^2} + \frac{\gamma_1^2}{b^2}}.$$

Thus we can conclude that Eqs. (4) admit in addition to first integrals (3) the energy type first integral

$$H = \frac{A}{2}\left(\omega_1^2 + \omega_2^2\right) + \frac{C}{2}\omega_3^2 - G\left(\gamma_1\right) = h = \text{const.}$$

The function $G(\gamma_1)$ is written differently depending on whether the ellipsoid, bounding the rigid body is prolate $(b > a)$ or oblate $(a > b)$. For a prolate ellipsoid of revolution $(b > a)$, the function $G(\gamma_1)$ has the form:

$$G(\gamma_1) = \frac{\rho v_0^2 \pi a^2 b l}{2} \gamma_1 \sqrt{\frac{1 - \gamma_1^2}{a^2} + \frac{\gamma_1^2}{b^2}} + \frac{\rho v_0^2 \pi b l}{2\sqrt{\frac{1}{a^2} - \frac{1}{b^2}}} \arctan\left(\frac{\sqrt{\frac{1}{a^2} - \frac{1}{b^2}} \gamma_1}{\sqrt{\frac{1 - \gamma_1^2}{a^2} + \frac{\gamma_1^2}{b^2}}}\right).$$

For an oblate ellipsoid of revolution $(a > b)$, the function $G(\gamma_1)$ has the form:

$$G(\gamma_1) = \frac{\rho v_0^2 \pi a^2 b l}{2} \gamma_1 \sqrt{\frac{1 - \gamma_1^2}{a^2} + \frac{\gamma_1^2}{b^2}} + \frac{\rho v_0^2 \pi b l}{2\sqrt{\frac{1}{b^2} - \frac{1}{a^2}}} \ln\left(a\sqrt{\frac{1}{b^2} - \frac{1}{a^2}} \gamma_1 + a\sqrt{\frac{1 - \gamma_1^2}{a^2} + \frac{\gamma_1^2}{b^2}}\right).$$

Further we will consider the case of a prolate ellipsoid of revolution (the case of an oblate ellipsoid of revolution is considered in a similar way and gives the same result). As generalized coordinates in this problem we introduce the standard Euler angles θ, ψ and φ. Then we have

$$\gamma_1 = \sin\theta \sin\varphi, \quad \gamma_2 = \sin\theta \cos\varphi, \quad \gamma_3 = \cos\theta$$

and the Hamilton function of the problem in standard notations has the form:

$$H = \frac{1}{2}\left(\frac{p_\theta^2}{A} + \frac{p_\varphi^2}{C} + \frac{(p_\psi - p_\varphi \cos\theta)^2}{A\sin^2\theta}\right) - \frac{\rho v_0^2 \pi a^2 b l}{2}\sin\theta \sin\varphi\sqrt{\frac{1 - \sin^2\theta \sin^2\varphi}{a^2} + \frac{\sin^2\theta \sin^2\varphi}{b^2}} -$$

$$- \frac{\rho v_0^2 \pi b l}{2\sqrt{\frac{1}{a^2} - \frac{1}{b^2}}} \arctan\left(\frac{\sqrt{\frac{1}{a^2} - \frac{1}{b^2}}\sin\theta \sin\varphi}{\sqrt{\frac{1 - \sin^2\theta \sin^2\varphi}{a^2} + \frac{\sin^2\theta \sin^2\varphi}{b^2}}}\right).$$

$$(5)$$

Obviously, the Hamilton function H does not depend on the generalized coordinate ψ, that is the generalized momentum p_ψ is a constant. The generalized momentum p_ψ is the area integral J_1 (see (3)). The equations of motion of the body have a hamiltonian form with the Hamilton function (5), in which p_ψ is a parameter. We will assume that the parameter p_ψ is the parameter that was mentioned in the statement of the V. V. Kozlov's Theorem 1. Let us obtain the necessary conditions for the existence of an additional first integral, analytic in p_ψ and independent of the Hamilton function H.

3 Application of V. V. Kozlov's Theorem 1

For any value of p_ψ the point

$$(p_\theta, p_\varphi, \theta, \varphi) = \left(0, 0, \frac{\pi}{2}, \frac{\pi}{2}\right) -$$

is the equilibrium of the considered Hamiltonian system. We denote

$$p_\theta = y_1, \quad p_\varphi = y_2, \quad \theta = \frac{\pi}{2} + x_1, \quad \varphi = \frac{\pi}{2} + x_2.$$

The units of measurement can always be chosen so, that

$$\pi \rho v_0^2 l a^2 = 1, \quad A = 1.$$

We introduce also the following parameters:

$$p_\psi = \sqrt{x}, \quad \frac{1}{C} = y, \quad \frac{b^2}{a^2} = z.$$

Then (x, y, z) are change in the domain $\mathbb{R}_+^3 = \{x, y, z : x > 0, y > 0, z > 0\}$. In a neighborhood of the equilibrium point $y_1 = 0$, $y_2 = 0$, $x_1 = 0$, $x_2 = 0$ the expansion of the Hamilton function (5) has the form:

$$H = H^{(2)} + H^{(3)} + H^{(4)} + \cdots,$$

$$H^{(2)}(y_1, y_2, x_1, x_2) = \frac{1}{2}y_1^2 + \frac{y}{2}y_2^2 + \sqrt{x}x_1y_2 + \frac{(1+x)}{2}x_1^2 + \frac{1}{2}x_2^2,$$

$$H^{(3)}(y_1, y_2, x_1, x_2) = 0,$$

$$H^{(4)}(y_1, y_2, x_1, x_2) = \frac{1}{2}x_1^2y_2^2 + \frac{5}{6}\sqrt{x}x_1^3y_2 + \left(\frac{z}{4} - \frac{1}{2}\right)x_1^2x_2^2 + \left(\frac{x}{3} + \frac{z}{8} - \frac{1}{6}\right)x_1^4 + \left(\frac{z}{8} - \frac{1}{6}\right)x_2^4.$$

Note that in the case of $z = 1$, i.e. when the rigid body is bounded by the sphere, the expressions $H^{(2)}(y_1, y_2, x_1, x_2)$ and $H^{(4)}(y_1, y_2, x_1, x_2)$ exactly coincide with the corresponding expressions obtained by V. V. Kozlov [1–3] when studying the problem of motion of a heavy dynamically symmetric rigid body with a fixed point, with the center of mass situated in the equatorial plane of the ellipsoid of inertia.

Equations of motion of the system with the Hamilton function $H^{(2)}$ has the form of the linearized equations of the system, namely

$$\begin{pmatrix} \dot{p}_1 \\ \dot{p}_2 \\ \dot{q}_1 \\ \dot{q}_2 \end{pmatrix} = \begin{pmatrix} 0 & -\sqrt{x} & -(x+1) & 0 \\ 0 & 0 & 0 & -1 \\ 1 & 0 & 0 & 0 \\ 0 & y & \sqrt{x} & 0 \end{pmatrix} \begin{pmatrix} p_1 \\ p_2 \\ q_1 \\ q_2 \end{pmatrix}. \tag{6}$$

The characteristic equation for determining the natural frequencies of the linear system (6) with the Hamilton function $H = H^{(2)}$ is written as follows:

$$\lambda^4 + (1 + x + y)\lambda^2 + y(1 + x) - x = 0. \tag{7}$$

Obviously, the roots of the characteristic equation are purely imaginary if

$$y > \frac{x}{1+x}.$$

Let us denote by E the subset of $\mathbb{R}_+^2$, where this inequality is satisfied. The characteristic Eq. (7) is biquadratic, therefore, if the frequency ratio is three, then the ratio of the squares of the frequencies should be nine. Calculating the squares of the frequencies and equating their ratio to nine, we obtain the following condition for the parameters x and y:

$$4\left(1 + x + y\right) = 5\sqrt{1 + 6x - 2y + x^2 - 2xy + y^2}. \tag{8}$$

Therefore, squaring both sides of this equation and subtracting the left side from the right side, we find that the ratio of the frequencies $\lambda_1/\lambda_2 = 3$ if the parameters x and y are connected by the following equation

$$9x^2 - 82xy + 9y^2 + 118x - 82y + 9 = 0. \tag{9}$$

This is the equation of a hyperbola; for $x > 0$ and $y > 0$ its branches are entirely in E.

From the triangle inequality for the moments of inertia $(A_1 + A_2 \geq A_3)$ it follows, that $y \geq 1/2$. For any fixed $y_0 \geq 1/2$, there exists $x_0 > 0$, such that the point (x_0, y_0) satisfies Eq. (9). Consider a small interval (a, b) of variation of the parameter x, including the point x_0. For $x \in (a, b)$, $y = y_0$ the roots of the characteristic equation are purely imaginary and distinct. When $x = x_0$, then the frequencies λ_1 and λ_2 are connected by the equation $\lambda_1 - 3\lambda_2 = 0$. It remains to find out, when the secular coefficient $h_{1,-3}^{(4)}$ is zero.

To calculate the coefficient $h_{1,-3}^{(4)}$ let us make the canonical change of variables $(y_1, y_2, x_1, x_2) \to (p_1, p_2, q_1, q_2)$ such, that in the new variables the quadratic part $H^{(2)}$ of the Hamilton function H is represented in the form:

$$H^{(2)} = \frac{B_1}{2}p_1^2 + \frac{K_1}{2}q_1^2 + \frac{B_2}{2}p_2^2 + \frac{K_2}{2}q_2^2,$$

where B_i and K_i, $(i = 1, 2)$ are coefficients to be determined.

The required change of variables in linear with respect to the variables p_1, p_2, q_1, q_2. Let us represent it in the most general form, namely:

$$y_1 = \alpha_1 p_1 + \beta_1 p_2 + \xi_1 q_1 + \eta_1 q_2, \quad y_2 = \alpha_2 p_1 + \beta_2 p_2 + \xi_2 q_1 + \eta_2 q_2,$$

$$x_1 = \alpha_3 p_1 + \beta_3 p_2 + \xi_3 q_1 + \eta_3 q_2, \quad x_2 = \alpha_4 p_1 + \beta_4 p_2 + \xi_4 q_1 + \eta_4 q_2. \tag{10}$$

This change of variables must satisfy two properties:

1. it should be a canonical transformation;
2. in the new variables the expression $H^{(2)}$ do not contain the mixed products $p_1 p_2$, $p_1 q_1$, $p_1 q_2$, $p_2 q_1$, $p_2 q_2$, $q_1 q_2$.

Using the standard condition of the canonicity of the change of variables in the Hamiltonian system (see, for example, [14, 15])

$$p_1 dq_1 + p_2 dq_2 - y_1 dx_1 - y_2 dx_2 = -dF$$

it can be shown that a linear change of variables (10) will be canonical transformation if the following conditions are satisfied:

$$\beta_1\alpha_3 + \beta_2\alpha_4 - \beta_3\alpha_1 - \beta_4\alpha_2 = 0, \quad \xi_1\alpha_3 + \xi_2\alpha_4 - \xi_3\alpha_1 - \xi_4\alpha_2 + 1 = 0,$$

$$\eta_1\alpha_3 + \eta_2\alpha_4 - \eta_3\alpha_1 - \eta_4\alpha_2 = 0, \quad \xi_1\beta_3 + \xi_2\beta_4 - \xi_3\beta_1 - \xi_4\beta_2 = 0, \qquad (11)$$

$$\eta_1\beta_3 + \eta_2\beta_4 - \eta_3\beta_1 - \eta_4\beta_2 + 1 = 0, \quad \eta_1\xi_3 + \eta_2\xi_4 - \eta_3\xi_1 - \eta_4\xi_2 = 0.$$

In addition to these six equations, we should write down the condition for the vanishing of the coefficients of the mixed terms in the Hamilton function $H^{(2)}$, written in the variables p_1, p_2, q_1, q_2 (there also be six such mixed members: p_1p_2, p_1q_1, p_1q_2, p_2q_1, p_2q_2, q_1q_2). These conditions are as follows:

$$\xi_1\eta_1 + \xi_3\eta_3 + \xi_4\eta_4 + \sqrt{x}\left(\xi_2\eta_3 + \xi_3\eta_2\right) + x\xi_3\eta_3 + y\xi_2\eta_2 = 0,$$

$$\alpha_1\xi_1 + \alpha_3\xi_3 + \alpha_4\xi_4 + \sqrt{x}\left(\alpha_2\xi_3 + \alpha_3\xi_2\right) + x\alpha_3\xi_3 + y\alpha_2\xi_2 = 0,$$

$$\beta_1\xi_1 + \beta_3\xi_3 + \beta_4\xi_4 + \sqrt{x}\left(\beta_2\xi_3 + \beta_3\xi_2\right) + x\beta_3\xi_3 + y\beta_2\xi_2 = 0,$$

$$\alpha_1\eta_1 + \alpha_3\eta_3 + \alpha_4\eta_4 + \sqrt{x}\left(\alpha_2\eta_3 + \alpha_3\eta_2\right) + x\alpha_3\eta_3 + y\alpha_2\eta_2 = 0, \qquad (12)$$

$$\alpha_1\beta_1 + \alpha_3\beta_3 + \alpha_4\beta_4 + \sqrt{x}\left(\alpha_2\beta_3 + \alpha_3\beta_2\right) + x\alpha_3\beta_3 + y\alpha_2\beta_2 = 0,$$

$$\beta_1\eta_1 + \beta_3\eta_3 + \beta_4\eta_4 + \sqrt{x}\left(\beta_2\eta_3 + \beta_3\eta_2\right) + x\beta_3\eta_3 + y\beta_2\eta_2 = 0.$$

Thus we have 12 Eqs. (11)–(12) on the 16 unknown coefficients α_i, β_i, ξ_i and η_i, $i = 1, \ldots, 4$. In order for the number of equations to be equal to the number of unknown coefficients, we assume from the very beginning that

$$\beta_1 = 0, \quad \alpha_2 = 0, \quad \eta_3 = 0, \quad \xi_4 = 0.$$

The solution of the obtained system of 12 Eqs. (11)–(12) with respect to 12 unknown coefficients α_1, α_3, α_4, β_2, β_3, β_4, ξ_1, ξ_2, ξ_3 and η_1, η_2, η_4 was found using the software for symbolic computations MAPLE 7. It turned out, that the solution has the form:

$$\xi_1 = 0, \quad \eta_2 = 0, \quad \alpha_3 = 0, \quad \beta_4 = 0, \quad \xi_2 = \Delta\xi_3,$$

$$\alpha_1 = \frac{\sqrt{x}}{\xi_3\left(2\sqrt{x} + (y-1-x)\Delta\right)}, \quad \eta_1 = \frac{\Delta\sqrt{x}}{\beta_2\left(2\sqrt{x} + (y-1-x)\Delta\right)},$$

$$\beta_3 = -\frac{\beta_2\left(\sqrt{x} + (y-1-x)\Delta\right)}{\Delta\sqrt{x}}, \quad \alpha_4 = -\frac{\sqrt{x} + (y-1-x)\Delta}{\xi_3\Delta\left(2\sqrt{x} + (y-1-x)\Delta\right)},$$

$$\eta_4 = \frac{\sqrt{x}}{\beta_2\left(2\sqrt{x} + (y-1-x)\Delta\right)},$$

where β_2 and ξ_3 are free parameters, and Δ is the positive root of the quadratic equation:

$$\sqrt{x}\Delta^2 + (x + 1 - y)\Delta - \sqrt{x} = 0$$

We will assume that the free parameters take the following values:

$$\beta_2 = \frac{\Delta\sqrt{x}}{(\sqrt{x} + (y - 1 - x)\Delta)}, \quad \xi_3 = 1.$$

For these values of the free parameters, the linear canonical transformation $(y_1, y_2, x_1, x_2) \rightarrow (p_1, p_2, q_1, q_2)$ takes the most simple form

$$y_1 = \frac{1}{1 + \Delta^2}p_1 + \frac{\Delta^2}{1 + \Delta^2}q_2, \quad y_2 = \frac{1}{\Delta}p_2 + \Delta q_1, \quad x_1 = q_1 - p_2, \quad x_2 = \frac{\Delta}{1 + \Delta^2}(q_2 - p_1)$$

The quadratic part $H^{(2)}$ of the Hamilton function H is represented as follows:

$$H^{(2)} = \frac{B_1}{2}p_1^2 + \frac{K_1}{2}q_1^2 + \frac{B_2}{2}p_2^2 + \frac{K_2}{2}q_2^2,$$

$$B_1 = \frac{1}{1 + \Delta^2}, \quad B_2 = \frac{y - 2\Delta\sqrt{x} + (1 + x)\Delta^2}{\Delta^2} = \frac{(1 + \Delta^2)(y - \sqrt{x}\Delta)}{\Delta^2},$$

$$K_1 = \Delta^2 y + 2\Delta\sqrt{x} + 1 + x = (1 + \Delta^2)\left(y + \frac{\sqrt{x}}{\Delta}\right), \quad K_2 = \frac{\Delta^2}{1 + \Delta^2}.$$

Now we introduce action – angle variables $(\boldsymbol{I}, \boldsymbol{\varphi})$ by the formulas:

$$q_1 = i\sqrt{\frac{I_1}{2}}\sqrt{\frac{B_1}{K_1}}\left(\exp(-i\varphi_1) - \exp(i\varphi_1)\right), \quad p_1 = \sqrt{\frac{I_1}{2}}\sqrt{\frac{K_1}{B_1}}\left(\exp(i\varphi_1) + \exp(-i\varphi_1)\right),$$

$$q_2 = i\sqrt{\frac{I_2}{2}}\sqrt{\frac{B_2}{K_2}}\left(\exp(-i\varphi_2) - \exp(i\varphi_2)\right), \quad p_2 = \sqrt{\frac{I_2}{2}}\sqrt{\frac{K_2}{B_2}}\left(\exp(i\varphi_2) + \exp(-i\varphi_2)\right).$$

Here i is the unit imaginary number. In the new variables the form $H^{(4)}$ will be written as follows:

$$H^{(4)} = \sum_{0 \leq |m_1| + |m_2| \leq 4} h^{(4)}_{m_1,m_2} \exp(i(m_1\varphi_1 + m_2\varphi_2)).$$

Let us calculate now the coefficient $h^{(4)}_{1,-3}$ explicitly. Note, that the exponent $\exp(i(\varphi_1 - 3\varphi_2))$ can only appear in the following expressions: $p_1p_2^3$, $p_1p_2^2q_2$, $p_1p_2q_2^2$, $p_1q_2^3$, $q_1p_2^3$, $q_1p_2^2q_2$, $q_1p_2q_2^2$, $q_1q_2^3$.

This remark greatly simplifies the process of calculating the coefficient $h^{(4)}_{1,-3}$. The condition for this coefficient to be zero can be written as follows:

$$5\sqrt{x}\Delta^3 + (3z + 8x - 10)\Delta^2 + 3(z - 7)\sqrt{x}\Delta + 6y - 3zy + 6 =$$
$$= ((4 - 3z)(y - \Delta\sqrt{x}) + 3(z - 2)\Delta^2)\sqrt{xy + y - x}.$$

(13)

Further simplifications of the Eq. (13) are based on the Eqs. (8)–(9) and also on the equations

$$\sqrt{xy + y - x} = \frac{3}{10}\left(1 + x + y\right), \quad \Delta = \frac{9y - x - 1}{10\sqrt{x}},$$

which can be derived by direct calculations from Eqs. (8)–(9) and from the definition of the parameter Δ.

Finally, the condition for vanishing of the coefficient $h_{1,-3}^{(4)}$ in the expansion of the function $H^{(4)}$ can be reduced to the following form:

$$27x^3z + 111x^2yz - 159xy^2z - 243y^3z - 9x^3 - 617x^2y - 39x^2z + 2093xy^2 - 118xyz + 1701y^3 +$$

$$+621y^2z + 653x^2 - 4374xy - 59xz - 2727y^2 - 129yz + 2633x + 543y + 7z - 29 = 0.$$

$$(14)$$

Thus, the following theorem is valid.

Theorem 2. *Necessary conditions for the existence of an additional integral, analytic in canonical variables and the parameter x and independent with the Hamilton function H, in the problem of motion in the flow of particles of a dynamically symmetric rigid body with a fixed point, bounded by the surface of an ellipsoid of revolution, whose center lies in the equatorial plane of the ellipsoid of inertia, have the form of Eqs. (9), (14).*

Remark 2. For $z = 1$ i.e. in the case when the rigid body is bounded by a sphere, the conditions (9), (14) take the form

$$9x^2 - 82xy + 9y^2 + 118x - 82y + 9 = 0, \tag{15}$$

$$18x^3 - 506x^2y + 1934xy^2 + 1458y^3 + 614x^2 - 4492xy - 2106y^2 + 2574x + 414y - 22 = 0, \tag{16}$$

and coincide with the necessary conditions for the existence of an additional integral in the problem of motion of a heavy dynamically symmetric rigid body with a fixed point and with the center of mass situated in the equatorial plane of the ellipsoid of inertia, obtained by V. V. Kozlov [1–3,6]. Algebraic curves (15) and (16) intersect at two points (x, y):

$$\left(\frac{4}{3}, 1\right) \quad \text{and} \quad (7, 2),$$

which correspond to the Lagrange integrable case $(A = C)$ and Kovalevskaya integrable case $(A = 2C)$. $\qquad\square$

Let us put in the conditions (9), (14) $y = 2$, i.e. consider a rigid body with the mass distribution corresponding to the Kovalevskaya integrable case in the problem of motion of a heavy rigid body with a fixed point. Then the condition (9) takes the form:

$$(9x + 17)\left(x - 7\right) = 0,$$

and can only be valid if $x = 7$. Substituting the values $x = 7$ and $y = 2$ into the condition (14) gives

$$12000\,(z - 1) = 0.$$

Thus, for a rigid body with a mass distribution corresponding to the Kovalevskaya case, an additional first integral, independent of the energy integral, can exist only when the rigid body is bounded by a sphere. In the case, when a rigid body, exposed by the flow of particles, is bounded by the ellipsoid, there is no additional first integral.

Analysis of Eqs. (9), (14), performed using MAPLE 7 symbolic computations software, shows that this system has solutions

$$x = 0,\ y = \frac{1}{9};\quad x = -\frac{16}{3},\ y = 1;\quad x = \frac{4}{3},\ y = 1. \tag{17}$$

existing for any value of the parameter z. The first two of the solutions (17) do not satisfy the conditions

$$x > 0,\quad y \ge \frac{1}{2}$$

and therefore they have no physical meaning. As for the third solution, it corresponds to the Lagrange integrable case ($A = C$). Thus, in this problem, for any shape of the ellipsoid (both when it is prolate and when it is oblate), there is an integrable case, corresponding to the Lagrange case.

In addition to the three solutions (17), Eqs. (9), (14) admit a z - dependent solution, in which y is a root of the quadratic equation with coefficients, depending on z, and x is expressed in terms of y and z:

$$(3z - 4)\,(7z - 52)\,y^2 - \left(76z^2 - 632z + 736\right) y + 20z^2 - 432z + 592 = 0,$$

$$x = \frac{\left(4048z^2 - 471z^3 - 3200 - 2672z\right) y + 3252z^2 - 54z^3 - 17424z + 18816}{2\,(3z - 4)\,(7z - 52)\,((23z - 32)\,y - 38z + 56)}.$$

Among the parameters (x, y, z) that belong to this solution, one can find such parameters, that have a physical meaning. These are, for example, the parameters

$$x = \frac{57}{23},\ y = \frac{30}{23},\ z = \frac{1}{5}.$$

Thus for some values of parameters, the necessary conditions for the existence of an additional first integral in the problem of motion of a rigid body with a fixed point in the flow of particles are satisfied. The study of existence of an additional first integral for such values of parameters is a problem, which we will try to investigate in the future.

4 Conclusions

In this paper we presented necessary conditions for the existence of an additional analytic first integral independent of the energy integral in the problem of motion of a rigid body with a fixed point in the flow of particles. The obtained necessary conditions is always fulfilled in the case of motion of a dynamically symmetric rigid body with the center of mass situated on the axis of dynamical symmetry of the body (the case similar to the Lagrange integrable case of the classical problem of motion of a heavy rigid body with a fixed point) and these conditions is not fulfilled for the dynamically symmetric rigid body with the center of mass situated in the equatorial plane of the ellipsoid of inertia (the mass distribution similar to the Kovalevskaya integrable case in the classical problem of motion of a heavy rigid body with a fixed point). Thereby we proved the nonexistence of the integrable case similar to the Kovalevskaya integrable case in the problem of motion in the flow of particles of a rigid body with a fixed point.

References

1. Kozlov, V.V.: The nonexistence of analytic integrals near equilibrium positions of Hamiltonian systems. Vestnik Moskov. Univ. Ser. 1. Maths. Mechs. (1), 110–115 (1976). (in Russian)
2. Kozlov, V.V.: Integrability and non-integrability in Hamiltonian mechanics. Russ. Math. Surv. **38**(1), 1–76 (1983)
3. Kozlov, V.V.: Symmetries, Topology and Resonances in Hamiltonian Mechanics. Springer, Heidelberg (1996)
4. Birkhoff, G.: Dynamical Systems, vol. IX. American Mathematical Society Colloquium Publications, Providence (1927)
5. Deprit, A., Henrard, J., Price, J.F., Rom, A.: Birkhoff's normalization. Celest. Mech. **1**(2), 222–251 (1969)
6. Guseva, N.A., Kuleshov, A.S.: On the application of V.V. Kozlov's theorem to prove the nonexistence of analytic integrals for some problems in mechanics. In: Problems in the Investigation of the Stability and Stabilization of Motion, pp. 84–109. Ross. Akad. Nauk, Vychisl. Tsentr im A. A. Dorodnitsyna, Moscow (2009). (in Russian)
7. Burov, A.A.: On the non - existence of a supplementary integral in the problem of a heavy two - link plane pendulum. J. Appl. Math. Mech. **50**(1), 123–125 (1986)
8. Burov, A.A., Nechaev, A.N.: On nonintegrability in the problem of the motion of a heavy double pendulum. In: Problems in the Investigation of the Stability and Stabilization of Motion, pp. 128–135. Ross. Akad. Nauk, Vychisl. Tsentr im A. A. Dorodnitsyna, Moscow (2002). (in Russian)
9. Ivochkin, M.Y.: The necessary conditions of the existence of an additional integral in the problem of the motion of a heavy ellipsoid on a smooth horizontal plane. J. Appl. Math. Mech. **73**(5), 616–620 (2009)
10. Kozlov, V.V., Onishchenko, D.A.: Nonintegrability of Kirchhoff's equations. Soviet Math. Dokl. **26**(2), 495–498 (1983)
11. Onishchenko, D.A.: Normalization of canonical systems depending on a parameter. Vestnik Moskov. Univ. Ser. 1. Math. Mech. (3), 78–81 (1982). (in Russian)
12. Gadzhiev, M.M., Kuleshov, A.S.: On the motion of a rigid body with a fixed point in a flow of particles. Moscow Univ. Mechs. Bull. **77**(3), 75–86 (2022)
13. Kuleshov, A.S., Gadzhiev, M.M.: Problem of motion of a rigid body with a fixed point in a particle flow. Vest. St. Petersburg Univ. Math. **55**(3), 353–360 (2022)
14. Vilasi, G.: Hamiltonian Dynamics. World Scientific, Singapore (2001)
15. Dittrich, W., Reuter, M.: Classical and Quantum Dynamics. Springer, Heidelberg (2017)

Numerical and Analytical Investigation of the Dynamics of a Body Under the Action of a Periodic Piecewise Constant External Force

Irina V. Nikiforova$^{(\boxtimes)}$ ⓘ, Vladimir S. Metrikin ⓘ, and Leonid A. Igumnov ⓘ

Lobachevsky University, Gagarin Avenue, 23 bld,
603022 Nizhny Novgorod, Russia
`tsii@list.ru`

Abstract. The paper investigates the dynamics of a body under the action of a piecewise constant periodic force with an arbitrary duty cycle and an oscillation limiter. Analytical relations for point mappings are presented for the first time using the Poincaré point mapping method. These relations allow one to study arbitrarily complex periodic motions both with a finite and infinite number of fixed points on the Poincaré surfaces. As a result, exact equations are presented in an analytical form that determine in the parameter space the existence and stability domains of periodic motions with an infinite number of fixed points on the Poincaré surfaces. The constructed with the help of a software product developed in the C++ language, bifurcation diagrams demonstrate, for some parameter values, the existence of chaotic regimes of body motion. Thus, the scenario for chaos origin is given. The comparison of numerical and analytical calculations is presented for different sets of parameters of the dynamical system.

Keywords: Non-linear dynamics · Point mapping method · Chaotic regimes

1 Introduction

Mechanisms whose motion is accompanied by impact interaction of the mechanisms components are widely used in practice. It turned out that the efficiency of many technological processes associated with the use of vibrational actions can be increased significantly by including collisions in the operating regime (vibroplatforms for consolidation of concrete mixtures, vibrating tools, vibrotransporting devices, etc.). There is another important reason for the keen interest in studying vibroimpact systems. The matter is that the enhanced specific speed of various machines and their sophisticated structure leads inevitably to

Supported by organization National Research Lobachevsky State University of Nizhny Novgorod.

an increase in dynamic stresses at the junctions of their units. This makes it necessary to allow for unwanted side effects caused by arising collisions. The overwhelming majority of the problems describing the impact interaction directly includes Newton's concept of the proportional dependence between relative pre- and post-impact velocities of translational bodies. It is assume that the proportionality factor depends on the property of the impactors and is independent of the impact velocity. The most interesting dynamic feature of vibroimpact systems, which can be detected using Newton's impact model, is the existence of an infinite impact process in a finite time interval. It is known that in collisional systems, motions with any finite number of impacts per period are possible theoretically. Therefore, studies of finite-impact and infinite-impact periodic motion regimes [1–4], as well as the conditions of creation of chaotic motions [5] are of great interest in the vibration engineering problems and the theory of vibroimpact systems. The conditions for existence of the above-specified motions are studied, as a rule, within an assumption of the harmonic character of the external force [5–8] using approximate consideration. In this case, the equations for the boundaries of existence of infinite-impact periodic motions with different multiplicity in the parameter space are presented mainly as approximate relationships [4,5]. For example, the authors of [9–15] consider a non-autonomous system of bouncing balls, which consists of a mass point in a constant gravity field, which bounces inelastically on a flat vibration bed. It is proved that there exists a finite time, within which the solution converges exactly on the equilibrium state. The top boundary of the gravitation time is presented. In this work, we study the body dynamics affected by a piecewise constant periodic force with a random duty factor and an oscillation limiter. Similar real technological processes, such as vibrotamping, pile driving, soil compaction in cramped production conditions, are widely used in many fields of science and technology [16–18]. The mathematical model is a strongly nonlinear dynamic system with a phase space truncated with respect to the phase coordinate. The mathematical apparatus of the method of point mapping of Poincare surfaces is used for the first time to present analytical relationships for point maps, which allow one to study arbitrary complex motions, both periodical, with a finite number of stationary points on Poincare surfaces, and with an infinite number of stationary points. As a result, exact equations, which determine the regions of existence and stability of periodic motions with an infinite number of stationary points on Poincare surfaces, are presented analytically. The specially developed C++ language code is used to construct bifurcation diagrams that demonstrate the existence of chaotic body motion regimes at certain values of the parameters. The scenario of chaos origination is described. The results of comparing the numerical calculations with analytical data for different sets of the parameters of the dynamic system are presented.

2 Problem Setting

The dimensionless equations of motion of a body hitting a fixed plane with a velocity recovery factor in the event of an impact $R \in [0,1]$ (according to Newton's assumption) can be written as

$$\dot{x} = y, \dot{y} = wf(t) - 1(x > 0), y_+ = -Ry_-(x = 0). \qquad (1)$$

Here, $\dot{x} = \frac{dx}{dt}, \dot{y} = \frac{dy}{dt}$, y_- and y_+ are the pre- and postimpact velocities of the particle, w is the overload parameter, and the pulsed external periodic force affecting the particle is specified by the relationship

$$f(t) = \begin{cases} 1, 0 \leq t \leq \gamma \\ -\gamma(2 - \gamma)^{-1}, \gamma < t < 2 \end{cases} \qquad (2)$$

within the determination period [0,2) and has the zero average component.

Equation (1) describe the dynamics of various vibro-impact systems (vibro-impact driving of piles, various medical devices, etc.), in particular, for the scheme shown in Fig. 1, [19] and Fig. 2, [20].

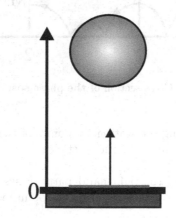

Fig. 1. Ball bouncing on an actuated surface.

3 Solution Method

The phase space x, y, t of system (1) is three-dimensional, cylindrical with respect to t, and truncated with respect to x (Fig. 3 presents the cross section of the phase space).

Therefore, if the mapping T_{j_1} of the points $M_0(x_0, y_0)$ on the plane $t = 0$ to the points $M_1(x_1, y_1)$ on the plane $t = \gamma$ and the mapping T_{j_2} of the points M_1 to the points $M_2(x_2, y_2)$ on the plane $t = 2$ are known, the motions of system (1)

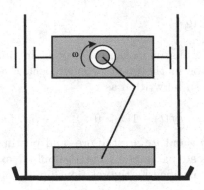

Fig. 2. Scheme of a single-piston vibro-impact mechanism.

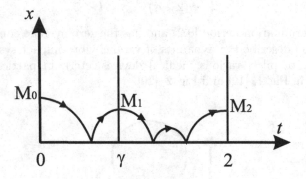

Fig. 3. Cross section of the phase space.

can be studied by considering the stationary points of the mapping $T = T_{j_1} T_{j_2}$, where j_1 and j_2 are the numbers of the particle's impact on the plane in the intervals $0 \leq t < \gamma$ and $\gamma < t < 2$.

Using the invariability of the right-hand parts of system (1) in each of the above-specified intervals, one can easily obtain the point mapping in the following form:

T_{j_1} at $j_1 = 0$

$$y_1 = y_0 + (w - 1)\gamma; x_1 = x_0 + \gamma y_0 + (w - 1)\frac{\gamma^2}{2}; (x_1 \geq 0); \qquad (3)$$

T_{j_1} at $j_1 = 1$

$$
\begin{aligned}
&y_1 = R\eta_0 + (w - 1)(\gamma - \tau); \eta_0 = (y^2 - 2(w - 1)x_0)^{1/2}, \\
&x_1 = (R\eta_0 + 0,5(w - 1)(\gamma - \tau))(\gamma - \tau), (x_1 \geq 0), \\
&\tau = \frac{y_0 + \eta_0}{1 - w}, (0 \leq \tau \leq \gamma).
\end{aligned}
\qquad (4)
$$

T_{j_2} at $j_2 = 0$

$$x_2 = x_1 + (2 - \gamma)y_1 - 0,5(\gamma(w - 1) + 2)(2 - \gamma);$$
$$y_2 = y_1 - (w - 1)\gamma - 2, (x_2 \geq 0). \tag{5}$$

T_{j_2} at $j_2 = m \geq 1$

$$y_2 = R^m \eta_1 - (2 - \gamma - \tau_m)(\gamma w(2 - \gamma)^{-1} + 1);$$
$$x_2 = (R^m \eta_1 - 0,5(2 - \gamma - \tau_m))(\gamma w(2 - \gamma)^{-1} + 1)(2 - \gamma - \tau_m);$$
$$\tau_m = (2 - \gamma)(\gamma(w - 1) + 2)^{-1}(y_1 + (1 - R)^{-1}\eta_1(1 + R - 2R^m)); \tag{6}$$
$$\eta_1 = (y_1^2 + 2(\gamma w(2 - \gamma)^{-1} + 1)x_1)^{1/2}; (0 \leq \tau_m \leq 2 - \gamma, x_2 \geq 0).$$

The typical feature of the simplest finite-impact and infinite-impact particle motions with the period being an n-multiple $(n = 1, 2, ...)$ of the period of the external force is a nonimpact pass during the $(n - 1)$th period and an impact process of the interaction with the plane for the nth period of the external force.

Using relationships (3) and (5), one can easily see that the coordinates $x_{2,n-1}, y_{2,n-1}$ of the point $M_{2,n-1} = T^{n-1}M_0$ are determined by the formulas

$$y_{2,n-1} = y_0 - 2(n - 1);$$
$$x_{2,n-1} = x_0 + 2(n - 1)y_0 + (n - 1)(\gamma w - 2(n - 1)). \tag{7}$$

The qualitative form of the regions of existence and stability of the periodic motions for $n = 1, 2, 3$ on the plane w, R at various values of the duty factor γ is shown in Fig. 4 and Fig. 5.

The simplest infinite-impact n-fold periodic motion regime is characterized by a non-impact passage of the particle over the plane in the time interval $(2(n - 1) + \gamma)$ and an infinite-impact exhaustive process during the second half-interval of the nth period of the external force. In this case, either the development of the process $(\tau_{m=\infty} \geq 0)$, or its disappearance $(\tau_{m=\infty} \leq 2 - \gamma)$ correspond to the boundaries of the region of regime existence. Assuming that $x_0 = y_0 = 0$ in (7), we get the values $x_{2,n-1} = y_{2,n-1}$, which we substitute into (3) as x_0, y_0 and find the values of x_1, y_1. Then, substituting into (6) the found values x_1, y_1 and $m = \infty$, we find that for the indicated boundary processes

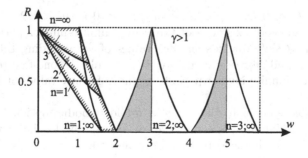

Fig. 4. Regions of existence and stability of the periodic motions for $\gamma > 1$.

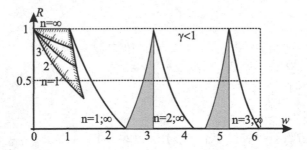

Fig. 5. Regions of existence and stability of the periodic motions for $\gamma < 1$.

$$\tau_\infty = \frac{2-\gamma}{\gamma(w-1)+2}(y_1 + \frac{1+R}{1-R}(y_1^2 + 2(\frac{\gamma w}{2-\gamma}+1)x_1)^{1/2}). \qquad (8)$$

Here,

$$y_1 = (w-1)\gamma - 2(n-1);$$
$$x_1 = 0,5y_1(\gamma + 2(n-1)). \qquad (9)$$

Allowing for (8) and (9), the condition $\tau_\infty \geq 0$ takes on the form

$$w \geq w_1 = 1 + 2(n-1)\gamma^{-1}, (y_1 \geq 0); \qquad (10)$$

and the condition $\tau_\infty \leq 2 - \gamma$ will look as

$$w \leq w_2 = \left(2n-2+\gamma+\left((2n-2+\gamma)^2+8n(2-\gamma)\left(\frac{1-R}{1+R}\right)^2\right)^{1/2}\right)(2\gamma)^{-1}; \quad (11)$$

Inequalities (10) and (11) allow one to estimate the influence of the parameter γ on the size and shape of the regime existence region with sufficient ease.

One can easily check that at any admissible values of γ the value $\Delta(R,\gamma) = w_2 - w_1 \geq 0$, and $\Delta(R=1,\gamma) = \Delta(R,\gamma=2) = 0$, while at $n \to \infty$

$$\Delta \to \Delta_\infty = (2-\gamma)(1-R)^2(1+R)^{-2}\gamma^{-1}. \qquad (12)$$

Since $\partial w_1/\partial\gamma < 0$ at $n \geq 2$ and $\partial w_2/\partial\gamma < 0$ at $n = 1, 2, ...$, then, as γ increases, the regions of existence of the n-fold infinite-impact periodic regime on the plane w, R shift towards smaller values of w, excluding the stationary boundary $w = 1$, and decrease in size. At $\gamma \to 0$, only the one-fold regime exists of the considered infinite-impact periodic regimes for any values of R and finite $w \geq 1$.

At $\gamma \to 2$ (as opposed to the one-fold n-fold periodic motions) the regions of existence of the infinite-impact regime disappear due to the merging of the boundaries w_2, w_1 with the corresponding intervals $w = n, 0 \leq R \leq 1(n = 1, 2, ...)$.

From (10) and (11), at $\gamma \to 0, w \to \infty$ and $w\gamma = C = const$ we obtain the asymptotic representation of the boundaries w_2, w_1 in the form

$$C_1 = 2(n-1);$$
$$C_2 = (n-1) + ((n-1)^2 + 4n(1-R)^2(1+R)^{-2})^{1/2}).$$

(13)

In this case, on the plane C, R we have a denumerable number of successive regions of existence of the infinite-impact periodic regime, which correspond to different multiplicity of the motion.

The points $C = n, R = 0$ are common for the neighboring regions (Fig. 6).

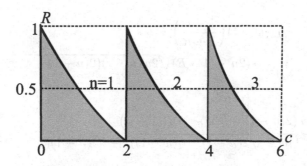

Fig. 6. Regions of existence of the infinite-impact periodic regime.

The infinite-impact n-fold periodic motion of a particle with a non-impact passage during the $(n-1)$ period of the external force, one impact in the interval $2(n-1) \leq t \leq 2(n-1) + \gamma$, and an infinite-impact exhaustive process in the time interval $2(n-1) + \gamma \leq t \leq 2(n-1)$ according to (4), (6), and (7) exists, if the inequalities

$$0 \leq \tau^{'} \leq \gamma; 0 \leq \tau^{'}_{\infty} \leq 2 - \gamma.$$

(14)

are fulfilled, as well as the conditions $x_0 = y_0 = 0, x_2 = y_2 = 0$ of the motion periodicity. In (14),

$$\tau^{'} = (1-w)^{-1}(y_{2,n-1} + (y_{2,n-1}^2 - 2(w-1)x_{2,n-1})^{1/2});$$

(15)

$$\tau^{'}_{\infty} = \frac{2-\gamma}{\gamma(w-1)+2}\left(y_1 + \frac{1+R}{1-R}\left(y_1^2 + 2\left(\frac{\gamma w}{2-\gamma}+1\right)x_1\right)^{1/2}\right).$$

(16)

In this case, the coordinates $x_{2,n-1}, y_{2,n-1}$ and x_1, y_1 satisfy the system

$$y_{2,n-1} = -2(n-1); x_{2,n-1} = (\gamma w - 2(n-1))(n-1);$$

(17)

$$y_1 = R(y_{2,n-1}^2 - 2(w-1)x_{2,n-1})^{1/2} + (w-1)(\gamma - \tau^{'});$$
$$x_1 = (R(y_{2,n-1}^2 - 2(w-1)x_{2,n-1})^{1/2}) + 0,5(w-1)(\gamma - \tau^{'}))(\gamma - \tau^{'}).$$

(18)

From (15) and (16), one can easily make sure that inequalities (14) take one, respectively, the form

$$2(n-1)\gamma^{-1} \le w \le w_1, \tag{19}$$

$$w_3 \le w \le w_1, \tag{20}$$

where w_1 is determined by relationship (10), and w_3 at $\gamma, R = const$ is the maximum, and smaller than w_1, root of the equation

$$H_1(w) = H_2(w), \tag{21}$$

in which

$$H_1(w) = (2-\gamma)\ (w-1)\left(\frac{1-R}{1+R}\right)^2$$
$$\cdot (2n - (1+R)\sqrt{2w(n-1)(2(n-1) - \gamma(w-1))})^2; \tag{22}$$

$$H_2(w) = 2w(2(n-1) - \gamma(w-1))(((1+R)$$
$$\cdot \sqrt{2(n-1)w} - \sqrt{2(n-1) - \gamma(w-1)})^2 - R^2(n-1)(2 + \gamma(w-1))). \tag{23}$$

It should be noted that $H_2(w = w_1) < H_1(w = w_1)$, and at $w = 2(n-1)\gamma^{-1}$ $H_2 > H_1$. Therefore, the value of $w_3 > 2(n-1)\gamma^{-1}$ and the region of existence of the considered infinite-impact periodic motion is determined by the inequality

$$w_3 \le w \le w_1. \tag{24}$$

The properties of the boundary $w = w_1$ have been considered above. Let us study the properties of the boundary $w = w_3$.

It follows from (21) that at $R = 1, w_3 = w_1$ and $\partial w_3/\partial R = 0$, i.e., the boundary $w = w_3$ on the plane w, R passes through the point $(w_1, 1)$ having the vertical tangent.

In the specific case at $\gamma \to 0, w \to \infty$ and $\gamma w = C = const$ $w_3 \to w_1$ according to Eqs. (20)–(23). Therefore, the region of existence of the considered infinite-impact regime collapses to zero.

At $R = 0$, the partitioning of the parameter plane w, γ into the regions of existence of single-impact periodic regimes, both with instantaneous, and long-term stops [9] can be performed using the previously found relationships and the method of constructing the formulas for point mapping T in a period of the external force. For example, the regions of existence of the single-impact periodic motion with instantaneous stops are determined by the inequality

$$\frac{2}{\gamma} \le w \le 2, (1 \le \gamma \le 2), \tag{25}$$

and the regions of existence of single-impact periodic motions with higher multiplicity do not exist.

According to (10) and (11), the region of existence of a single-fold periodic motion with a long-term stop is determined by the inequality

$$1 \leq w \leq \frac{2}{\gamma}(0 < \gamma \leq 1). \tag{26}$$

It follows directly from conditions (25) and (26) that the left-hand boundary of the region of existence of the single-impact motion on the plane w, γ is a part of the right-hand boundary of the region of existence of the single-fold periodic motion (Fig. 7).

The equations of the boundaries, at which the single-fold regime with an instantaneous stop and the single-fold regime with a long-term stop are transformed, as they cross this boundary, into the regime with a long-term stop and an additional instantaneous impact are determined, respectively, by the relationships

$$w = 2; 1 \leq \gamma \leq 2; \tag{27}$$

$$w = 2/\gamma; 0 < \gamma \leq 1. \tag{28}$$

As the parameter w increases, the above-stated two-fold periodic regime is transformed continuously into the simplest two-fold regime. The boundary of disappearance of the latter, according to (11), is determined as follows:

$$w = \frac{4}{\gamma}. \tag{29}$$

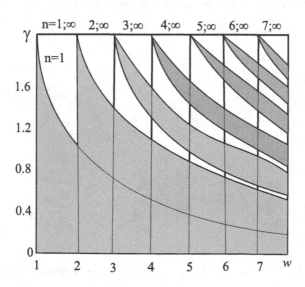

Fig. 7. Regions of existence of the single-impact motion with instantaneous stops and the periodic motion with long-term stops for various n.

The regions of existence of the simplest periodic motions and motions with an additional instantaneous impact having the multiplicity $n \geq 3$, according to Eqs. (10), (11), and (24), are specified by the inequalities

$$\gamma^* \leq \gamma \leq \frac{2n}{w}, (w \geq n). \tag{30}$$

Here,

$$\gamma^* = 2(n-1)\ (w-1)^{-1} + 2(n-w)((n-2)^2(w-1)w)^{-1}$$
$$\cdot (2w(n-1) - n^2 - 2((n-1)(w-n)(w(n-1)-n))^{1/2}). \tag{31}$$

It follows from the analysis of relationships (30) that the regions of existence of the above-specified periodic regimes do not intersect on the plane w, γ and at $\gamma = const$, the motions with great multiplicity can be realized only at great values of w.

4   Numerical Study of the Dynamics of a Body

Bifurcation diagrams for the overload parameter w for different values of the velocity recovery factor during impact R and the duty cycle parameter γ were obtained using the original software. In the Figs. 8 9, 10, 11, 12, 13, 14 and 15 the values of the overload parameter are plotted along the abscissa axis, and the values of the particle velocities on the Poincaré surfaces at the moments $\tau = \gamma$ and $\tau = 2$ are plotted along the ordinate axis, when constructing a point mapping of the Poincaré surfaces into themselves (3)–(6).

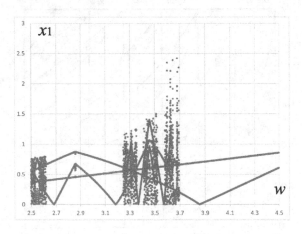

Fig. 8. Bifurcation diagrams for the parameter on the Poincaré surface $\tau = \gamma$ at $R = 0.3; \gamma = 0.7$.

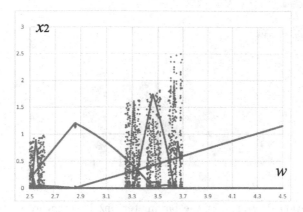

Fig. 9. Bifurcation diagrams for the parameter on the Poincaré surface $\tau = 2$ at $R = 0.3; \gamma = 0.7$.

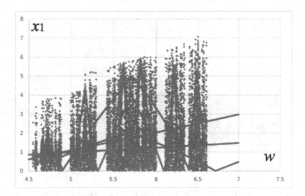

Fig. 10. Bifurcation diagrams for the parameter on the Poincaré surface $\tau = \gamma$ at $R = 0.3; \gamma = 0.7$.

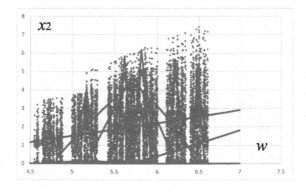

Fig. 11. Bifurcation diagrams for the parameter on the Poincaré surface $\tau = 2$ at $R = 0.3; \gamma = 0.7$.

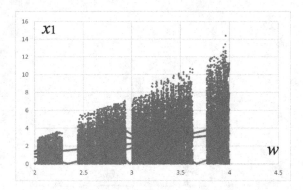

Fig. 12. Bifurcation diagrams for the parameter on the Poincaré surface $\tau = \gamma$ at $R = 0.3; \gamma = 1.5$.

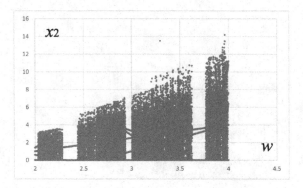

Fig. 13. Bifurcation diagrams for the parameter on the Poincaré surface $\tau = 2$ at $R = 0.3; \gamma = 1.5$.

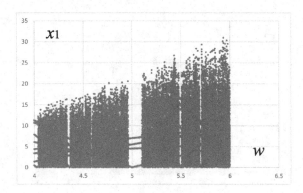

Fig. 14. Bifurcation diagrams for the parameter on the Poincaré surface $\tau = \gamma$ at $R = 0.3; \gamma = 1.5$.

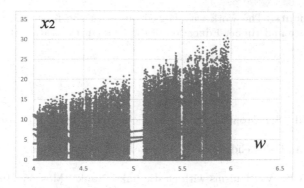

Fig. 15. Bifurcation diagrams for the parameter on the Poincaré surface $\tau = 2$ at $R = 0.3; \gamma = 1.5$.

Figures 8, 9, 10, 11, 12, 13, 14 and 15 for $R = 0,3; \gamma = 0.7$ show that the intervals of existence of periodic motions (with respect to the parameter w) with a finite number of impacts ($j_1 = 1, j_2 \geq 1$) belong to the segments $w_1 \in [2.6; 3.3], w_2 \in [3.7; 4.5]$ and with an increase in the overload parameter, the sizes of the segments of the existence of periodic motions decrease significantly. In this case, the size of the intervals in terms of the duty cycle parameter of the existence of chaotic motions increases significantly.

5   Conclusion

- The equations of point mappings of Poincare surfaces in a mathematical model describing the dynamics of a particle colliding with a fixed plane under the action of a periodic piecewise continuous force with an arbitrary duty cycle are given. This type of periodic force covers various symmetrical and asymmetric effects, including a function of the form $\delta(t)$.
- Analytical boundary equations for the domains of existence of periodic motions with an infinite and finite number of particle impacts on a fixed plane are obtained for the first time. It should be noted that such motion modes are effectively used in the technological process of compaction of various media (soil, sand, cement, etc.).
- The developed software product in the C++ high-level programming language made it possible to present bifurcation diagrams that illustrate the scenario chaotic motions origin, and the bifurcation parameters - the birth of motions with a finite and infinite number of particle impacts.
- The given in the paper methodology of studying the particle dynamics using the mathematical apparatus of the point mappings method has shown its extreme efficiency.
- Bifurcation diagrams made it possible to trace the influence of the off-duty and overload parameters on the scenario of transition to chaotic particle motions.

Acknowledgements. The work was carried out with the financial support of the Ministry of Science and Higher Education of the Russian Federation (task 0729-2020-0054).

References

1. Blekhman, I.I.: Synchronization of dynamic systems. Fizmatgiz, Moscow (1971). (in Russian)
2. Blekhman, I.I., Dzhanelidze, Yu.G.: Vibratory movement. Nauka, Moscow (1964). (in Russian)
3. Kobrinsky, A.E.: Mechanisms with elastic links. Nauka, Moscow (1964). (in Russian)
4. Kobrinsky, A.E., Kobrinsky, A.A.: Vibroimpact systems (dynamics and stability). Nauka, Moscow (1973). (in Russian)
5. Ksendzov, A.A., Nagaev, R.F.: Infinite-impact periodic modes in the problem of vibrotransport with tossing. Izv.AN SSSR, MTT, no. 5, pp. 29–35 (1971)
6. Nagaev, R.F.: General problem of quasi-plastic impact. Izv. USSR Academy of Sciences, MTT, no. 3. pp. 94–103 (1971)
7. Feigin, M.I.: Forced oscillations of systems with discontinuous nonlinearities. Nauka, Moscow (1994). (in Russian)
8. Zheleztsov, N.A.: The method of point transformations and the problem of forced oscillations of an oscillator with combined friction. Prikl. Mat. I Mekh. **13**(1), 30–40 (1949)
9. Leine, R.I., Heimsch, T.F.: Global uniform symptotic attractive stability of the non-autonomous bouncing ball system. Phys. D **241**, 2029–2041 (2012)
10. Luck, J.M., Mehta, A.: Bouncing ball with a finite restitution: chattering, locking, and chaos. Phys. Rev. E **48**(5), 3988–3997 (1993)
11. Everson, R.M.: Chaotic dynamics of a bouncing ball. Phys. D **19**(3), 355–383 (1986)
12. Giusepponi, S., Marchesoni, F.: The chattering dynamics of an ideal bouncing ball. Europhys. Lett. **64**, 36–42 (2003)
13. Klages, R., Barna, I.F., Matyas, L.: Spiral modes in the diffusion of a granular particle on a vibrating surface. Phys. Lett. A **333**, 79–84 (2002)
14. Or, Y., Teel, A.R.: Zeno stability of the set-valued bouncing ball. IEEE Trans. Autom. Control **56**(2), 447–452 (2011)
15. Biemond, J.B., van de Wouw, N., Heemels, W.P.M.H., Sanfelice, R.G., Nijmeijer, H.: Tracking control of mechanical systems with a unilateral position constraint inducing dissipative impacts. In: 51st IEEE Conference on Decision and Control, 10–13 December 2012, Maui, Hawaii, USA, pp. 4223–4228 (2012)
16. Nikiforova, I.V., Metrikin, V.S., Igumnov, L.A.: Mathematical modeling of multidimensional strongly nonlinear dynamic systems. In: Balandin, D., Barkalov, K., Gergel, V., Meyerov, I. (eds.) MMST 2020. CCIS, vol. 1413, pp. 63–76. Springer, Cham (2021). https://doi.org/10.1007/978-3-030-78759-2_5
17. Pavlovskaia, E., Hendry, D.C., Wiercigroch, M.: Modelling of high frequency vibroimpact drilling. Int. J. Mech. Sci. **91**, 110–119 (2015)
18. Woo, K.-C., Rodger, A.A., Neilson, R.D., Wiercigroch, M.: Application of the harmonic balance method to ground moling machines operating in periodic regimes. Chaos Solitons Fract. **11**(15), 2515 (2000)

19. Naldi, R., Sanfelice, R.G.: Passivity-based control for hybrid systems with applications to mechanical systems exhibiting impacts. Automatica **49**, 1104–1116 (2013). https://doi.org/10.1016/j.automatica.2013.01.018
20. Igumnov L.A., Metrikin, V.S., Nikiforova, I.V.: The dynamics of eccentric vibration mechanism (part 1). J. Vibroeng. **19**(7) (2017). ISSN 1392-8716

Motif of Two Coupled Phase Equations with Inhibitory Couplings as a Simple Model of the Half-Center Oscillator

Artyom Emelin[ID], Alexander Korotkov[ID], Tatiana Levanova[(✉)][ID], and Grigory Osipov[ID]

Lobachevsky University, Nizhny Novgorod 603950, Russia
tatiana.levanova@itmm.unn.ru

Abstract. We propose a new simple model of the half-center oscillator (HCO) consists of two oscillatory neurons interacting via the inhibitory coupling. We found the regions of dynamics, typical for central pattern generators, in the parameter space of the model. Various bifurcation transitions between all these states are in the focus of the proposed study.

Keywords: Adler equation · Chemical synaptic coupling · Half-center oscillator · In-phase and anti-phase synchronization · Bifurcations

1 Introduction

Locomotor dynamics of animals is based on the rhythmic limb movements. It is well known that the basic rhythmic pattern of flexion-extension alternation can be generated in absence of any inputs by neural circuits known as central pattern generators (CPGs) [3,9,10]. It is widely accepted that such rhythmic activity involves reciprocal inhibitory couplings between neuronal ensembles. Nevertheless, the precise topology of the CPG circuits in many animals and in humans, as well as the mechanisms of rhythmogenesis and control of locomotor pattern, are not fully understood [2]. A large amount of experimental studies devoted to the organization of CPGs in experimental models allowed to formulate several general concepts of rhythm generation.

The most widely used hypothesis on the organization of locomotor CPG is based on the classical half-center model that was proposed by Brown [3] and further elaborated by others [11,16]. This concept suggests a quasi-symmetric organization of two half-center oscillators. Also, mutual inhibition for flexor-extensor alternation is critical for rhythmogenesis. A great amount of data collected from experimental studies also support the half-center hypothesis [5–18].

In this study we propose a new simple model of the half-center oscillator, which consists of two identical neurons coupled by chemical inhibitory synapses. The proposed mathematical model is described by two phase oscillators, each of them, without coupling, demonstrates regular oscillatory dynamics (spiking). We use the phase oscillator as a single unit because reproduction of temporal patterns, not the dynamics of an individual neuron, plays a crucial role [17] in the

D. Balandin et al. (Eds.): MMST 2022, CCIS 1750, pp. 82–94, 2022.
https://doi.org/10.1007/978-3-031-24145-1_7

paradigm of the half-centers. This approach pays tribute to the early modeling of animal locomotor CPG [4,6]. The elements are inhibitory coupled.

The paper is organized as follows. In Sect. 2 we introduce a model of the half-center oscillator which consists of two phase oscillators with inhibitory couplings. We investigate in Sect. 3 temporal patterns that can be observed in the system under study depending on different values of control parameters. Bifurcation transitions between these patters are also in the focus of the study. In Sect. 4 we discuss our findings, before we draw our conclusions.

2   The Model

Let us consider the model of a minimal neural ensemble consisting of two neuron-like elements coupled by inhibitory synaptic couplings as shown in the Fig. 1.

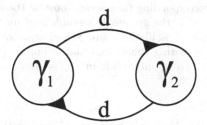

Fig. 1. Model of two oscillatory neurons with mutual inhibitory couplings.

To describe an individual element of an ensemble, we use the Adler equation [1]:

$$\dot{\phi} = \gamma - \sin\phi, \tag{1}$$

where the variable ϕ corresponds to the phase of an individual element, and γ is a parameter that determines the type of neuron behavior. For example, for $\gamma < 1$ in the phase space of the system, which is a unit circle, there are two equilibria: stable and unstable ones, which corresponds to the unexcited state of the neuron (Fig. 2(a)). When $\gamma = 1$, a saddle-node bifurcation occurs: both equilibria merge into one. In this case a neuron still remains unexcited, but now it can generate a single response on the external stimulus (Fig. 2(b)). Finally, for $\gamma > 1$, there are no equilibria in the phase space of the system, due to which the phase point begins to move counterclockwise along this circle. In this case if γ is slightly greater than 1, e.g. $\gamma = 1.01$, the neuron begins to generate spiking activity (Fig. 2(c)). Also, it should be noted, that (1) can be transformed to theta-neuron equation [8].

The connection between the elements will be specified using the $I(\phi)$ function, which, in accordance with biological principles, is specified in the following way:

$$I(\phi) = \frac{1}{1 + e^{k(\cos(\sigma) - \sin(\phi))}}. \tag{2}$$

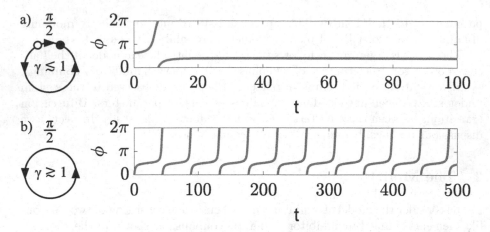

Fig. 2. Phase space and corresponding time series showing the saddle-node bifurcation in the Eq. (1). (a) For $\gamma < 1$, the system has stable and unstable equilibria, which corresponds to the absence of oscillations (unexcited state of the neuron) with the ability to generate activity with a small external stimulus. (b) If the value of the parameter $\gamma > 1$, there are no equilibria left in the system and the neuron generates spiking activity.

A coupling function of this kind was first introduced in [13] and then tested in [12,14,15]. Here we use its inhibitory analogue. This function simulates signal transmission from the presynaptic element to the postsynaptic element as follows. When the phase ϕ of the active presynaptic element reaches the value $\frac{\pi}{2} - \sigma$, this element stops inhibitory effect on the postsynaptic element. The duration of the cessation of the inhibitory effect is determined by the σ parameter and is 2σ. The dependence of the link function $I(\phi)$ on the phase of the presynaptic element ϕ is shown in the Fig. 3(a). It is important to note that this connection function takes into account the basic principles of the chemical interaction of neurons: (i) the presence or absence of activity of the postsynaptic element depends on the level of activity of the presynaptic element; (ii) all interactions between neuron cells are inertial due to the fact that signal transmission is not instantaneous. Using the σ parameter, which is responsible for the duration of the effect, we can simulate different types of couplings.

Thus, the system of two neuron-like phase elements with mutual synaptic inhibitory couplings is described by the following system of ordinary differential equations:

$$\begin{cases} \dot{\phi}_1 = \gamma_1 - \sin\phi_1 - d \cdot I(\phi_2) \\ \dot{\phi}_2 = \gamma_2 - \sin\phi_2 - d \cdot I(\phi_1) \end{cases} \tag{3}$$

where parameter d corresponds to the strength of inhibitory coupling $I(\phi)$. The phase space of (3) is torus (ϕ_1, ϕ_2). In this case, the regions where the inhibitory effect of the corresponding neurons stops are marked with blue and green areas in the Fig. 3(b).

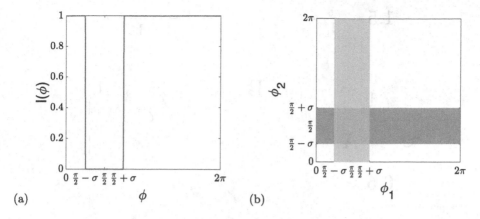

Fig. 3. (a) Type of inhibitory coupling function. (b) Regions on the phase torus in which the inhibitory effect of one element on another stops. In the green area, the first element ceases inhibit the second one. Similarly, in the blue area, the second element stops inhibitory effect on the first one. In the blue-green area, the mutual influence of elements on each other completely stops. (Color figure online)

3   The Results

In this section we study the case of identical elements. Each element is initially in a oscillatory activity, so we choose the values of natural frequencies slightly greater than one. To do this, we fix the parameters $\gamma_1 = \gamma_2 = 1.01$. We also fix the parameter $k = -500$, which is responsible for the switching speed in the coupling function (2). We will study the dynamics of the half-center oscillator by changing the parameter σ, which is responsible for the duration of the inhibitory effect, as well as the parameter d, which is responsible for the strength of the influence of elements on each other. It follows from the method of choosing the coupling function (2) that the parameter σ can take values from 0 to $\frac{\pi}{2}$. The values of the parameter of coupling strength d for biological reasons should not be chosen too large, since in this case the simulation will not be biologically relevant. Also, the parameter d obviously cannot take negative values. For the convenience of modeling, we assume that d can vary in the range from 0 to 1.5.

3.1   Maps of Temporal Patterns

To study and classify the dynamics in the system under consideration on the parameter plane $P = (\sigma; d)$, where $\sigma \in [0; \frac{\pi}{2}]$, $d \in [0; 1.5]$, two-parameter maps of neuron-like activity were obtained.

Let us note that initially, without couplings, both neuron-like elements demonstrated spiking activity. This situation corresponds to the parameter $d = 0$. Next, we consider how the dynamics of the system changes for different values of couplings parameters. To do this, we will study and describe each of the areas of neuron-like activity in the Fig. 4.

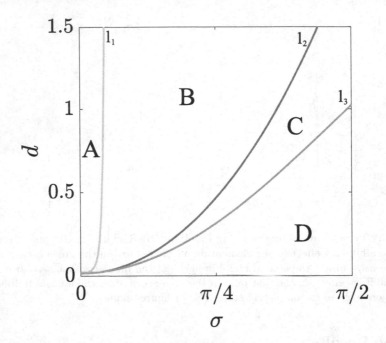

Fig. 4. Map of neuron-like temporal patterns in the system (3). Curves of different colors mark different bifurcation transitions. Curves l_1, l_3 correspond to a saddle-node bifurcation, and l_2 to a heteroclinic one. Regions A and B doesn't show any spiking activity and demonstrate only silence temporal pattern, sub-threshold oscillations or both. The next region C is characterized by the fact that the regime of anti-phase oscillations is added to the previously observed temporal patterns. In the case of D, only two temporal patterns remain in phase space: anti-phase spiking pattern and in-phase spiking pattern.

Next, some implementations of phase portraits and their corresponding time series will be demonstrated. Let us define some notation. Thus, on the phase portraits, dots of different colors mark the states of equilibria, namely:

- The red dot with a red outline is the saddle state of equilibrium.
- A pink dot with a purple outline is a state of equilibrium of the center type.
- The blue dot with a dark blue outline is a stable node.
- The blue dot with a red outline is an unstable node.
- A gray dot with a black outline is a complex equilibrium state.

Further, the pale red curves represent the vector field. In the phase portrait, the solid or dashed blue trajectories marks the trajectory calculated according to some initial conditions and correspond to blue (phase ϕ_1) and red (phase ϕ_2) solid or blue and red dashed ones on the time series respectively. The beginning of such a trajectory is marked with a dark blue dot. The time series of the corresponding trajectory is also shown to the right of the phase portrait. Finally, the areas where the inhibitory effect on the corresponding elements ceases are marked in green.

Let us list the types of neuron-like activity that can be observed in different regions from Fig. 4. The dynamics in region A is rather simple and corresponds to silence regime (absence of spikes). All other regions in the map Fig. 4 are regions of multistability, i.e. in this case several attractors coexist in the phase space of the system (3). In region B in addition to silence regime one can observe sub-threshold oscillations. Region C is characterized by three types of neuron-like temporal patterns: absence of oscillations, sub-threshold oscillations and anti-phase spiking. The region D features both anti-phase and in-phase spiking regimes.

Let us have a closer look at each of described regions in Fig. 4 and transitions between observed types of temporal activity patterns.

In the region A, the system (3) is characterized by the smallest area of interaction between elements in the ensemble. In this case, the system is able to demonstrate only the absence of activity (silence temporal pattern, see Fig. 5 (a, b)), since there are only four equilibria in its phase space, namely: a stable and unstable nodes, as well as two saddles.

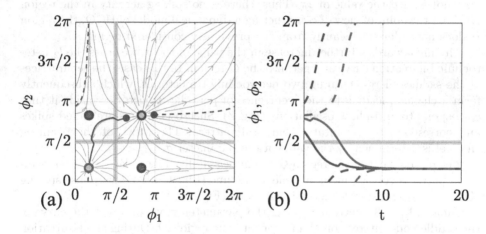

Fig. 5. (a) Phase portrait for the region A, demonstrating the silence temporal pattern. Here $\sigma = 0.08$, $d = 0.5$. (b) Time series for the first trajectory (blue and red solid lines represent phases ϕ_1 and ϕ_2 respectively) starting from $\phi_1^0 = 2$, $\phi_2^0 = 2.5$ and for the second trajectory (blue and red dashed lines represent phases ϕ_1 and ϕ_2 respectively) starting from $\phi_1^0 = 3.2$, $\phi_2^0 = 2.7$. (Color figure online)

The transition from region A to the region B occurs when crossing the curve l_1. When the parameter σ reaches the bifurcation value, a complex equilibrium state is formed with zero eigenvalues at the point with coordinates $(\phi_1, \phi_2) = (\frac{\pi}{2}; \frac{\pi}{2})$. Which the value of σ increases further, complex one splits into two equilibria of the type center and two saddle ones (Fig. 6).

As a result, in addition to the silence regime (Fig. 7(a, b)), a sub-threshold oscillations also appear (Fig. 7(c, d)). Note, however, that the amplitude of these

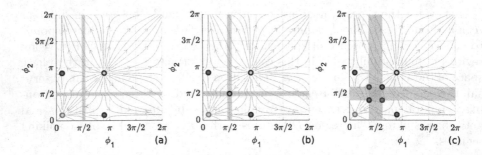

Fig. 6. Phase portraits showing the transition through the l_1 curve corresponding to the saddle-node bifurcation. The phase portraits are plotted for a fixed parameter $d = 0.6896$ and different values of the parameter σ: (a) $\sigma = 0.05$, (b) $\sigma = 0.13$, (c) $\sigma = 0.5$. (Color figure online)

sub-threshold oscillations is significantly less than the spike amplitude, i.e., it does not exceed the value of 2π. Thus, there is no spiking activity in the region B. From the point of view of constructing a dynamical model of HCO, the region B does not differ significantly from the previously considered region A.

Moving towards a further increase in the value of the σ, we cross the l_2 heteroclinic bifurcation curve and get into the C region. In this case, the separatrices of the saddles merge, forming two heteroclinic trajectories, which subsequently form a channel, leading to the occurrence of periodic trajectories inside it that correspond to anti-phase oscillations (Fig. 8). Such oscillations are called spikes and correspond to the rotator temporal pattern. The width of the resulting channel is determined by the value of the parameter σ.

Let us note that in the region C the multistability takes place: different temporal patterns coexist, including silence regime (Fig. 9(a, b)), anti-phase spiking (Fig. 9(c, d)) and sub-threshold oscillations (Fig. 9(e, f)).

Finally, by sufficiently increasing the parameter σ, we intersect the curve of the saddle-node bifurcation l_3 and get into the region D. During the bifurcation at the l_3 boundary, two centers, a stable and an unstable node with saddles, merge pairwise, resulting in the formation of four complex equilibrium states, which subsequently disappear (Fig. 10).

The region D is also a region of multisatbility, and the system strongly depending on the initial conditions. In that case only two temporal patterns remain in phase space: anti-phase spiking pattern (Fig. 11(a, b)) and in-phase spiking pattern (Fig. 11(c, d)).

3.2   Analytical Study of Bifurcation Transitions

Let us describe the bifurcation scenarios for the appearance and disappearance of various temporal patterns of neuron-like activity that are observed at the boundaries between regions in Fig. 4.

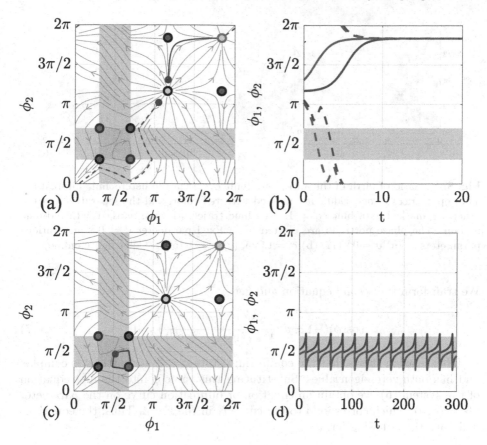

Fig. 7. Phase portraits and time series for the region B, demonstrating the silence regime (absence of oscillations) (a), (b) and sub-threshold oscillations (c), (d). Here $\sigma = 0.6171$, $d = 1.5$. In the phase portrait, the solid or dashed blue trajectories correspond to blue (phase ϕ_1) and red (phase ϕ_2) solid or blue and red dashed ones on the time series respectively. (Color figure online)

The equilibrium states of the system can be found from the following system of equations:

$$\begin{cases} \gamma_1 - \sin\phi_1 - d \cdot I(\phi_2) = 0 \\ \gamma_2 - \sin\phi_2 - d \cdot I(\phi_1) = 0 \end{cases} \tag{4}$$

Eigenvalues of equilibrium states can be found from the equation:

$$\lambda_{1,2} = \frac{-(\cos\phi_1 + \cos\phi_2) \pm \sqrt{(\cos\phi_1 - \cos\phi_2) + 4d^2 I'(\phi_1) I'(\phi_2)}}{2} \tag{5}$$

Equilibrium states of the first type $\phi_1 = \phi_2 = \phi$. Let us write down the conditions under which a saddle-node bifurcation occurs for such equilibrium states:

$$\begin{cases} \gamma - \sin(\phi) - I(\phi) = 0 \\ -\cos(\phi) \pm dI'(\phi) = 0 \end{cases} \tag{6}$$

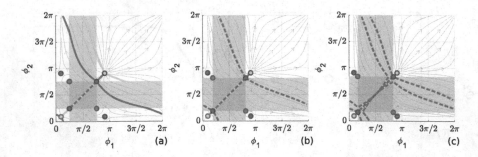

Fig. 8. Scenario of birth of the anti-phase limit cycle via the heteroclinic bifurcation. The separatrices of one saddle are marked with red color, and the separatrices of the other are marked with blue color. Heteroclinic trajectories are marked with red-blue marker. The phase portraits are plotted for a fixed parameter $d = 0.8$ and various parameters σ: (a) $\sigma = 0.9177$, (b) $\sigma = 1.025$, (c) $\sigma = 1.1507$. (Color figure online)

We transform the second equation and get

$$\cos(\phi)(-1 \pm dk \frac{e^{k(\cos(\sigma)-\sin(\phi))}}{1 + e^{k(\cos(\sigma)-\sin(\phi))}}) = 0 \tag{7}$$

This relation shows that the equilibrium state $\phi = \frac{\pi}{2} + \pi k, k \in Z$ is complex and has both zero eigenvalues. Substituting this value ϕ into the first equation of the system (6), we obtain an equation of bifurcation curve on the parameter plane (σ, d), which correspond to the curve l_1 in the Fig. 4. Thus, the equations will have the following form

$$d = (\gamma \pm 1)(1 + e^{k(\cos \sigma \pm 1)}) \tag{8}$$

Now consider the expression in the parentheses. For the values of the parameters $d \geq 0$ and $k << 0$, it is satisfied only for the case

$$-1 - dk \frac{e^{k(\cos(\sigma)-\sin(\phi))}}{(1 + e^{k(\cos(\sigma)-\sin(\phi))})^2} = 0 \tag{9}$$

Using the replacement $A = e^{k(\cos(\sigma)-\sin(\phi))}$, provided $e^{k(\cos(\sigma)+1)} \leq A \leq e^{k(\cos(\sigma)-1)}$, we get

$$e^{k(\cos(\sigma)-\sin(\phi))} = \frac{-(2 + dk) \pm \sqrt{(2 + dk)^2 - 4}}{2} \tag{10}$$

Expressing $\sin \phi$ from this equation and substituting it into the first equation of the system (6), we get

$$\begin{cases} \gamma - \cos \sigma + \frac{1}{k} \ln \frac{-(2+dk) \pm \sqrt{(2+dk)^2-4}}{2} - \frac{2d}{-dk \pm \sqrt{(2+dk)^2-4}} = 0 \\ e^{k(\cos(\sigma)+1)} \leq \frac{-(2+dk) \pm \sqrt{(2+dk)^2-4}}{2} \leq e^{k(\cos(\sigma)-1)} \end{cases} \tag{11}$$

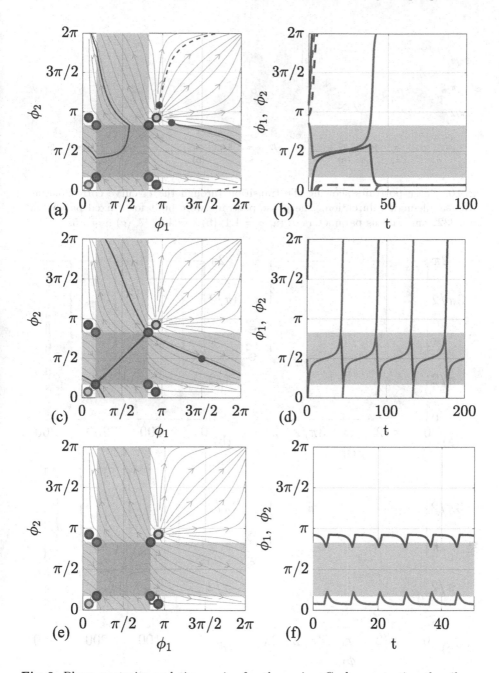

Fig. 9. Phase portraits and time series for the region C, demonstrating the silence regime (a), (b), anti-phase spiking (c), (d) and sub-threshold oscillations (e), (f). Here $\sigma = 1.0306$, $d = 0.7738$. In the phase portrait, the solid or dashed blue trajectories correspond to blue (phase ϕ_1) and red (phase ϕ_2) solid or blue and red dashed ones on the time series respectively. (Color figure online)

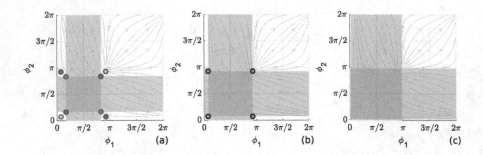

Fig. 10. Phase portraits showing the transition through the l_3 curve corresponding to the saddle-node bifurcation. The phase portraits were built for a fixed parameter $d = 0.8322$ and various parameters σ: (a) $\sigma = 1.1$, (b) $\sigma = 1.3777$, (c) $\sigma = 1.5$.

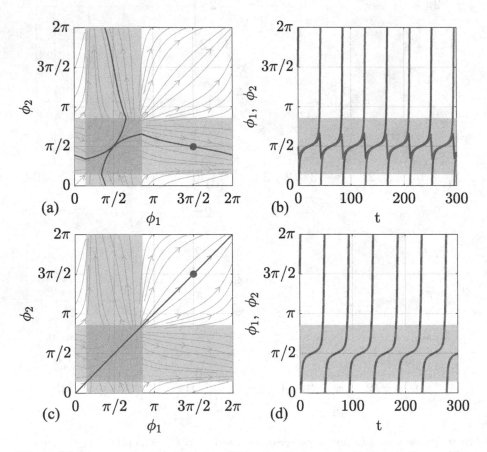

Fig. 11. Phase portraits and time series for region D, demonstrating anti-phase spiking (a), (b) and in-phase spiking (c), (d). Here $\sigma = 1.1175$, $d = 0.369$. In the phase portrait, the solid blue trajectory correspond to blue (phase ϕ_1) and red (phase ϕ_2) solid one on the time series. (Color figure online)

It can be verified numerically that the described system with given parameters is valid only for the case

$$
\begin{cases}
\gamma - \cos\sigma + \frac{1}{k}\ln\frac{-(2+dk)-\sqrt{(2+dk)^2-4}}{2} - \frac{2d}{-dk-\sqrt{(2+dk)^2-4}} = 0 \\
e^{k(\cos(\sigma)+1)} \leq \frac{-(2+dk)-\sqrt{(2+dk)^2-4}}{2} \leq e^{k(\cos(\sigma)-1)}
\end{cases}
\tag{12}
$$

The resulting system of equations describes the curve l_3 in Fig. 4.

The l_2 curve was constructed numerically using the MatCont [7] mathematical package and corresponds to a heteroclinic bifurcation (Fig. 8).

In this case, before the bifurcation, stable separatrices come to the saddle from an unstable node, and unstable separatrices leaving the saddle tend to a stable node (Fig. 8(a)). As the parameter σ increases until it meets the curve l_2, a heteroclinic bifurcation occurs, during which unstable separatrices of one of the saddles become stable separatrices of the other saddle (Fig. 8(b)). Finally, after the σ parameter passes the l_2 curve and further increases, the heteroclinic trajectories between the saddles are preserved forming a channel (Fig. 8(c)).

4 Conclusion

In this study we have proposed a new phenomenological model of the HCO, consists of two oscillatory neurons coupled by chemical inhibitory synapses. On the one hand, it is simple and thus allows one to conduct analytical studies; on the other hand, despite its simplicity, it reflects the main properties of the biological HCO and reproduces all typical temporal patterns, including silent state, in-phase and anti-phase spiking. We have identified regions in the control parameters space that correspond to multistability, which support the hypothesis that the same pattern generator circuit can generate several types of neuron-like activity. We also have analyzed bifurcation transitions that lead to the occurrence of the specified temporal patterns.

This study may help one to gain new insights into the nature of the locomotor CPG and its functioning under different conditions. Namely, suggested model can be used as a building block of specific complex CPGs in studies of animal and robot locomotion.

Acknowledgements. This work was supported by the Ministry of Science and Education of Russian Federation (Project No. 0729-2020-0036).

References

1. Adler, R.: A study of locking phenomena in oscillators. Proc. IEEE **61**(10), 1380–1385 (1973)
2. Ausborn, J., Snyder, A.C., Shevtsova, N.A., Rybak, I.A., Rubin, J.E.: State-dependent rhythmogenesis and frequency control in a half-center locomotor CPG. J. Neurophysiol. **119**(1), 96–117 (2018)

3. Brown, T.G.: The intrinsic factors in the act of progression in the mammal. Proc. Roy. Soc. London Ser. B Containing Pap. Biol. Character **84**(572), 308–319 (1911)
4. Buono, P.L., Golubitsky, M.: Models of central pattern generators for quadruped locomotion I. Primary gaits. J. Math. Biol. **42**(4), 291–326 (2001)
5. Burke, R., Degtyarenko, A., Simon, E.: Patterns of locomotor drive to motoneurons and last-order interneurons: clues to the structure of the CPG. J. Neurophysiol. **86**(1), 447–462 (2001)
6. Cohen, A.H., Holmes, P.J., Rand, R.H.: The nature of the coupling between segmental oscillators of the lamprey spinal generator for locomotion: a mathematical model. J. Math. Biol. **13**(3), 345–369 (1982)
7. Dhooge, A., Govaerts, W., Kuznetsov, Y.A.: MATCONT: a MATLAB package for numerical bifurcation analysis of ODEs. ACM Trans. Math. Softw. (TOMS) **29**(2), 141–164 (2003)
8. Ermentrout, G.B., Kopell, N.: Parabolic bursting in an excitable system coupled with a slow oscillation. SIAM J. Appl. Math. **46**(2), 233–253 (1986)
9. Grillner, S.: Neurobiological bases of rhythmic motor acts in vertebrates. Science **228**(4696), 143–149 (1985)
10. Grillner, S.: Biological pattern generation: the cellular and computational logic of networks in motion. Neuron **52**(5), 751–766 (2006)
11. Jankowska, E., Jukes, M., Lund, S., Lundberg, A.: The effect of DOPA on the Spinal Cord 5. Reciprocal organization of pathways transmitting excitatory action to alpha motoneurones of flexors and extensors. Acta Physiol. Scand. **70**(3–4), 369–388 (1967)
12. Korotkov, A.G., Kazakov, A.O., Levanova, T.A.: Effects of memristor-based coupling in the ensemble of FitzHugh-Nagumo elements. Eur. Phys. J. Spec. Top. **228**(10), 2325–2337 (2019)
13. Korotkov, A.G., Kazakov, A.O., Levanova, T.A., Osipov, G.V.: Chaotic regimes in the ensemble of FitzhHugh-Nagumo elements with weak couplings. IFAC-PapersOnLine **51**(33), 241–245 (2018)
14. Korotkov, A.G., Kazakov, A.O., Levanova, T.A., Osipov, G.V.: The dynamics of ensemble of neuron-like elements with excitatory couplings. Commun. Nonlinear Sci. Numer. Simul. **71**, 38–49 (2019)
15. Korotkov, A.G., Levanova, T.A., Zaks, M.A., Maksimov, A.G., Osipov, G.V.: Dynamics in a phase model of half-center oscillator: two neurons with excitatory coupling. Commun. Nonlinear Sci. Numer. Simul. **104**, 106045 (2022)
16. Lundberg, A.: Half-centres revisited. In: Regulatory Functions of the CNS Principles of Motion and Organization, pp. 155–167. Elsevier (1981)
17. Sakurai, A., Newcomb, J.M., Lillvis, J.L., Katz, P.S.: Different roles for homologous interneurons in species exhibiting similar rhythmic behaviors. Curr. Biol. **21**(12), 1036–1043 (2011)
18. Yakovenko, S., McCrea, D., Stecina, K., Prochazka, A.: Control of locomotor cycle durations. J. Neurophysiol. **94**(2), 1057–1065 (2005)

Solutions of Multidimensional Hydrodynamic Evolution Equations Using the Fast Legendre Transformation

A. E. Spivak(✉) [iD], S. N. Gurbatov [iD], and I. Yu. Demin [iD]

Lobachevsky State University of Nizhny Novgorod, 23 Gagarin Avenue, Nizhny Novgorod 603950, Russia

spivak@rf.unn.ru

Abstract. The numerical solution of multidimensional nonlinear evolution equations, in which the field depends on several spatial coordinates, is considered. The work was based on the Fast Legendre Transform algorithm. A model for the numerical calculation of the processes of surface growth and the evolution of a two-dimensional velocity field, implemented in the MATLAB software environment, is presented. With the help of this model, the growth of regular surfaces and surfaces with a random initial shape (two-dimensional Gaussian noise with zero mean and unit variance) was considered, and the level lines and lines of discontinuities of the growing surfaces were also shown. The result of numerical simulation of a localized random velocity field with an initial field in the form of two-dimensional Gaussian noise with zero mean and unit variance is presented.

Keywords: Nonlinear acoustics · Fast Legendre Transform · Burgers equation · MATLAB

1 Introduction

The equation of nonlinear diffusion, presented by J. Burgers in 1939 [1], is a model of hydrodynamic turbulence and describes two main effects: nonlinear redistribution of energy over the spectrum and dissipation in the region of small scales. This equation has much in common with the Navier-Stokes equation: the type of nonlinearity, invariance groups, the energy-dissipation relation [2, 3].

In the absence of external forces, the Burgers equation describes the degeneration of turbulence, i.e. nonlinear Transformation of a random initial perturbation. However, despite the fact that this equation has an exact solution - the Hopf-Cole solution, the study of the statistical properties of this equation is a complex mathematical problem. In this case, the initial conditions significantly affect the turbulence degeneracy regime.

In nonlinear acoustics, the BE is derived from a system of hydrodynamic equations taking into account the viscosity and thermal conductivity of the medium [2–5]. In particular, under random initial conditions, this equation describes the evolution of intense acoustic noise (such solutions are called one-dimensional acoustic turbulence).

D. Balandin et al. (Eds.): MMST 2022, CCIS 1750, pp. 95–105, 2022.
https://doi.org/10.1007/978-3-031-24145-1_8

The variation of this equation in the multidimensional case with external random forces is widely used as a pressureless Navier-Stokes hydrodynamic turbulence model.

In 1986 M. Kardar, G. Parisi and Yu.K. Zhang was the first to propose a non-linear equation with a random source describing the non-equilibrium evolution of a surface (the Kardara-Parisi-Zhang equation or KPZ equation) [5, 6]. This equation coincides with the nonlinear equation for the potential of the velocity field and describes the growth of the surface, the deposition of impurities, and the propagation of the flame front. In these cases, the velocity potential corresponds to the surface profile, and the equation describing its evolution is equivalent to the KPZ equation.

The Burgers vector equation, which is a natural generalization of the one-dimensional equation, is used to model the evolution of the large-scale structure (LSS) of the Universe [7–9]. Elements of this structure are objects with a higher concentration of galaxies— galaxy groups and clusters, filaments (fiber-like groups of galaxies connecting galaxy clusters) and "walls". It is believed that the large-scale formations existing today were formed from small initial density perturbations due to gravitational instability.

In astrophysics, the solution of this equation in the case of vanishingly small viscosity is commonly known as the adhesion model. In this case, the Burgers equation describes a "skeleton" of the large-scale matter distribution. As it follows from the properties of the three-dimensional Burgers equation, at long times, there appears a cellular structure of matter: regions with a density much smaller than the mean density, surfaces with an elevated concentration of matter separating these dark regions, surface intersections— lines and, finally, line intersections—clusters.

From the point of view of numerical simulation, the Burgers equation is of great interest, because it is possible to construct sufficiently simple and fast algorithms for the numerical solution of this equation [2].

The purpose of this work is to consider the problem of changing the shape of the surface, which is of the statistical nature of the equations described above using the Fast Legendre Transform method.

2 Evolution Equations of Multidimensional Flows and Growth of Surfaces

The equation for the velocity field of a multidimensional flow of uniformly moving particles is written as follows [2, 3]:

$$\frac{\partial \mathbf{v}}{\partial t} + (\mathbf{v}, \nabla)\mathbf{v} = 0, \ \mathbf{v}(\mathbf{x}, 0) = \mathbf{v}_0(\mathbf{x}). \tag{1}$$

In this case, the velocity field potential $s(x, t)$ satisfies the following equation:

$$\frac{\partial s}{\partial t} = \frac{1}{2}(\nabla s(\mathbf{x}, t))^2, \ s(\mathbf{x}, 0) = s_0(\mathbf{x}). \tag{2}$$

In the Lagrangian description, the above nonlinear partial differential equations are reduced to ordinary differential equations.

For the Euler coordinates, the Lagrangian velocity field and its potential, the equations are as follows:

$$\frac{d\mathbf{X}}{dt} = \mathbf{V}, \ \mathbf{X}(\mathbf{y}, 0) = \mathbf{y}, \tag{3}$$

$$\frac{d\mathbf{V}}{dt} = 0, \ \mathbf{V}(\mathbf{y}, 0) = \mathbf{v}_0(\mathbf{y}), \tag{4}$$

$$\frac{dS}{dt} = \frac{1}{2}\mathbf{V}^2, \ S(\mathbf{y}, 0) = s_0(\mathbf{y}). \tag{5}$$

Solving Eqs. (3) and (4) together, we obtain a mapping of Lagrangian coordinates into Euler coordinates:

$$\mathbf{x} = \mathbf{X}(\mathbf{y},t) = \mathbf{y} + \mathbf{v}_0(\mathbf{y})t. \tag{6}$$

The solution of Eqs. (4) and (5) gives the Lagrangian velocity fields and its potential:

$$\mathbf{V}(\mathbf{y}, t) = \mathbf{v}_0(\mathbf{y}), \ S(\mathbf{y}, t) = s_0(\mathbf{y}) + \frac{1}{2}\mathbf{v}_0^2(\mathbf{y})t \tag{7}$$

Now, to define the Euler fields, we need to find the inverse mapping (6)

$$\mathbf{y} = \mathbf{y}(\mathbf{x}, t) \tag{8}$$

If mapping (8) is known, then the velocity and potential fields are determined by the equalities:

$$\mathbf{v}(\mathbf{x}, t) = \mathbf{v}_0(\mathbf{y}(\mathbf{x}, t)), \ s(\mathbf{x}, t) = s_0(\mathbf{y}(\mathbf{x}, t)) + \frac{1}{2}\mathbf{v}_0^2(\mathbf{y}(\mathbf{x}, t))t \tag{9}$$

or in explicit notation through the map of Lagrangian coordinates to Euler coordinates (8):

$$\mathbf{v}(\mathbf{x}, t) = \frac{\mathbf{x} - \mathbf{y}(\mathbf{x}, t)}{t}, \tag{10}$$

$$s(\mathbf{x}, t) = s_0(\mathbf{y}) + \frac{(\mathbf{y} - \mathbf{x})^2}{2t}. \tag{11}$$

We will analyze the growth of the surface using the example of the propagation of a volumetric fire. Let us consider the leading edge of a bulk flame propagating at unit speed in a direction perpendicular to the front, while the front surface $z = h(\mathbf{x}, t)$ satisfies the following [2–5]:

$$\frac{\partial h}{\partial t} = \sqrt{1 + (\nabla h)^2}. \tag{12}$$

If the front propagates predominantly along the z axis, then the inequality $\nabla h \ll 1$.

The right side of Eq. (12) can be expanded in a Taylor series, restricting ourselves to only the first two terms of the expansion, and we get the following equation for h:

$$\frac{\partial h}{\partial t} = \frac{1}{2}(\nabla h(\mathbf{x}, t))^2, \; h(\mathbf{x}, 0) = h_0(\mathbf{x}). \tag{13}$$

Let us introduce a field $\mathbf{u}(\mathbf{x}, t)$, whose modulus characterizes the tangent of the slope angle between the front normal and the z axis, and obeys the equation

$$\frac{\partial \mathbf{u}}{\partial t} + (\mathbf{u}, \nabla)\mathbf{u} = 0, \; \mathbf{u}(\mathbf{x}, t) = -\nabla h(\mathbf{x}, t) \tag{14}$$

To solve Eqs. (13) and (14), we pass to the system of characteristic equations:

$$\frac{d\mathbf{X}}{dt} = \mathbf{U}, \; \frac{d\mathbf{U}}{dt} = 0, \; \frac{dH}{dt} + \frac{1}{2}\mathbf{U}^2 = 0, \tag{15}$$

whose solutions look like:

$$\mathbf{X}(\mathbf{y}, t) = \mathbf{y} - \nabla h_0(\mathbf{y})t, \; \mathbf{U}(\mathbf{y}, t) = -\nabla h_0(\mathbf{y}), \tag{16}$$

$$H(\mathbf{y}, t) = h_0(\mathbf{y}) - \frac{1}{2}(\nabla h_0(\mathbf{y}))^2 t \tag{17}$$

Moreover, if the mapping $\mathbf{y}(\mathbf{x}, t)$ is single-valued, then the expressions for the desired fields $h(\mathbf{x}, t)$ and $\mathbf{u}(\mathbf{x}, t)$ are written as (9) and (10).

In the case when the mapping becomes multi-valued, the real appearance of the surface will be determined by the absolute maximum principle, i.e. from all branches of the multivalued function $h(\mathbf{x}, t)$, one should choose the branch that has the largest value at a given point x. Thus, the solution of the surface growth equation can be represented as:

$$h(\mathbf{x}, t) = \max_{\mathbf{y}} \left[h_0(\mathbf{y}(\mathbf{x}, t)) - \frac{(\mathbf{y}(\mathbf{x}, t) - \mathbf{x})^2}{2t} \right] \tag{18}$$

3   Results of Numerical Simulation of Surface Growth and Multidimensional Velocity Field Using the Two-Dimensional Fast Legendre Transform Algorithm

3.1   Using the Fast Legendre Transform for the Numerical Calculation of Evolutionary Nonlinear Equations

In this paper, for numerical experiments, we used the Fast Legendre Transform algorithm, which allows us to significantly reduce the number of required operations compared to standard methods [10]. The Legendre Transform of a scalar function $\Phi(a)$ is

$$\phi(x) = \min_{a}[\Phi(a) + x \cdot a] \tag{19}$$

provided that the second derivative of $\Phi(a)$ with respect to a exists.

An increase in the speed of the algorithm is possible due to the fact that $y(x)$ is a non-decreasing function of the argument x at constant time. Mathematically, this will be written as follows:

$$[y(x) = y(x')](x - x') \geq 0 \tag{20}$$

Below is a graphical interpretation of the described algorithm for sixteen points (Fig. 1).

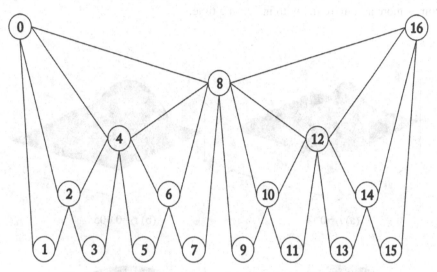

Fig. 1. Graphical interpretation of the Fast Legendre Transform.

As applied to the Burgers equation in the case of vanishingly low viscosity

$$\frac{\partial v}{\partial t} + v \frac{\partial v}{\partial x} = 0 \tag{21}$$

we get the expression:

$$\min_y \Phi(y(x, t), x, t) = \min_y \left[s_0(y) + \frac{(x - y)^2}{2t} \right] \tag{22}$$

which is essentially also the Legendre Transform. This is what underlies the algorithm for solving the problem [11].

Using (22) we find $y(x)$ corresponding to the minimum of the function $S(y(x, t), x, t)$ at a given time t and coordinate x, we determine velocity field (10).

3.2 Results of Numerical Simulation

For numerical calculation, the MATLAB software environment was used [12, 13]. For the 2D Fast Legendre Transform, a grid of 129×129 pixels was defined. As you can see, as a result of nonlinear interaction, its surface shape changes. In particular, in Fig. 2(b) one can see that the troughs are sharpened and the vertices are smoothed. Further on Fig. 2(c) one can see a clear "collision" of the vertices with each other, and it is also clearly seen that at the junctions the field $u(x,t)$ is broken, while small vertices are "absorbed" by large ones. In the end (Fig. 2(d)) there will be only one peak, which at the initial moment of time was the highest. Further, it can be observed that the peak becomes more and more flat with increasing time.

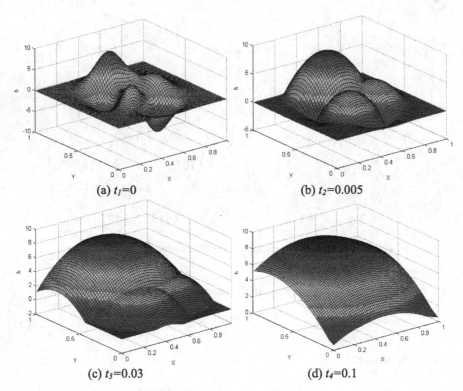

(a) $t_1=0$ (b) $t_2=0.005$

(c) $t_3=0.03$ (d) $t_4=0.1$

Fig. 2. The growth of a regular surface.

For the same regular surface in Fig. 3 shows the evolution of level lines and lines of discontinuities in the slope field $u(x,t)$ at different times. The internal parameters of the original program during the numerical experiment corresponded to the previous consideration.

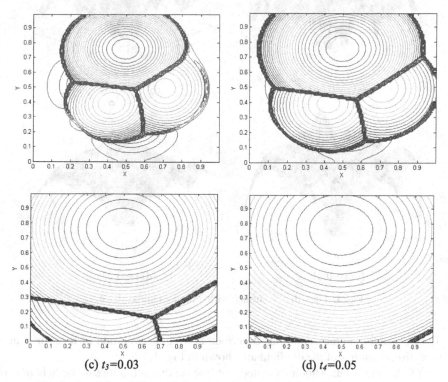

(c) $t_3=0.03$

(d) $t_4=0.05$

Fig. 3. Level lines (thin) and discontinuity lines (bold) of the growing surface.

It is worth noting that in Fig. 3(b, c) the phenomenon of "colliding" of the vertices on each other is clearly visible, the breaks of the level lines are also clearly visible, which are located along the lines of discontinuities in the field of surface slopes.

Next, we will numerically simulate the problem of surface growth with a randomly given initial shape (two-dimensional Gaussian noise with zero mean and unit variance). We also save the previous settings of the program using the FLT and use the numerical analysis scheme as in the case of a regular surface.

On Fig. 4 shows the result of numerical simulation of the growth of a surface with a random initial shape.

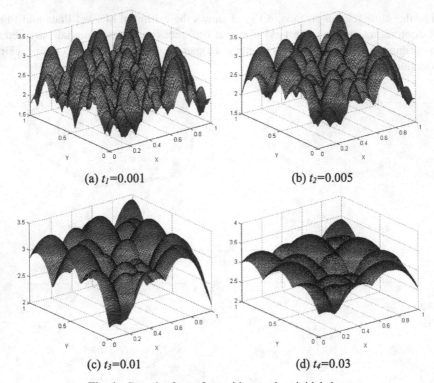

(a) t_1=0.001   (b) t_2=0.005

(c) t_3=0.01   (d) t_4=0.03

Fig. 4. Growth of a surface with a random initial shape.

For the random initial conditions described above, the evolution of level lines and lines of discontinuities of the dip field are shown in Fig. 5.

The following feature should be noted: the lines of discontinuities in the field $u(x, t)$ have a pronounced cellular structure, the scale of which increases with time. Thus, as a result of the nonlinear interaction of the discontinuities of the field $u(x, t)$, they merge, and hence the regularization of the structure, which is formed by the discontinuity lines.

Let us turn to the consideration of the multidimensional Burgers equation. In general, it looks like [1, 14]:

$$\frac{\partial \mathbf{v}}{\partial t} + (\mathbf{v}, \nabla)\mathbf{v} = \mu \Delta \mathbf{v}, \ \mathbf{v}(\mathbf{x}, 0) = \mathbf{v}_0(\mathbf{x}) \tag{23}$$

In the case of potentiality of the initial velocity field, i.e. if it can be represented as a gradient of some scalar function:

$$\mathbf{v}_0(\mathbf{x}) = \nabla s_0(\mathbf{x}) \tag{24}$$

solution (23) remains potential at all points for any $t > 0$, and can be described by the expression:

$$\mathbf{v}(\mathbf{x}, t) = \frac{\mathbf{x} - \{\mathbf{y}\}(\mathbf{x}, t)}{t}, \tag{25}$$

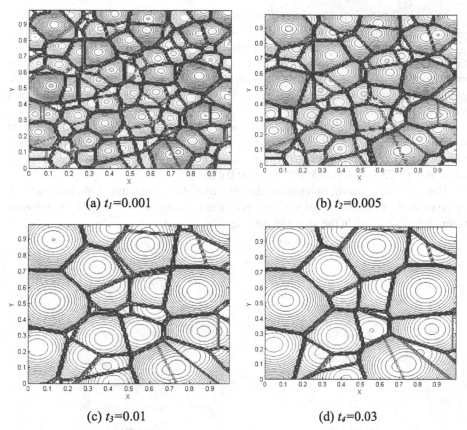

(a) $t_1=0.001$

(b) $t_2=0.005$

(c) $t_3=0.01$

(d) $t_4=0.03$

Fig. 5. Level lines (thin) and discontinuity lines (bold) of a growing surface with a random initial shape.

curly brackets in (25) denote spatial averaging using the function

$$f(\mathbf{y}(\mathbf{x}, t)) = \frac{\exp\left[-\frac{1}{2\mu t}\Phi(\mathbf{y}(\mathbf{x}, t))\right]}{\int \exp\left[-\frac{1}{2\mu t}\Phi(\mathbf{y}(\mathbf{x}, t))\right]d^n y}, \quad \Phi(\mathbf{y}(\mathbf{x}, t)) = s_0(\mathbf{y})t + \frac{(\mathbf{y} - \mathbf{x})^2}{2t} \quad (26)$$

As in the one-dimensional case, applying the Hopf-Cole change [1]

$$s(\mathbf{x}, t) = 2\mu \ln U(\mathbf{x}, t), \quad \mathbf{v}(\mathbf{x}, t) = 2\mu \ln U(\mathbf{x}, t) \quad (27)$$

it is possible to reduce the Eq. (23) to the linear diffusion equation

$$\frac{\partial U(\mathbf{x}, t)}{\partial t} = \mu \Delta U(\mathbf{x}, t), \quad (28)$$

$$U(\mathbf{x}, 0) = U_0(\mathbf{x}) = \exp\left[-\frac{s_0(\mathbf{x})}{2\mu}\right]. \quad (29)$$

Considering the solution of the problem in the limit of infinitely small viscosity μ → 0, we obtain the following expressions for the potential of the velocity field (9–11).

Let us present the results of numerical simulation of the evolution of the velocity field. For a random velocity field, the qualitative form of evolution for lines of discontinuities and lines of the potential level does not qualitatively differ from the results given for the growth of a random surface; therefore, we will give an example of the evolution of a localized initial perturbation.

In Fig. 6. a picture of the evolution of a localized random velocity field is presented for the following parameters: the size of the FLT grid is 257×257; initial field - two-dimensional Gaussian noise with zero mean and unit variance; the field is localized in the region $0.375 < x < 0.625$; $0.375 < y < 0.625$.

Due to the fact that the potential of the velocity field in a medium without dispersion in the case of a vanishingly low viscosity obeys an equation similar to (13), its evolution is similar to a change in the growing surface.

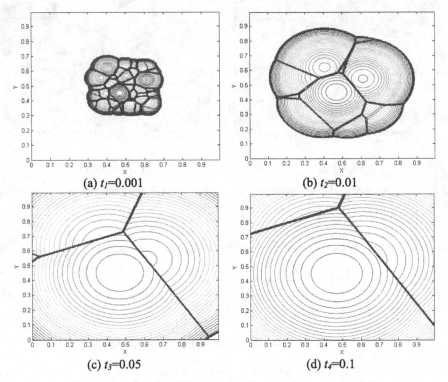

(a) $t_1=0.001$ (b) $t_2=0.01$

(c) $t_3=0.05$ (d) $t_4=0.1$

Fig. 6. Potential level lines of a localized random velocity field.

An important feature of the evolution of a localized field is the fact that its delocalization occurs, that is, an increase in the region of localization of the velocity field.

Acknowledgement. This work has been supported by the grants the Russian Science Foundation, RSF 19-12-00256.

References

1. Burgers, J.M.: Mathematical examples illustrating relations occurring in the theory of turbulent fluid motion. In: Nieuwstadt, F.T.M., Steketee, J.A. (eds.). Selected Papers of J.M. Burgers, pp. 281–334. Springer, Dordrecht (1995). https://doi.org/10.1007/978-94-011-0195-0_10

2. Gurbatov, S.N., Rudenko, O.V., Saichev, A.I.: Waves and Structures in Nonlinear Nondispersive Media. 1st edn. Springer, Heidelberg (2011). https://doi.org/10.1007/978-3-642-23617-4

3. Saichev, A.I., Woyczyński, W.A.: Distributions in the Physical and Engineering Sciences, vol. 2, 1st edn. Birkhäuser, New York (2013). https://doi.org/10.1007/978-0-8176-4652-3

4. Rudenko, O.V., Soluyan, S.I.: Theoretical Foundations of Nonlinear Acoustics. Consultants Bureau, New York (1977)

5. Saichev, A.I., Woyczyński, W.A.: Distributions in the Physical and Engineering Sciences, vol. 3, 1st edn. Birkhäuser, Cham (2008). https://doi.org/10.1007/978-3-319-97958-8

6. Mehran, K.M., Giorgio, P.G., Zhang, Y.-C.: Dynamic scaling of growing interfaces. Phys. Rev. Lett. **56**(9), 889 (1986). https://doi.org/10.1103/PhysRevLett.56.889

7. Gurbatov, S.N., Saichev, A.I., Shandarin, S.F.: The large-scale structure of the Universe in the frame of the model equation of non-linear diffusion. Mon. Not. R. Astron. Soc. **236**(2), 385–402 (1989)

8. Vergassola, M., Dubrulle, B., Frisch, U., Noullez, A.: Burgers' equation, Devil's staircases and the mass distribution for large-scale structures. Astron. Astrophys. **289**, 325–356 (1994)

9. Gurbatov, S.N., Saichev, A.I., Shandarin, S.F.: Large-scale structure of the Universe. The Zeldovich approximation and the adhesion model. Phys. Usp. **55**(3), 223–249 (2012). https://doi.org/10.3367/UFNe.0182.201203a.0233

10. Noullez, A., Vergassola, M.: A fast Legendre transform algorithm and applications to the adhesion model. J. Sci. Comput. **9**, 259–281 (1994). https://doi.org/10.1007/BF01575032

11. Spivak, A.E., Gurbatov, S.N., Demin, I.Yu.: Numerical solution of the Burgers equation in the case of vanishingly small viscosity using the fast Legendre transformation. Memoirs of the Faculty of Physics (2022). (in press)

12. Murugan, S.D., Frisch, U., Nazarenko, S., Besse, N., Ray, S.S.: Suppressing thermalization and constructing weak solutions in truncated inviscid equations of hydrodynamics: lessons from the Burgers equation. Phys. Rev. Res. **2**(3), 033202 (2020). https://doi.org/10.1103/PhysRevResearch.2.033202

13. Bequin, P.: Burgers equation and convexification method: propagation of wave packets and narrowband noise. J. Acoust. Soc. Am. **151**(2), 1223 (2022). https://doi.org/10.1121/10.0009578

14. Gurbatov, S.N., Moshkov, A., Noullez, A.: The evolution of anisotropic structures and turbulence in the multi-dimensional Burgers equation. Phys. Rev. E **81**(4), 046312 (2010). https://doi.org/10.1103/PhysRevE.81.046312

An Iterative Method for Solving a Nonlinear System of the Theory of Radiation Transfer and Statistical Equilibrium in a Plane-Parallel Layer

Aleksey Kalinin[1,2] , Alla Tyukhtina[1(✉)] , and Aleksey Busalov[1]

[1] N.I. Lobachevsky State University, 23 Gagarin Av., 603022 Nizny Novgorod, Russia
tyukhtina@iee.unn.ru

[2] Institute of Applied Physics, Russian Academy of Sciences,
46 Ulyanova St., 603950 Nizhny Novgorod, Russia

Abstract. A nonlinear system of integro-differential equations of radiation transfer and statistical equilibrium in a plane-parallel layer is studied within the framework of a two-level atom model under the assumption of a complete redistribution of radiation in frequency. A boundary value problem for the kinetic transport equation with a condition corresponding to the absence of an external particle flux incident on the boundary of the region is considered. The results on the existence and uniqueness of the solution of the problem are presented. To find this solution, an iterative linearizing algorithm is proposed and justified. A finite-difference scheme of the integro-interpolation method is used for the numerical solution of the problem. Its main properties - the stability condition, the approximation order, the conservativeness condition of the scheme - are investigated. The efficiency of the algorithm is numerically illustrated on model problems for specific media under various assumptions about the optical density of the matter.

Keywords: System of radiation transfer equations · Nonlinear integro-differential equations · Iterative method · Finite-difference scheme · DSn method

1 Introduction

The mathematical foundations and issues of numerical solution of linear problems of the theory of radiation transfer are discussed in [1–7]. Taking into account the interaction of radiation with the medium in the absence of local thermodynamic equilibrium leads to rather complex nonlinear systems of integro-differential equations [7–11]. The principal aspects of these nonlinear problems can be described by a system of integro-differential equations of radiation transfer and statistical equilibrium [9,10].

Supported by the Scientific and Education Mathematical Center "Mathematics for Future Technologies" (Project No. 075-02-2022-883).

The issues of mathematical correctness and properties of solutions of the system of radiation transfer equations and statistical equilibrium in bounded domains were studied in [12–15]. In the present work, we study a nonlinear system of integro-differential equations of radiative transfer and statistical equilibrium for the model of a two-level atom in a plane-parallel layer in the spatially one-dimensional case. The problem under consideration has specific features associated with the unboundedness of the domain and requires independent research [2].

The paper proves the theorem on the existence and uniqueness of the solution of the boundary value problem for the system under consideration, a linearizing iterative algorithm for its search is proposed and justified. The proposed iterative algorithm is numerically implemented. The issues of stability and approximation of the finite-difference scheme of the integro-interpolation method [7] are investigated. The efficiency of the algorithm is numerically illustrated on model problems for specific media under various assumptions about the optical density of matter.

2   Problem Statement and Main Results

We consider a stationary nonlinear system of integro-differential equations of radiation transfer and statistical equilibrium corresponding to the model of a two-level atom under the assumption of a complete redistribution of radiation in frequency [9,10] in a plane-parallel layer:

$$\mu \frac{\partial}{\partial x}\varphi(x,\nu,\mu) + h\nu_{12}\frac{\kappa(\nu)}{2}[B_{12}C_1(x) - B_{21}C_2(x)]\varphi(x,\nu,\mu)$$

$$= h\nu_{12}\frac{\kappa(\nu)}{2}A_{21}C_2(x), \tag{1}$$

$$\left[C_{12}n_e(x) + B_{12}\int_I\int_{-1}^1 \frac{\kappa(\nu)}{2}\varphi(x,\nu,\mu)d\mu d\nu\right]C_1(x)$$

$$= \left[A_{21} + C_{21}n_e(x) + \int_I\int_{-1}^1 \frac{\kappa(\nu)}{2}\varphi(x,\nu,\mu)d\mu d\nu\right]C_2(x), \tag{2}$$

$$C_1(x) + C_2(x) = f(x), \tag{3}$$

$$\varphi(x_1,\nu,\mu) = 0, \ \mu > 0, \ \varphi(x_2,\nu,\mu) = 0, \ \mu < 0. \tag{4}$$

Here $x \in [x_1, x_2]$, $x_2 - x_1 = d > 0$, $\mu \in [-1, 1]$, $\nu \in I = [0, \nu_0]$. The function φ is the specific radiation intensity, C_1 and C_2 are the spatial densities of the atoms of the medium in the ground and excited states respectively. The boundary conditions (4) mean the absence of an external particle flux incident on the boundary of the region.

It is assumed that h, ν_{12}, ν_0, A_{21}, B_{12}, B_{21}, C_{12}, C_{21} are given positive numbers satisfying the condition $B_{12}C_{21} - B_{21}C_{12} > 0$. Functions $n_e(x)$, $f(x)$,

$x \in (x_1, x_2)$, $\kappa(\nu)$, $\nu \in I$ are given, measurable and non-negative almost everywhere in their domains,

$$\text{esssup}_{x \in (x_1, x_2)} n_e(x) = n_e^* < \infty, \text{esssup}_{x \in (x_1, x_2)} f(x) = f^* < \infty, \qquad (5)$$

$$\text{esssup}_{\nu \in I} \kappa(\nu) = \kappa^* < \infty, \int_I \kappa(\nu) d\nu = 1. \qquad (6)$$

Detailed information about the physical meaning of these functions and coefficients is given in [9, 10].

For an arbitrary subset Π of Euclidean space we denote by $K_\infty(\Pi)$ the cone of non-negative functions in $L_\infty(\Pi)$. Let $D = [x_1, x_2] \times I \times [-1, 1]$, $\mathcal{D}_\infty(D)$ is the class of functions $\varphi \in L_\infty(D)$ that absolutely continuous on $[x_1, x_2]$ with almost all fixed $\nu \in I$, $\mu \in [-1, 1]$, satisfy the boundary conditions (4) and such that $\mu \partial \varphi / \partial x \in L_\infty(D)$; $\mathcal{D}_\infty^+(D) = \mathcal{D}_\infty(D) \cap K_\infty(D)$. Similar functional spaces were generally considered in [1].

A solution of problem (1)–(4) is a function

$$\Phi(x, \nu, \mu) = \{\varphi(x, \nu, \mu), C_1(x), C_2(x)\} \in \mathcal{D}_\infty^+(D) \times K_\infty(x_1, x_2) \times K_\infty(x_1, x_2),$$

that satisfies (1)–(3) almost everywhere.

From (2), (3) the equalities

$$C_1(x) = \frac{A_{21} + C_{21} n_e(x) + B_{21} J(\varphi)(x)}{A_{21} + (C_{12} + C_{21}) n_e(x) + (B_{12} + B_{21}) J(\varphi)(x)} f(x), \qquad (7)$$

$$C_2(x) = \frac{C_{12} n_e(x) + B_{12} J(\varphi)(x)}{A_{21} + (C_{12} + C_{21}) n_e(x) + (B_{12} + B_{21}) J(\varphi)(x)} f(x) \qquad (8)$$

follow, where

$$J(\varphi)(x) = \int_I \int_{-1}^1 \frac{\kappa(\nu)}{2} \varphi(x, \nu, \mu) d\mu d\nu.$$

Equation (1) takes the form

$$\mu \frac{\partial}{\partial x} \varphi(x, \nu, \mu) + h \nu_{12} \frac{\kappa(\nu)}{2} \frac{(A_{21} B_{12} + (B_{12} C_{21} - B_{21} C_{12}) n_e(x)) f(x)}{A_{21} + (C_{12} + C_{21}) n_e(x) + (B_{12} + B_{21}) J(\varphi)(x)}$$

$$= h \nu_{12} \frac{\kappa(\nu)}{2} A_{21} \frac{C_{12} n_e(x) + B_{12} J(\varphi)(x)}{A_{21} + (C_{12} + C_{21}) n_e(x) + (B_{12} + B_{21}) J(\varphi)(x)} f(x). \qquad (9)$$

Thus $\Phi(x, \nu, \mu) = \{\varphi(x, \nu, \mu), C_1(x), C_2(x)\}$ if and only if is a solution of system (1)–(3) when it satisfies almost everywhere (7)–(9). Therefore, to study the solvability of problem (1)–(4), it is enough to study the solvability of a nonlinear integro-differential equation (9) in $\mathcal{D}_\infty^+(D)$. After that, using (7), (8), functions C_1, $C_2 \in K_\infty(x_1, x_2)$ can be found.

Theorem 1. *Let formulated conditions on the coefficients of the system (1)–(3) are satisfied. Then there exists a unique solution of problem (7)–(9).*

We consider the following iterative linearizing algorithm for solving problem (1)–(4). Let $\varphi^0 \in K_\infty(D)$. For $k = 0, 1, \ldots$ define the functions C_1^k, C_2^k by the right hands of (7), (8) respectively, where φ^k is substituted instead φ, and let $\varphi^{k+1} \in \mathcal{D}_\infty^+(D)$ is a solution of the linear differential equation

$$\mu \frac{\partial}{\partial x} \varphi^{k+1} + h\nu_{12} \frac{\kappa(\nu)}{2} [B_{12}C_1^k - B_{21}C_2^k]\varphi^{k+1} = h\nu_{12} \frac{\kappa(\nu)}{2} A_{21}C_2^k. \qquad (10)$$

Theorem 2. *There exists $N > 0$, that for any $\varphi^0 \in K_\infty(D)$, $\|\varphi^0\|_{L_\infty(D)} \leq N$, there is a limit*

$$\varphi = \lim_{k \to \infty} \varphi^k$$

of the iterative process (10) by the norm of $L_p(D)$, $1 \leq p < \infty$. Moreover $\varphi \in \mathcal{D}_\infty^+(D)$ and $\{\varphi, C_1, C_2\} \in \mathcal{D}_\infty^+(D) \times K_\infty(x_1, x_2) \times K_\infty(x_1, x_2)$, where C_1, C_2 determined by the relations (7) and (8), is a solution of problem (1)–(4).

Let $\psi \in K_\infty(D)$. We denote

$$F(\psi)(x) = h\nu_{12}f(x) \frac{A_{21}B_{12} + (B_{12}C_{21} - B_{21}C_{12})n_e(x)}{A_{21} + (C_{12} + C_{21})n_e(x) + (B_{12} + B_{21})J(\psi)(x)}; \qquad (11)$$

$$P(\psi)(x) = h\nu_{12}f(x)A_{21} \frac{C_{12}n_e(x) + B_{12}J(\psi)(x)}{A_{21} + (C_{12} + C_{21})n_e(z) + (B_{12} + B_{21})J(\psi)(x)}. \qquad (12)$$

Equation (9) take the form

$$\mu \frac{\partial}{\partial x} \varphi(x, \nu, \mu) + \frac{\kappa(\nu)}{2} F(\varphi)(x)\varphi(x, \nu, \mu) = \frac{\kappa(\nu)}{2} P(\varphi)(x). \qquad (13)$$

The Theorems 1, 2 are a consequence of a more general result. Let "$\succ$" is the order relation generated by the cone $K_\infty(D)$ in $L_\infty(D)$. Consider differential-operator equation (13), where $F, P : K_\infty(D) \to K_\infty(x_1, x_2)$ are operators satisfying the following conditions for all $\psi, \psi_1, \psi_2 \in K_\infty(D)$, $\psi_1 \succ \psi_2$, and almost all $x \in (x_1, x_2)$.

1) $P(\psi)(x) \leq M$, $F(\psi)(x) \leq M$ for some $M \geq 0$.
2) $P(\psi)/F(\psi) \in K_\infty(x_1, x_2)$ and there is $N \geq 0$ that if $\|\psi\|_{L_\infty(D)} \leq N$ then

$$\frac{P(\psi)(x)}{F(\psi)(x)} \leq N.$$

3) $P(\psi_1)(x) - P(\psi_2)(x) \geq 0$, $F(\psi_1)(x) - F(\psi_2)(x) \leq 0$.
4) $F(\psi_1)(x) - F(\psi_2)(x) = 0$ if $P(\psi_1)(x) - P(\psi_2)(x) = 0$.
5)

$$F(\psi_1)(x)J(\psi_1)(x) - F(\psi_2)(x)J(\psi_2)(x) \geq P(\psi_1)(x) - P(\psi_2)(x).$$

6)
$$\left\|\frac{F(\psi_2) - F(\psi_1}{F(\psi_1)}\right\|_{L_p(x_1,x_2)} \le C\|\psi_1 - \psi_2\|_{L_p(D)},$$

$$\left\|\frac{P(\psi_1) - P(\psi_2)}{F(\psi_1)}\right\|_{L_p(x_1,x_2)} \le C\|\psi_1 - \psi_2\|_{L_p(D)}$$

for some $C > 0$.

Theorem 3. *If conditions 1)–3) are satisfied, there exist a solution $\varphi \in \mathcal{D}_\infty^+(D)$ of equation (13); if conditions 1)–5) are satisfied, the solution of (13) is unique. Under conditions 1)–6), the solution of Eq. (13) can be obtained as the limit*

$$\varphi = \lim_{k \to \infty} \varphi^k$$

by the norm of space $L_p(D)$ ($1 \le p < \infty$), where the functions $\varphi^k \in \mathcal{D}_\infty^+(D)$ ($k = 1, 2, \ldots$) are defined recursively as a result of solving the sequence of problems

$$\mu\frac{\partial}{\partial x}\varphi^{k+1} + \frac{\kappa(\nu)}{2}F(\varphi^k)\varphi^{k+1} = \frac{\kappa(\nu)}{2}P(\varphi^k), \tag{14}$$

$\varphi^0 \in K_\infty(D)$ an arbitrary element such that $\|\varphi^0\|_{L_\infty(D)} \le N$.

3 Proofs of Theorems

3.1 Preliminary Statements

To prove Theorem 3, preliminary statements formulated in the following lemmas will be required.

Lemma 1. *Let $a, b \in K_\infty(D)$, $b/a \in K_\infty(D)$. There is a unique solution $\varphi \in \mathcal{D}_\infty(D)$ of the equation*

$$\mu\frac{\partial}{\partial x}\varphi(x,\nu,\mu) + a(x,\nu,\mu)\varphi(x,\nu,\mu) = b(x,\nu,\mu). \tag{15}$$

Moreover $\varphi \in \mathcal{D}_\infty^+(D)$ and

$$\|\varphi\|_{L_\infty(D)} \le \|b/a\|_{L_\infty(D)}. \tag{16}$$

For any $1 \le s < p < \infty$ there is a constant $C(s, p) > 0$ that

$$\|\varphi\|_{L_s(D)} \le C(s,p)\|a\|_{L_\infty(D)}^{1/p}\|b/a\|_{L_p(D)}. \tag{17}$$

Proof. The solution of (15), (4) has the form

$$\varphi(x,\nu,\mu) = \begin{cases} \varphi_-(x,\nu,\mu), & \mu < 0, \\ \varphi_+(x,\nu,\mu), & \mu > 0, \end{cases}$$

where

$$\varphi_-(x,\nu,\mu) = \frac{1}{\mu}\int_{x_2}^{x} b(\xi,\nu,\mu)\exp\{\frac{1}{\mu}\int_{x}^{\xi} a(\xi',\nu,\mu)d\xi'\}d\xi,$$

$$\varphi_+(x,\nu,\mu) = \frac{1}{\mu}\int_{x_1}^{x} b(\xi,\nu,\mu)\exp\{\frac{1}{\mu}\int_{x}^{\xi} a(\xi',\nu,\mu)d\xi'\}d\xi.$$

Obviously, $\varphi \in \mathcal{D}_\infty^\pm(D)$. We have for all $x \in [x_1, x_2]$ and almost all $\nu \in I,\ \mu > 0$

$$\varphi_+(x,\nu,\mu) \le \|b/a\|_{L_\infty(D)}\left(1 - \exp\{-\frac{1}{\mu}\int_{x_1}^{x} a(\xi,\nu,\mu)d\xi\}\right) \le \|b/a\|_{L_\infty(D)}.$$

Let $1 \le s < p < \infty,\ q = p/(p-1)$. Then

$$\varphi_+(x,\nu,\mu) = \int_{x_1}^{x} \mu^{-\frac{1}{q}-\frac{1}{p}}\frac{b(\xi,\nu,\mu)}{a(\xi,\nu,\mu)}a^{\frac{1}{q}+\frac{1}{p}}(\xi,\nu,\mu)\exp\{\frac{1}{\mu}\int_{x}^{\xi} a(\xi',\nu,\mu)d\xi'\}d\xi$$

$$\le \mu^{-1/p}\|a\|_{L_\infty(D)}^{1/p}\|b/a\|_{L_p(D)}q^{-1/q},$$

$$\int_0^1 \varphi_+^s(x,\nu,\mu)d\mu \le q^{-s/q}\frac{p}{p-s}\|a\|_{L_\infty(D)}^{s/p}\|\frac{b}{a}\|_{L_p(D)}^s.$$

Similar estimates are fulfilled for $\varphi_-(x,\nu,\mu)$. Thus, inequalities (16) and (17) are valid, $C(s,p) = (2d\nu_0)^{1/s}(p-1)^{1-1/p}(p-s)^{-1/s}p^{1/p+1/s-1}$.

Lemma 2. *Let $a,\ b \in K_\infty(x_1,x_2),\ a(x) > 0,\ x \in [x_1,x_2],\ \kappa \in K_\infty(I)$ and conditions (6) are met, a function $\varphi \in \mathcal{D}_\infty^+(D)$ satisfies the inequality*

$$\mu\frac{\partial}{\partial x}\varphi(x,\nu,\mu) + \frac{\kappa(\nu)}{2}a(x)\varphi(x,\nu,\mu) \ge \frac{\kappa(\nu)}{2}b(x).$$

Then for some $\alpha > 0$

$$\int_{x_1}^{x_2}\int_I\int_{-1}^1 \mu\frac{\partial}{\partial x}\varphi(x,\nu,\mu)d\mu d\nu dx \ge \alpha\int_{x_1}^{x_2} b(x)dx.$$

Lemma 3. *Let $a_n,\ b_n \in K_\infty(x_1,x_2),\ n = 1,2,\ a_1(x) \le a_2(x),\ b_1(x) \ge b_2(x)$ almost everywhere; $\varphi_n \in \mathcal{D}(D),\ n = 1,2$, are solutions of corresponding equations*

$$\mu\frac{\partial}{\partial x}\varphi_n(x,\nu,\mu) + \frac{\kappa(\nu)}{2}a_n(x)\varphi_n(x,\nu,\mu) = \frac{\kappa(\nu)}{2}b_n(x). \tag{18}$$

Then $\varphi_1 \succ \varphi_2$.

Lemmas 2, 3 are proved in [16].

3.2   Proof of Theorem 3

We define the operator $A : K_\infty(D) \to \mathcal{D}_\infty^+(D)$, which assigns each function $\psi \in K_\infty(D)$ the function $\varphi = A(\psi)$, defined as a solution from $\mathcal{D}_\infty^+(D)$ of equation

$$\mu\frac{\partial}{\partial x}\varphi(x,\nu,\mu) + \frac{\kappa(\nu)}{2}F(\psi)(x)\varphi(x,\nu,\mu) = \frac{\kappa(\nu)}{2}P(\psi)(x). \tag{19}$$

It follows from Lemma 1 that a solution $\varphi \in \mathcal{D}_\infty^+(D)$ exists, is unique and, according to condition 2), $\|\varphi\|_{L_\infty(D)} \le N$ if $\|\psi\|_{L_\infty(D)} \le N$. Therefore the operator A leaves the cone segment $\langle 0, U \rangle = \{\psi \in K_\infty(D) : \psi \prec U\}$ invariant, where $U(x,\nu,\mu) = N$ almost everywhere. The cone segment $\langle 0, U \rangle$ is a complete sublattice [17] of a conditionally complete lattice $\mathcal{D}_\infty^+(D)$.

Thus, the problem of solvability of Eq. (13) in the class $\mathcal{D}_\infty^+(D)$ reduces to the problem of determining fixed points of the operator A', which is the restriction of A to the complete sublattice $\langle 0, U \rangle$.

Let $\psi_1, \psi_2 \in\, < 0, U >,\, \psi_1 \succ \psi_2$. Then, by condition 3), $P(\psi_1)(x) \ge P(\psi_2)(x)$, $F(\psi_1)(x) \le F(\psi_2)(x)$. Applying Lemma 3, we conclude that $A' : \langle 0, U \rangle \to \langle 0, U \rangle$ is an isotone operator. According to Tarski's theorem [17] and from results of [12], the set of its fixed points is not empty and contains its infimum ψ_* and supremum ψ^*.

Assuming now that conditions 4), 5) are satisfied, we show that the operator A' has at most one fixed point. Suppose that functions $\psi_*, \psi^* \in \mathcal{D}_\infty^+(D)$ satisfy the Eq. (13). Then we get that $\varphi = \psi^* - \psi_*$ satisfies the equation

$$\mu\frac{\partial}{\partial x}\varphi + \frac{\kappa(\nu)}{2}[F(\psi^*)\psi^* - F(\psi_*)\psi_*] = \frac{\kappa(\nu)}{2}[P(\psi^*) - P(\psi_*)]. \tag{20}$$

It follows that

$$\int_{x_1}^{x_2}\int_I\int_{-1}^{1}\mu\frac{\partial\varphi}{\partial x}(x,\nu,\mu)d\mu d\nu dx - \int_{x_1}^{x_2}[P(\psi^*)(x) - P(\psi_*)(x)]dx$$

$$+ \int_{x_1}^{x_2}[F(\psi^*)(x)J(\psi^*)(x) - F(\psi_*)(x)J(\psi_*)(x)]\,dx = 0.$$

Using condition 5), we get

$$\int_{x_1}^{x_2}\int_{-1}^{1}\int_0^{\nu_0}\mu\frac{\partial\varphi}{\partial x}(x,\nu,\mu)d\nu d\mu dx \le 0. \tag{21}$$

On the other hand, from condition 3) we conclude that the inequality

$$\mu\frac{\partial}{\partial x}\varphi(x,\nu,\mu) + \frac{\kappa(\nu)}{2}F(\psi^*)(x)\varphi(x,\nu,\mu) \ge \frac{\kappa(\nu)}{2}[P(\psi^*)(x) - P(\psi_*)(x)]$$

is true. Applying Lemma 2, we have

$$\int_{x_1}^{x_2}\int_{-1}^{1}\int_I\mu\frac{\partial}{\partial x}\varphi(x,\nu,\mu)d\nu d\mu dx \ge \alpha\int_{x_1}^{x_2}[P(\psi^*)(x) - P(\psi_*)(x)]dx, \tag{22}$$

where $\alpha = \int_0^1 \exp[-\kappa^* M d/(2\mu)]d\mu > 0$. Comparing (21) and (22) and taking into account condition 3) we obtain that $P(\psi^*)(x) = P(\psi_*)(x)$ almost everywhere. Therefore, by condition 4) $F(\psi^*) = F(\psi_*)$. Thus, ψ^*, ψ_* satisfy the same equation

$$\mu\frac{\partial}{\partial x}\psi(x,\nu,\mu) + \frac{\kappa(\nu)}{2}F(\psi_*)(x)\psi(x,\nu,\mu) = \frac{\kappa(\nu)}{2}P(\psi_*)(x).$$

Since by Lemma 1 the solution of this equation is unique, so $\psi^* = \psi_*$. Thus, the uniqueness of the solution of Eq. (13) in the class $\mathcal{D}_\infty^+(D)$ is established.

The convergence of the iterative method formulated in Theorem 3, by virtue of the definition of the operator A, is obviously equivalent to the convergence for any $\varphi^0 \in \langle 0, U \rangle \subset K_\infty(D)$ of successive approximations

$$\varphi^{k+1} = A\varphi^k. \tag{23}$$

We define the sequences $\{\xi_n\}_{n=0}^\infty$, $\{\eta_n\}_{n=0}^\infty$ of elements of the complete lattice $\langle 0, U \rangle$, taking $\xi_0 = 0$, $\xi_{n+1} = A\xi_n$, $\eta_0 = U$, $\eta_{n+1} = A\eta_n$, $n = 0, 1, \ldots$. Obviously,

$$\xi_0 \prec \xi_1 \prec \ldots, \quad \eta_0 \succ \eta_1 \succ \ldots.$$

Denote $\xi = \sup\{\xi_n\}$, $\eta = \inf\{\eta_n\}$. Since the order convergence in $L_p(D)$ implies convergence according to the norm [18], $\|\xi - \xi_n\|_{L_p(D)} \to 0$, $\|\eta - \eta_n\|_{L_p(D)} \to 0$.

Let $1 \le s < p < \infty$. Applying Lemma 1 and condition 6), we obtain for any $\psi_1, \psi_2 \in \langle 0, U \rangle$, $\psi_1 \succ \psi_1$

$$\|A\psi_1 - A\psi_2\|_{L_s(D)} \le C(s,p)(M\kappa^*/2)^{1/p}C(1+N)\|\psi_1 - \psi_2\|_{L_p(D)}.$$

Thus, $\|A\xi - A\xi_n\|_{L_s(D)} \to 0$, $\|A\eta - A\eta_n\|_{L_s(D)} \to 0$.

As the norm $\|\cdot\|_{L_s(D)}$ is monotone, we conclude that

$$\|A\xi - \xi\|_{L_s(D)} \le \|A\xi - A\xi_n\|_{L_s(D)} \to 0, \quad \|A\eta - \eta\|_{L_s(D)} \le \|A\eta - A\eta_n\|_{L_s(D)} \to 0,$$

so ξ, η are fixed points of the operator A', and $\xi = \eta$ due to the uniqueness of the fixed point of A'. Therefore, $\lim_{n\to\infty}\|\xi_n - \eta_n\|_{L_p(D)} = 0$.

Let us now take the sequence $\{\varphi^k\}_{k=0}^\infty$ defined by (23) for $\varphi^0 \in \langle 0, U \rangle$. How easy to see, $\xi_n \prec A^{(n+1)}\varphi^0 \prec \eta_n$, from which it follows that

$$\|A^{(n+1)}\varphi^0 - \xi\|_{L_p(D)} \le \|A^{(n+1)}\varphi^0 - \xi_n\|_{L_p(D)} + \|\xi - \xi_n\|_{L_p(D)} \le 2\|\xi_n - \eta_n\|_{L_p(D)}.$$

Theorem 3 is proved.

3.3 Proof of Theorems 1, 2

To prove Theorems 1, 2, it is enough to show that the operators F, P taken in a concrete form (11), (12) satisfy conditions 1)–6).

The positivity of the operators follows from the conditions on the coefficients of the problem, condition 1) is fulfilled for $M = \nu_{12}hf^* \max\{A_{21}, B_{12}\}$.

Further, for all $\psi \in K_\infty(D)$ and for almost all $x \in (x_1, x_2)$

$$\frac{P(\psi)(x)}{F(\psi)(x)} \leq \frac{A_{21}[C_{12}n_e(x) + B_{21}\|\psi\|_{L_\infty(D)}]}{A_{21}B_{12} + (B_{12}C_{21} - B_{21}C_{12})n_e(x)}.$$

Condition 2) is met, therefore, for $N \geq A_{21}C_{12}/(B_{12}C_{21} - B_{21}C_{12})$.

Let $\psi_1, \psi_2 \in K_\infty(D)$, $\psi_1 \succ \psi_2$.

$$F(\psi_2) - F(\psi_1) = h\nu_{12}f(B_{12} + B_{21})\frac{A_{21}B_{12} + (B_{12}C_{21} - B_{21}C_{12})n_e}{R(\psi_1)R(\psi_2)}J(\psi_1 - \psi_2),$$

$$P(\psi_1) - P(\psi_2) = h\nu_{12}fA_{21}\frac{A_{21}B_{12} + (B_{12}C_{21} - B_{21}C_{12})n_e}{R(\psi_1)R(\psi_2)}J(\psi_1 - \psi_2),$$

where

$$R(\psi_i) = A_{21} + (C_{12} + C_{21})n_e + (B_{12} + B_{21})J(\psi_i), \ i = 1, 2.$$

Thus, contitions 3) is true.

$P(\psi_1)(x) - P(\psi_2)(x) = 0$ if and only if $f(x)J(\psi_1 - \psi_2)(x) = 0$, hence the condition 4) is fulfilled.

$$F(\psi_1)(x)J(\psi_1) - F(\psi_2)(x)J(\psi_2) - [P(\psi_1)(x) - P(\psi_2)(x)]$$

$$= h\nu_{12}f(x)(C_{12} + C_{21})n_e(x)\frac{A_{21}B_{12} + (B_{12}C_{21} - B_{21}C_{12})n_e(x)}{R(\psi_1)(x)R(\psi_2)(x)}J(\psi_1 - \psi_2)(x),$$

that is, inequality 5) is true.

Since

$$\frac{F(\psi_2) - F(\psi_1)}{F(\psi_1)} = \frac{B_{12} + B_{21}}{R(\psi_2)}J(\psi_1 - \psi_2), \quad \frac{P(\psi_1) - P(\psi_2)}{F(\psi_1)} = \frac{A_{21}}{R(\psi_2)}J(\psi_1 - \psi_2),$$

condition 6) is satisfied, where $C = \kappa^* 2^{-1/p}\nu_0^{1-1/p}\max\{1, (B_{12} + B_{21})/A_{21}\}$.

The validity of the assertions of Theorems 1, 2 follows from Theorem 3.

4   Numerical Solution of the Transfer Equation

4.1   Difference Approximation of the Kinetic Equation

We apply the considered iterative algorithm to solve the nonstationary transport equation

$$\frac{1}{c}\frac{\partial\varphi(x,\nu,\mu,t)}{\partial t} + \mu\frac{\partial}{\partial x}\varphi(x,\nu,\mu,t) + \frac{\kappa(\nu)}{2}F(\varphi)(x,t)\varphi(x,\nu,\mu,t) = \frac{\kappa(\nu)}{2}P(\varphi)(x,t). \tag{24}$$

Let $\varphi^0 \in \langle 0, U\rangle \subset K_\infty(D)$. For $k = 0, 1, \ldots$

$$C_1^k(x,t) = \frac{A_{21} + C_{21}n_e(x) + B_{21}J(\varphi^k)(x,t)}{A_{21} + (C_{12} + C_{21})n_e(x) + (B_{12} + B_{21})J(\varphi^k)(x,t)}f(x), \tag{25}$$

$$C_2^k(x,t) = \frac{C_{12}n_e(x) + B_{12}J(\varphi^k)(x,t)}{A_{21} + (C_{12} + C_{21})n_e(x) + (B_{12} + B_{21})J(\varphi^k)(x,t)} f(x), \qquad (26)$$

$$a^k(x,\nu,t) = \frac{\kappa(\nu)}{2}F(\varphi^k)(x,t) = h\nu_{12}\frac{\kappa(\nu)}{2}[B_{12}C_1^k(x,t) - B_{21}C_2^k(x,t)],$$

$$b^k(x,\nu,t) = \frac{\kappa(\nu)}{2}P(\varphi^k)(x,t) = h\nu_{12}\frac{\kappa(\nu)}{2}A_{21}C_2^k(x,t).$$

The function φ^{k+1} is a solution of the equation

$$\frac{1}{c}\frac{\partial\varphi^{k+1}}{\partial t} + \mu\frac{\partial}{\partial x}\varphi^{k+1} + a^k\varphi^{k+1} = b^k. \qquad (27)$$

We introduce a difference grid for the variables t (index n), x (index i) and μ (index j), and approximate Eq. (27) at step $k+1$ using the integro-interpolation method:

$$\frac{1}{c}\frac{\varphi_{i+1/2,j+1/2}^{n+1} - \varphi_{i+1/2,j+1/2}^{n}}{\Delta tt} + \mu_{j+1/2}\frac{\varphi_{i+1,j+1/2}^{n+1} - \varphi_{i,j+1/2}^{n+1}}{\Delta x}$$

$$+ a_{i+1/2}^{n+1}\varphi_{i+1/2,j+1/2}^{n+1} = b_{i+1/2}^{n+1}, \qquad (28)$$

where $\Delta t = t_{n+1} - t_n$, $\Delta x = x_{i+1} - x_i$.

To close the system of grid equations (28) with respect to the spatial variable x, we use an additional approximation relation of the WDD scheme [7]

$$\begin{cases} \varphi_{i+1,j+1/2}^{n+1} = (\delta + 1)\varphi_{i+1/2,j+1/2}^{n+1} - \delta\varphi_{i,j+1/2}^{n+1}, \; \mu_{j+1/2} > 0, \\ \varphi_{i,j+1/2}^{n+1} = (\delta + 1)\varphi_{i+1/2,j+1/2}^{n+1} - \delta\varphi_{i+1,j+1/2}^{n+1}, \; \mu_{j+1/2} < 0, \end{cases}$$

where weight $\delta \in [0,1]$ determines the order of approximation in space: for $\delta = 1$ we get a DD-scheme of the second order of approximation, for $\delta = 0$ we get a ST-scheme of the first order. In general, for $\delta < 1$, the WDD scheme is non-positive and non-monotonic.

The function of radiation intensity, due to "physicality" cannot take negative values. Correction algorithms are used to improve the properties of the difference scheme. They allow, at the cost of some deterioration in the accuracy of calculating integral expressions that depend on the solution, to improve its local characteristics. In this work, two correction algorithms were used – the zero fix up method and step-by-step correction (St-method) [7].

The essence of the zero fix up method is that if during the calculation according to the DD-scheme on the interval $[x_i, x_{i+1}]$ negative solutions are obtained, then this solution is forced to zero.

The correction according to the St-scheme consists in that when negative solutions are obtained during the solution according to the DD-scheme on the interval $[x_i, x_{i+1}]$, the parameter δ on this segment is chosen equal to zero, which implies that the DD-scheme goes into St, which is positive. After recalculation of the solution at this interval, further calculations continue according to the DD-scheme.

The resulting system of grid equations is solved by a simple iteration method in accordance with the linearizing algorithm:

$$\frac{1}{c}\frac{\varphi_{i+1/2,j+1/2}^{n+1,k+1} - \varphi_{i+1/2,j+1/2}^{n,k+1}}{\Delta tt} + \mu_{j+1/2}\frac{\varphi_{i+1,j+1/2}^{n+1,k+1} - \varphi_{i,j+1/2}^{n+1,k+1}}{\Delta x}$$

$$+ a_{i+1/2}^{n+1,k}\varphi_{i+1/2,j+1/2}^{n+1,k+1} = b_{i+1/2}^{n+1,k}, \tag{29}$$

where k is the iteration number.

4.2   Properties of the Difference Dcheme

The Order of Approximation

Consider an operator equation $Au = F$ and its difference analog $A_h u_h = f_h$. The expression $f_h - A_h u$, where u is an exact solution of the operator equation, is called the discrepancy of difference scheme [19].

Find the discrepancy for scheme (29). To do this, we decompose the solution into a Taylor series at the node $(x_{i+1/2}, t_n)$.

$$\varphi_{i+1/2}^{n+1} = \varphi_{i+1/2}^n + \Delta t\frac{\partial\varphi}{\partial t}|_{i+1/2,n} + \frac{\Delta t^2}{2}\frac{\partial^2\varphi}{\partial t^2}|_{i+1/2,n} + O(\Delta t^3),$$

$$\varphi_{i+1}^{n+1} = \varphi_{i+1/2}^n + \frac{\Delta x}{2}\frac{\partial\varphi}{\partial t}|_{i+1/2,n} + \frac{\Delta x^2}{8}\frac{\partial^2\varphi}{\partial x^2}|_{i+1/2,n} + \frac{\Delta x^3}{48}\frac{\partial^3\varphi}{\partial x^3}|_{i+1/2,n} + O(\Delta x^4),$$

$$\varphi_i^{n+1} = \varphi_{i+1/2}^n - \frac{\Delta x}{2}\frac{\partial\varphi}{\partial t}|_{i+1/2,n} + \frac{\Delta x^2}{8}\frac{\partial^2\varphi}{\partial x^2}|_{i+1/2,n} - \frac{\Delta x^3}{48}\frac{\partial^3\varphi}{\partial x^3}|_{i+1/2,n} + O(\Delta x^4).$$

Taking into account that

$$\frac{1}{c}\frac{\partial\varphi}{\partial t}|_{i+1/2,n} + \mu_{j+1/2}\frac{\partial\varphi}{\partial x}|_{i+1/2,n} + a_{i+1/2}^n\varphi|_{i+1/2,n} = b_{i+1/2}^n,$$

we get that the discrepancy is equal to

$$\psi = \frac{1}{c}\frac{\Delta t}{2}\frac{\partial^2\varphi}{\partial t^2}|_{i+1/2,n} + O(\Delta t^2) + \frac{\Delta x^2}{24}\frac{\partial^3\varphi}{\partial x^3}|_{i+1/2,n} + O(\Delta x^4) = O(\Delta t) + O(\Delta x^2).$$

Therefore, we obtain the first order of approximation in the time variable and the second in the spatial variable.

Stability of the Difference Scheme

Write the difference scheme (29) as

$$B\frac{\varphi^{k+1} - \varphi^k}{\Delta t} + A\varphi^k = F.$$

With a fixed right-hand side, the solution error satisfies the homogeneous equation

$$B\Delta\varphi^{k+1} = (B - \Delta t A)\Delta\varphi^k.$$

We represent the solution of this equation in the form of separable variables:

$$\Delta\varphi(x_i, t_n) = \lambda_q^n e^{iqx_i}, \quad q = 0, \pm1, \pm2, \ldots$$

Then $\Delta\varphi^{n+1} = \lambda_q \Delta\varphi^n$, so λ_q is a the growth multiplier of the q-th harmonic. According to the stability attribute [19], for all q the inequality $|\lambda_q| \leq 1 + C\Delta\tau$ must be satisfied. Applying this method to the difference scheme (29) we obtain:

$$\frac{\lambda_q - 1}{\Delta t} + \frac{\mu}{\Delta x}\lambda_q \left(e^{iq\Delta x/2} - e^{-iq\Delta x/2}\right) + a\lambda_q = 0,$$

$$|\lambda_q| = \left(1 + a\Delta t + \left(\frac{2\Delta t\mu}{\Delta x}\sin(q\frac{\Delta x}{2})\right)^2\right)^{-1/2} < 1.$$

Therefore, the difference scheme under consideration is stable.

Conservativeness of the Difference Scheme

When the conservation law regarding particle transfer is fulfilled, the number of particles entering the cell is equal to the number of particles that flew out of neighboring cells in the direction of the specified cell.

Write the scheme (29) for two adjacent cells $[x_i, x_{i+1}]$, $[x_{i+1}, x_{i+2}]$.

$$\frac{1}{c}\frac{\varphi_{i+1/2,j+1/2}^{n+1,k+1} - \varphi_{i+1/2,j+1/2}^{n,k+1}}{\Delta t} + \mu_{j+1/2}\frac{\varphi_{i+1,j+1/2}^{n+1,k+1} - \varphi_{i,j+1/2}^{n+1,k+1}}{\Delta x}$$

$$+ a_{i+1/2}^{n+1,k}\varphi_{i+1/2,j+1/2}^{n+1,k+1} = b_{i+1/2}^{n+1,k};$$

$$\frac{1}{c}\frac{\varphi_{i+3/2,j+1/2}^{n+1,k+1} - \varphi_{i+3/2,j+1/2}^{n,k+1}}{\Delta t} + \mu_{j+1/2}\frac{\varphi_{i+2,j+1/2}^{n+1,k+1} - \varphi_{i+1,j+1/2}^{n+1,k+1}}{\Delta x}$$

$$+ a_{i+3/2}^{n+1,k}\varphi_{i+3/2,j+1/2}^{n+1,k+1} = b_{i+3/2}^{n+1,k}.$$

The direction of particle transfer is determined by the $\mu_{j+1/2}$ sign: for positive values, this is movement in the positive direction of the x axis; when negative, vice versa. The number of particles passing through $x = x_{i+1}$ in both cases is equal to $\mu_{j+1/2}\varphi_{i+1,j+1/2}^{n+1,k+1}/\Delta x$.

Thus, the difference scheme (29) preserves the balance of movement across cell boundaries, which determines the conservativeness property of this scheme with respect to transfer.

5   Computational Results

Numerical studies of the performance of the proposed iterative algorithm for the selected difference scheme (WDD scheme) were carried out on two test problems with different optical properties (optically dense and transparent). The values of the Einstein coefficients of spontaneous and forced radiation and absorption A_{21}, B_{12}, B_{21}, C_{12}, C_{21} were taken from [20–22].

The transfer equation is solved in $\Pi = \{0 \leq x \leq 1\}$. The initial value of the total number of particles in the system is $\varphi_0 = 1$, the input flow $\varphi = 1$ is set on the left boundary of the region. A series of calculations were carried out in both stationary and non-stationary modes. The space step was taken uniform $\Delta x = 0,1$, the time step was taken $\Delta t = 0,0001$.

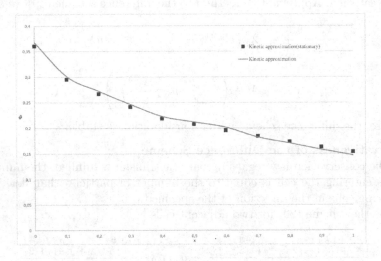

Fig. 1. Distribution of scalar flux density

The calculation in the non-stationary mode was carried out before the solution entered the stationary mode. The scalar flow of particles $\tilde{\varphi} = \int_{-1}^{1} \varphi d\mu$ is considered as a result.

The iterative process is repeated until the condition of achieving the specified accuracy according to the criterion

$$|\tilde{\varphi}^{k+1} - \tilde{\varphi}^{k}| \leq \varepsilon_0 \tilde{\varphi}^{k+1} + \varepsilon_1, \tag{30}$$

ε_0, ε_1 are given constants, we take $\varepsilon_0 = 10^{-4}$, $\varepsilon_1 = 10^{-16}$.

For the angular variable μ the 32 directions were selected.

Figure 1 shows a comparison of the scalar flux density (stationary and non-stationary).

In the second example the region is optically dense and filled with rubidium vapors. The 12 directions were selected for the angular variable μ.

Figure 2 shows a comparison of the scalar flux density (stationary and non-stationary). Figure 3 shows the scalar flow profiles for the 5th, 100, 350, 750 and the last 870 iterations, respectively.

Based on the results of numerical studies, the following conclusions can be made.

The proposed numerical iterative algorithm is convergent. In the optically dense problem the iterative process converged in 870 iterations, in the optically

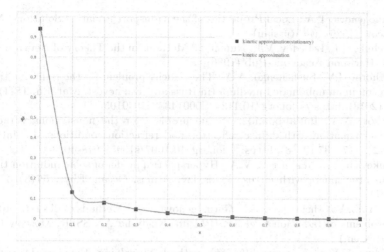

Fig. 2. Distribution of scalar flux density

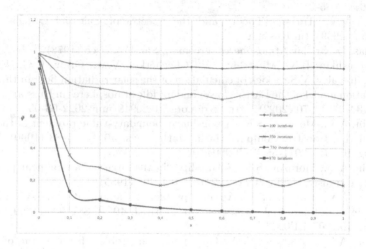

Fig. 3. Scalar flow profile

transparent problem, in 24 iterations with a given accuracy. The difference in the number of iterations is due to the properties of the medium [23].

When solving the kinetic equation in a non-stationary approximation, it is sufficient without loss of generality to make the time step sufficiently large.

References

1. Vladimirov, V.S.: Mathematical problems in the one-velocity theory of particle transport. Tr. MIAN SSSR 61. Publ. House of the Academy of Sciences of the USSR, Moscow (1961)

2. Germogenova, T.A.: Local Properties of the Transport Equation Solutions. Nauka, Moscow (1986). (in Russian)
3. Marchuk, G.I., Lebedev, V.I.: Numerical Methods in the Theory of Neutron Transport. Harwood Academic Pub (1986)
4. Prokhorov, I.V., Sushchenko, A.A.: The Cauchy problem for the radiative transfer equation in an unbounded medium (in Russian). Dal'nevost. Mat. Zh. **18**(1), 101–111 (2018). https://doi.org/10.1134/S0001433821030105
5. Amosov, A.A.: Initial-boundary value problem for the nonstationary radiative transfer equation with diffuse reflection and refraction conditions. J. Math. Sci. **235**(2), 117–137 (2018). https://doi.org/10.1007/s10958-018-4063-y
6. Sushkevich, T.A., Falaleeva, V.A.: Hyperspectral model of solar radiation transfer in an atmosphere with cirrus clouds Izv. Atmos. Ocean. Phys. **57**(3), 277–285 (2021)
7. Bass, L.P., Voloshchenko, A.M., Germogenova, T.A.: Methods of discrete ordinates in problems of radiation transfer. IPM im. Keldysha AN SSSR, Moscow (1986). (in Russian)
8. Gol'din, V.Ya.: A quasi-diffusion method of solving the kinetic equation. Comp. Math. Math. Phys. **4**(6), 136–149 (1964). https://doi.org/10.1016/0041-5553(64)90085-0
9. Ivanov, V.V.: Transfer Theory and the Spectra of Calestial Objects. Nauka, Moscow (1969). (in Russian)
10. Mihalas, D.: Stellar Atmospheres. Freeman, San Francisco (1978)
11. Chetverushkin, B.N., Mingalev, I.V., Chechetkin, V.M., Orlov, K.G., Fedotova, E.A., Mingalev, V.S.: Block of calculation of the solar radiation field in the general circulation model of the Earth's lower and middle atmosphere (in Russian). Matem. Mod. **34**(3), 43–70 (2022). https://doi.org/10.20948/mm-2022-03-03
12. Kalinin, A.V., Morozov, S.F.: A non-linear boundary-value problem in the theory of radiation transfer. Comput. Math. Math. Phys. **30**(4), 76–83 (1990). https://doi.org/10.1016/0041-5553(90)90046-U
13. Kalinin, A.V., Morozov, S.F.: Solvability "in the large" of a nonlinear problem of radiative transfer. Differ. Uravn. **21**(3), 484–494 (1985)
14. Kalinin, A.V., Morozov, S.F.: A mixed problem for a nonstationary system of non-linear integro-differential equations. Sib. Math. J. **40**(5), 887–900 (1999). https://doi.org/10.1007/BF02674718
15. Kalinin, A.V., Tyukhtina, A.A.: On a nonlinear problem for a system of integro-differential equations of radiative transfer theory. Comput. Math. Math. Phys. **62**(6), 933–944 (2022). https://doi.org/10.1134/S0965542522060094
16. Kalinin, A.V., Kozlov, A.Y.: On anonlinear problem of radiative transfer in a plane-parallel layer (in Russian). Vest. NNGU **4**(1), 140–145 (2011)
17. Birkhoff, G.: Lattice Theory. American Mathematical Society (1961)
18. Kantorovich, L.V., Akilov, G.P.: Functional Analysis. Pergamon Press (1982)
19. Kalitkin, N.N.: Numerical Methods. Nauka, Moscow (1978). (in Russian)
20. Frisch, S.E.: Optical Spectra of Atoms. Fizmatlit, Moscow (1963). (in Russian)
21. Elyashevich, M.A.: Atomic and Molecular Spectroscopy. Fizmatlit, Moscow (1962). (in Russian)
22. Allen, C.W.: Astrophysical Quantities. The Athlone Press, London (1973)
23. Zel'dovich, Ya.B., Raizer, Y.P.: Physics of Shock Waves and High-Temperature Hydrodynamic Phenomena. Academic Press, Cambridge (1967)

Role of Abnormal Calcium Signaling and Liver Tissue Structure in Glucose and Lipid Metabolism: Mathematical Modeling

Arina V. Martyshina[ID] and Irina V. Dokukina[(✉)][ID]

Sarov Physics and Technology Institute, National Research Nuclear University MEPhI, Sarov, Nizhniy Novgorod 607186, Russian Federation
IVDokukina@mephi.ru

Abstract. One of the main components of glucose and lipid metabolism in hepatocytes that provide liver's metabolic flexibility is the cell ability to temporarily store glucose in the form of glycogen. The glycogen storage and release processes are regulated by hormones insulin and glucagon and by intracellular calcium signaling. Correct calcium signaling strongly depends on proper intracellular structure, in particular on adequate functioning of mitochondria-associated membranes (MAMs). MAMs defects were shown to affect calcium signaling and expected to alter glucose metabolism and storage. Using mathematical modeling we research the role of both abnormal MAMs functioning and calcium release from endoplasmic reticulum in hepatocyte glucose and lipid metabolism. Also we estimate the consequences of decreased amount of hormones, that reach pericentral liver zone in comparison to periportal zone, for the amount of stored glycogen, TAG and glucose released by hepatocyte in the glycogenolytic mode.

Keywords: Hepatocyte · MAM · Metabolism · Calcium ions

1 Introduction

1.1 Main Processes

The main organ processing glucose is the liver. Hepatocytes can adjust their functioning depending on the organism dietary status to maintain stable plasma glucose level. Hepatocytes store excess glucose during and after feeding and release it into plasma in the fasting state. Glucose exchange between hepatocyte and extracellular space is provided by glucose-specific receptor channels GLUT2 located in the cell plasma membrane (PM) [45]. Due to abundance of GLUT2 receptors and their high glucose permeability glucose concentration levels both inside and outside the cell could be considered as the same [4,9].

In cell's cytosol glucose is phosphorylated by glucokinase and could be dephosphorylated back by glucose phosphatase [4,18]. Glucose-6-phosphate is

© The Author(s), under exclusive license to Springer Nature Switzerland AG 2022
D. Balandin et al. (Eds.): MMST 2022, CCIS 1750, pp. 121–135, 2022.
https://doi.org/10.1007/978-3-031-24145-1_10

the basic compound of early stages of glucose metabolism. It is utilized in several ways: in glycogen storage, energy production, and lipid synthesis. Glycogen storage process - glycogenesis - is multistage enzymatic process regulated by glycogen synthase [1]. The opposite process glycogenolysis is a two-step process regulated by glycogen phosphorylase [4, 18].

Conversion of glucose into energy starts with glucose breakdown by glycolysis. This is a complex multistage process, regulated among others by phosphofructokinase and pyruvate kinase, with pyruvate as the end product [30]. In the mitochondria pyruvate transforms into acetyl-CoA with the help of pyruvate dehydrogenase. After that acetyl-CoA molecules enter Krebs cycle to produce energy [4] and get transported to cytosol where they are used as the substrate for de novo lipogenesis with the fatty acids (FA) as the end product and as the substrate for further triglyceride (TAG) synthesis [19]. In addition to FA, TAG synthesis requires glycerol-3-phosphate produced during glycolysis. Glycerol is obtained from glycerol-3-phosphate. Both glycerol and FA could also come from the bloos plasma as a part of chylomicron and as a result of adipocyte lipolysis. Glycerol entering the cell gets phosphorylated by glycerol kinase. Hepatocyte glycerol and FA uptake from chylomicrons takes place with delay due to the food absorbtion by intestine and further delivery to the liver via blood flow. FA could be transformed into energy molecules via β-oxidation [37].

Opposite to glycolysis is the process of gluconeogenesis by which hepatocytes can produce glucose from glycerol, amino acids, lactate and other substrates [35]. In particular, glycerol-3-phosphate with the help of glycerol-3-phosphate dehydrogenase and fructose-1,6-bisphosphatase turns into glucose-6-phosphate [4].

TAG molecules form very low density lipoproteins (VLDL) that are released into the blood plasma to be delivered to muscles to fulfill their energy needs and adipose tissue for later storage [12].

Normal values of main metabolites in $mmol \cdot L^{-1}$ are:

– glucose 3.39–5.5 in fasting and less than 7.72 postprandial [16],
– glycogen 200–300 in fasting and about 500 postprandial [43],
– FA 0.4–0.7 [34],
– TAG about 10–30 [17, 32],
– plasma glycerol 0.1–0.2 [32],
– glucose-6-phosphate 0.2–0.3 [43].

1.2   Hormonal Regulation

Key regulators of glucose and lipid metabolism are hormones insulin and glucagon produced by pancreas. They provide metabolic flexibility of the liver. Hormonal regulation can stimulate some processes and suppress another [4]. In particular,

– when plasma glucose level is elevated insulin induces gene transcription of glucokinase, phosphofructokinase, pyruvate kinase, and glycogen synthase. At the same time insulin suppresses fructose-1,6-bisphosphatase, glucose phosphatase and glycogen phosphorylase. Glucagon acts opposite to insulin: it suppresses glucokinase, phosphofructokinase, pyruvate kinase, induces transcription of glucose phosphatase, increases activity of fructose-1,6-bisphosphatase and glycogen phosphorylase, and deactivates glycogen synthase;

- hormone-sensitive lipase regulating adipocyte lipolysis is activated by phosphorylation by protein kinase A in response to elevated plasma glucagon. Gluconeogenesis as well as lipolysis are inhibited by insulin;
- gluconeogenesis from glycerol ir regulated by insulin and glucagon via fructose-1,6-bisphosphatase. Both glycerol kinase and glycerol-3-phosphate dehydrogenase are not regulated by either insulin or glucagon;
- glycerol-3-phosphate production from glucose-6-phosphate is regulated by both insulin and glucagon in the same way as glycolysis, via phosphofructokinase, since early stages of both processes overlap;
- despite enzymatic suppression of gluconeogenesis by insulin experimental research confirms its low efficiency [37], so we do not include the regulation in our model;
- glucagon induces transcription of carnitine palmitoyltransferase I (CPT-1), which provides catabolism of long-chain FA transforming them into acyl-carnitines for further transportation into the mitochondria and β-oxidation [25,41];
- insulin suppresses apolipoprotein B synthesis needed for VLDL assembly. Insulin also activates phosphatidylinositol-3-kinase needed for phosphatidylinositol-3,4,5-trisphosphate (PIP3), which prevents merging nascent VLDL with TAG droplets [33,40]. In other words, insulin suppresses VLDL formation.

Metabolic zonation is an important structure property of the liver. Researchers single out three main zones inside each liver lobule: pericentral (PC), periportal (PP) and intermediate zone. For rodents changes in mitochondria size and morphology, graded depositions of glycogen, and gradients in certain important substrates, hormones and enzymes responsible for glycolysis, gluconeogenesis, FA oxidation, and others across the porto-central axis of the liver lobule are very well characterized [5,24]. Human hepatocyte heterogeneity is much less researched due to limited availability of healthy human liver lobule samples at physiological conditions [5,11]. Despite some differences between rodent and human liver zonation characteristics [11], many similarities are still present. For example, hormones glucagon and insulin decrease during blood passage through the liver from PP to PC zone by 50% in the fasting, whereas glucagon decreases by the same 50% and insulin decreases only by 15% in the postprandial state [22]. Note, that GLUT2 glucose transporters appear to be evenly distributed throughout the normal liver tissue [22]. It was shown that glucose release by hepatocytes in the fasting state is significantly reduced if stimulated by glucagon concentrations half below saturated [42]. Shortage in hormones in PC zone could be somewhat compensated through Ca^{2+} signaling with Ca^{2+}, inositol-1,4,5-triphosphate (IP$_3$) and glucose molecules being able to pass through gap junctions connecting adjacent hepatocytes [46]. It was also shown in human liver that accumulated glycogen amounts in the fasting state were higher in PP than in PC zone [10].

There are still very little and controversial data regarding lipid zonation in hepatocytes (for review see [38]). Several research works results suggest higher TAG synthesis and VLDL secretion in PC zone in both normal and pathological organ (non-alcoholic fatty liver disease), whereas others report the opposite results with prevalence of PP zone.

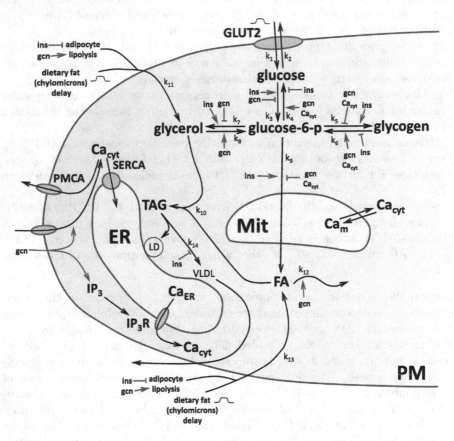

Fig. 1. The scheme of processes and their regulations included in the model. See text for details.

1.3   Calcium Regulation

Both glucagon and insulin regulate glucose and lipid metabolism in hepatocytes partially via Ca^{2+} signaling pathway [3,36]. In the fasting state even small amount of glucagon stimulate production of IP_3 in the hepatocyte cytosol [8].

Elevated IP_3 activates IP_3-receptors located in the endoplasmic reticulum (ER) membrane which serve as Ca^{2+} release channels from the ER store [20]. Higher glucagon concentration stimulates extracellular Ca^{2+} influx into the cytosol [29]. In hepatocyte there are three major mechanisms to eliminate toxic effect of high cytosolic Ca^{2+} [8]:

- sarco/endoplasmic reticulum Ca^{2+}-ATPase (SERCA) pumps excessive cytosolic Ca^{2+} back inside the ER,
- the mitochondria located close to the ER membrane form mitochondria-associated membranes (MAMs) to effectively uptake large amounts of cytosolic Ca^{2+} [44],
- plasma membrane Ca^{2+}-ATPase (PMCA) transports Ca^{2+} from the cytosol to the extracellular space.

Cytosolic Ca^{2+} oscillations stimulate a signaling cascade resulting in

- suppression of glycogen synthesis via glycogen synthase inhibition, and
- stimulation of glycogenolysis and gluconeogenesis (at the stage of glucose-6-phosphate dephosphorylation) via calcium/calmodulin-dependent kinase II (CaMKII) [4,18].

Increased cytosolic Ca^{2+} also decreases activity of pyruvate kinase and thereby suppresses glycolysis. All these processes eventually result in glucose release by hepatocyte [2].

In the postprandial state glucose level is increased and insulin decreases cytosolic Ca^{2+} concentration by phosphorylation and thereby activity suppression of IP_3-receptors [23]. As the result hepatocytes start to store glucose into glycogen.

Inability of the mitochondrial membrane to form MAMs and incorrect modulation of IP_3-receptors were shown by mathematical modeling to be associated with hepatocyte insulin resistance [13]. Combination of MAMs dysfunction and incorrect modulation of IP_3-receptors results in significant increase in cytosolic Ca^{2+} concentration. However the model [13] does not include description of plasma glucose levels and lipid metabolism components, and to our best knowledge there is no experimental research done in hepatocytes to check any correlation between MAMs dysfunction, the metabolic liver zonation, and hepatic lipid metabolism.

1.4   Mathematical Models

Hepatocyte metabolic pathways are very difficult to research both experimentally and theoretically. Mathematical models of hepatocyte metabolic pathways could be divided in two main classes: flux models and dynamic models. Flux models consider both fluxes of various substances through the cell and organelle membranes and fluxes as the result from chemical and enzyme kinetics with certain stoichiometry as steady-state processes [27,31]. Dynamical models are alternative to flux models. They describe metabolites chemical transformations and transport by means of differential equations using chemical kinetics approach and the law of mass action [15, 26, 28].

For example, model [26] gives description of hepatocyte role in glucose homeostasis including glycolysis, gluconeogenesis, and glycogen metabolism regulated by insulin and glucagon. However, the model does not include transcription regulation and hepatocyte lipid metabolism. Model [28] is the classic chemical kinetics model with automated formulation of model equations. The model takes into account glycolysis, gluconeogenesis, glycogen metabolism, pentose phosphate path, Krebs cycle, and FA, NAD and ATP metabolism. Nevertheless, high dimension of the model and large number of unknown parameters needed to be calibrated gets the model analysis complicated. The model [28] includes hepatocyte lipid metabolism only in part. In addition, to our best knowledge no models consider calcium regulation of glucose and lipid metabolism in full detail.

Somewhat simplified multicellular model proposed in [39] allows to research the role of nutrient and FA uptake gradient in TAG accumulation hepatic zonation. More complex multiscale models [6] and [7] allow to research lipid zonation in the liver at various conditions. However, these last three models do not take into account any Ca^{2+} signaling and its possible dysfunction.

Here we propose dynamical model of main processes of lipid and glucose metabolism in hepatocytes including their Ca^{2+} regulation. Our model posses all drawbacks typical for dynamical models and therefore can give only qualitative predictions and estimates. With our model we analyse the role of both cytosolic Ca^{2+} signaling abnormalities and hormone gradient along porto-central axis of the human liver in development of pathological condition of glycemic control system.

2   Model Description

The main processes of glucose and lipid metabolism in hepatocytes are shown in Fig. 1 as long as all model variables - concentrations of Ca^{2+}, glucose, glycogen, FA, TAG, glycerol, and glucose-6-phosphate (*glucose-6-p*). There are three compartments: cytosol (Cyt), mitochondria (Mit), and endoplasmic reticulum (ER). In Fig. 1 all black arrows show fluxes, and red arrows show either process stimulation or suppression. We take all equations for Ca^{2+} dynamics (see Eqs. 1–4) from our model [13]. Using the law of mass action we write down the system of ordinary differential equations describing all other processes (see Eqs. 5–11) shown in Fig. 1.

$$\frac{dCa_{cyt}}{dt} = \beta_{cyt} \cdot [\varepsilon \cdot (J_{in} - J_{out}) + J_{rel} - J_{SERCA} + J_{mo} - J_{mi}] \qquad (1)$$

$$\frac{dCa_{ER}}{dt} = \frac{\beta_{ER}}{\rho_{ER}} \cdot (J_{SERCA} - J_{rel}) \qquad (2)$$

$$\frac{dCa_m}{dt} = \frac{\beta_m}{\rho_m} \cdot (J_{mi} - J_{mo}) \qquad (3)$$

$$\frac{dh}{dt} = a \cdot (Ca_{cyt} + d_{inh}) \cdot [\frac{d_{inh}}{Ca_{cyt} + d_{inh}} - h] \tag{4}$$

$$\frac{d[glucose]}{dt} = k_1 - (k_2 + k_3(ins, gcn)) \cdot [glucose]$$
$$+ k_4(ins, gcn, Ca_{cyt}) \cdot [glucose\text{-}6\text{-}p] \tag{5}$$

$$\frac{d[glycogen_0]}{dt} = k_5(ins, gcn, Ca_{cyt}) \cdot [glucose\text{-}6\text{-}p]$$
$$- k_6(ins, gcn, Ca_{cyt}) \cdot [glycogen_0] \tag{6}$$

$$\frac{d[glycogen]}{dt} = \frac{[glycogen_0] - [glycogen]}{\tau} \tag{7}$$

$$\frac{d[FA]}{dt} = k_{13}(ins, gcn) + k_9(ins, gcn, Ca_{cyt}) \cdot [glucose\text{-}6\text{-}p]$$
$$- (k_{12}(gcn) + k_{10} \cdot [glycerol]) \cdot [FA] \tag{8}$$

$$\frac{d[TAG]}{dt} = k_{10} \cdot [FA] \cdot [glycerol] - k_{14}(ins) \cdot [TAG] \tag{9}$$

$$\frac{d[glycerol]}{dt} = k_{11}(ins, gcn) + k_7(ins, gcn) \cdot [glucose\text{-}6\text{-}p]-$$
$$- (k_8(gcn) + k_{10} \cdot [FA]) \cdot [glycerol] \tag{10}$$

$$\frac{d[glucose\text{-}6\text{-}p]}{dt} = k_3(ins, gcn) \cdot [glucose] + k_8(gcn) \cdot [glycerol]$$
$$+ k_6(ins, gcn, Ca_{cyt}) \cdot [glycogen] - (k_4(ins, gcn, Ca_{cyt}) \tag{11}$$
$$+ k_5(ins, gcn, Ca_{cyt}) + k_9(ins, gcn, Ca_{cyt}) + k_7(ins, gcn)) \cdot [glucose\text{-}6\text{-}p]$$

Note, that we introduce in the model a time delay ($\tau = 200$ s) in glycogenolysis via additional Eq. 7 as in [21].

To qualitatively describe insulin and glucagon dynamics during transitions between the fasting and postprandial states we use smoothed time-dependent step functions (Fig. 2a, b). Low insulin and high glucagon values are related to the fasting state, and high insulin and low glucagon values are related to the postprandial (fed) state. The dependence could be described as:

$$f(t) = (A - B) \cdot \frac{(t - p)^{ord}}{(t - p)^{ord} + l^{ord}} + B, \tag{12}$$

where A and B are high and low extreme values of hormone concentrations. For insulin we use: $A = 10$ $\mu IU \cdot mL^{-1}$, $B = 50$ $\mu IU \cdot mL^{-1}$, $ord = 6$, $p = 7200$ s, $l = 3000$ s, for glucagon: $A = 50$ $pg \cdot mL^{-1}$, $B = 10$ $pg \cdot mL^{-1}$, $ord = 6$, $p = 7200$ s, $l = 3000$ s. We describe the fasting/postprandial state-dependent dynamics of coefficients k_{11} and k_{13} for dietary fat uptake and k_1 for dietary glucose uptake with the same function type as in Eq. 12.

In order to incorporate hormone and Ca^{2+} regulation in the model we use sigmoid functions to describe coefficients $k_i(ins, gcn)$ (Fig. 2c, d) and both sigmoid and rational functions to describe coefficients $k_i(Ca_{cyt})$ (Fig. 2e) in Eqs. 5–11. The coefficient increase up to k_{sat} value refers to the process stimulation, whereas the decrease below k_{sat} refers to the process suppression. Here k_{sat} is the saturated value of the hormone activity. Activation function (Fig. 2c) is described as:

$$f_{act}(x) = k_{sat} \cdot \left(\frac{(1 + e^{-a \cdot b})}{1 + e^{-a \cdot (x-b)}} - \frac{(1 + e^{-a \cdot b})}{1 + e^{a \cdot b}} \right), \tag{13}$$

whereas suppression function (Fig. 2d) is:

$$f_{inh}(x) = k_{sat} - f_{act}(x). \tag{14}$$

The choice of sigmoid as a hormone regulatory function was made due to supposed threshold-type transition between two distinct states - fasting or postprandial, with either glucose release or glucose storage mode in hepatocyte.

The rational function (Fig. 2e) for coefficients $k_i(Ca_{cyt})$ of Ca^{2+} regulation has horizontal asymptote and could be described as:

$$f(Ca_{cyt}) = k_{sat} \cdot \frac{(Ca_{cyt})^c}{d + (Ca_{cyt})^c}. \tag{15}$$

The rational function describes Ca^{2+} regulation better than sigmoid because of oscillating Ca^{2+} behavior in hepatocyte with no actual threshold value and cell switching from one mode to another.

All coefficient functions (Eq. 12–15) have unique parameters as amplitude (the difference between the maximum and the minimum values), the slope, and the horizontal shift. Variation of these parameters allows to control the regulation process and to find the critical points of the signal scheme (see Fig. 1) where a pathology could arise. For the full list of parameters see Table 1.

Some processes have multiple regulators. For example, we describe the total regulatory coefficient $k_i(ins, gcn, Ca_{cyt})$ as a sum of the single regulatory coefficients:

$$k_i(ins, gcn, Ca_{cyt}) = k_i(ins) + k_i(gcn) + k_i(Ca_{cyt}). \tag{16}$$

3   Results

At first we validate our modeling results on available experimental data on normal and abnormal values of metabolite concentrations and on qualitative data on appropriate dynamics of concentrations at various conditions. All metabolite concentrations in our model are within healthy range at normal conditions (see Fig. 3). Decreased insulin level causes increased model glucose concentrations related to type I diabetes (results are not shown).

We model metabolic hepatic zonation in a simple way of mimicking insulin and glucagon gradient from PP to PC zone. Since we propose only a single cell model, we could consider two different furthest cells from PP and PC zones. To take hormone gradient into account we reduce:

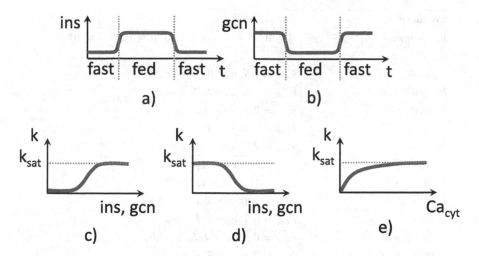

Fig. 2. Model description of dietary state-dependent concentrations of insulin (a) and glucagon (b). Sigmoid functions of hormone activation (c) and suppression (d) in the model. Rational function of Ca^{2+} regulation (e) in the model. See text for detailed description.

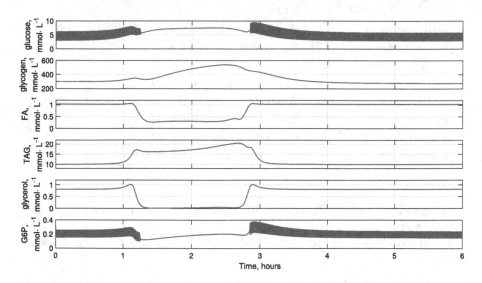

Fig. 3. Modeling results for normal cell metabolites in fasting and postprandial (from 1 h 10 min to 2 h 50 min) states.

Table 1. Model parameters.

Coefficient	Parameters
$k_1(t)$, $mmol \cdot s^{-1} \cdot L^{-1}$	$A = 0$, $B = 155$, $ord = 6$, $p = 7200$, $l = 3000$
k_2, s^{-1}	20
$k_3(ins)$, s^{-1}	$k_{sat} = 7.5$, $a = 0.05$, $b = 30$
$k_3(gcn)$, s^{-1}	$k_{sat} = 0.5$, $a = 0.05$, $b = 30$
$k_4(ins)$, s^{-1}	$k_{sat} = 220$, $a = 0.05$, $b = 30$
$k_4(gcn)$, s^{-1}	$k_{sat} = 220$, $a = 0.05$, $b = 30$
$k_4(Ca_{cyt})$, s^{-1}	$k_{sat} = 220$, $c = 0.9$, $d = 0.2$
$k_5(ins)$, s^{-1}	$k_{sat} = 50$, $a = 0.1$, $b = 30$
$k_5(gcn)$, s^{-1}	$k_{sat} = 10$, $a = 0.1$, $b = 30$
$k_5(Ca_{cyt})$, s^{-1}	$k_{sat} = 200$, $a = 0.1$, $b = 0.13$
$k_6(ins)$, s^{-1}	$k_{sat} = 0.2$, $a = 1$, $b = 30$
$k_6(gcn)$, s^{-1}	$k_{sat} = 0.16$, $a = 0.5$, $b = 30$
$k_6(Ca_{cyt})$, s^{-1}	$k_{sat} = 2$, $c = 0.9$, $d = 0.2$
$k_7(ins)$, s^{-1}	$k_{sat} = 0.03$, $a = 0.1$, $b = 30$
$k_7(gcn)$, s^{-1}	$k_{sat} = 0.01$, $a = 0.1$, $b = 30$
$k_8(gcn)$, s^{-1}	$k_{sat} = 100$, $a = 0.2$, $b = 30$
$k_9(ins)$, s^{-1}	$k_{sat} = 0.001$, $a = 0.1$, $b = 30$
$k_9(gcn)$, s^{-1}	$k_{sat} = 0.001$, $a = 0.1$, $b = 30$
$k_9(Ca_{cyt})$, s^{-1}	$k_{sat} = 0.0003$, $a = 0.1$, $b = 0.13$
k_{10}, $L \cdot s^{-1} \cdot mmol^{-1}$	0.12
$k_{11}(t)$, $mmol \cdot s^{-1} \cdot L^{-1}$	$A = 0$, $B = 0.1$, $ord = 4$, $p = 10800$, $l = 3000$
$k_{11}(ins)$, $mmol \cdot s^{-1} \cdot L^{-1}$	$k_{sat} = 39.6$, $a = 0.5$, $b = 30$
$k_{11}(gcn)$, $mmol \cdot s^{-1} \cdot L^{-1}$	$k_{sat} = 39.6$, $a = 0.5$, $b = 30$
$k_{12}(gcn)$, s^{-1}	$k_{sat} = 1.35$, $a = 0.4$, $b = 30$
$k_{13}(t)$, $mmol \cdot s^{-1} \cdot L^{-1}$	$A = 0$, $B = 0.002$, $ord = 4$, $p = 10800$, $l = 3000$
$k_{13}(ins)$, $mmol \cdot s^{-1} \cdot L^{-1}$	$k_{sat} = 0.75$, $a = 0.5$, $b = 30$
$k_{13}(gcn)$, $mmol \cdot s^{-1} \cdot L^{-1}$	$k_{sat} = 0.75$, $a = 0.5$, $b = 30$
$k_{14}(ins)$, s^{-1}	$k_{sat} = 0.01$, $a = 0.4$, $b = 30$

- glucagon level for the PC cell by 50% both in the fasting and postprandial state, and
- insulin level by 50% in the fasting and by 15% postprandial (as was shown in [22]).

Reduced hormone levels have almost no effect on the postprandial glucose levels, but result in hypoglycemia in the fasting state (see Table 2), consistent with available experimental data [42].

Table 2. Model results.

Condition	Glucose	Glycogen	TAG
	fed/fast, $mmol \cdot L^{-1}$	fed/fast, $mmol \cdot L^{-1}$	fed/fast, $mmol \cdot L^{-1}$
Normal Ca^{2+}, PP	7.4/3.0–5.7	540/274	18/10
Abnormal Ca^{2+}, PP	6.9–10.3/2.9–7.6	436/287	18/10
Normal Ca^{2+}, PC	7.5/1.6–3.2	456/191	59/44
Abnormal Ca^{2+}, PC	7.0–9.9/1.4–4.4	367/202	59/44

We see similar to the latter decreasing effect on glycogen concentrations regardless the dietary state when switching from PP to PC zone cell. Note, that the same results were obtained experimentally for the fasting state in human liver [10]. In addition, reduced hormone levels result in noticeable increase in synthesized TAG amounts in PC zone model cell, supporting one of the hypothesis reviewed in [38].

We also checked the effect of abnormal Ca^{2+} signaling, researched in [13], on glucose, glycogen and TAG levels in hepatocyte in both the fasting and postprandial states (see Table 2). Our modeling results suggest glucose increase in both the fasting and postprandial states in response to abnormally elevated cytosolic Ca^{2+} concentration. At the same time postprandial glycogen level gets decreased whereas its fasting level gets somewhat increased, in total suggesting less effective glycogen accumulation and release process. Abnormal Ca^{2+} increase though has no effect on hepatic TAG levels. This dynamics is observed for both PP and PC zones.

4   Discussion

The results of our modeling are mostly related to medical biophysics and are not focused on mathematics due to several reasons. To begin with, the model includes 11 equations with the large number of parameters many of which are not numerical constants, but nonlinear dynamic functions of other variables what depend on the cell changing feeding state. The model structure makes any kind of mathematical analysis, e.g. stability analysis, very complicated and lengthy. In addition, the focus on getting biologically relevant results makes parameter ranges very narrow and, so, search of possible modes of system behavior uninteresting. Preliminary analysis of model sensitivity to slight change of parameters values results in either very little effect or significant increase of some variables values way out of physiological limits. Since variable's shift to unphysiological value makes the model biologically irrelevant we decided not to proceed any further with stability analysis of the model.

We would like to emphasize that our model is the model of the part of living organism with many individual characteristics. It is well known from experiment that there are not alike cells with the same dynamics. Amplitudes and times characteristics vary significantly from cell to cell and from person to person, that's why researchers mostly look for trends and thresholds between healthy behavior and pathology. We do the same in our model. We are unable to compare our results with any experimental dependency since there is no such a dependency available for the system we model. We have to rely on normal ranges of metabolites concentrations and vague textual descriptions of corresponding trends from experimental research papers to validate our model. Thus our model should be considered qualitative rather than quantitative since it only could and it does provide some new insights in possible reasons in pathology development and some estimates of its severity.

In particular, we consider two feeding states (fast and fed) in norm and pathology. Here we check only one possible reason of pathology development, namely abnormal calcium signaling for different zones of liver lobule. To the best of our knowledge there are no models incorporating calcium regulation of hepatocyte lipid metabolism. Also there is no such experimental research done to this day. That's why we believe that our results could be considered new and original and could improve understanding of hepatocyte lipid metabolism pathology development.

Thus we propose the valid model of glucose and lipid metabolism process in hepatocytes which correctly takes into account hormone gradient between PP and PC zone and its effect on hepatic glucose, glycogen and TAG concentration. The model also allows to check the impact of abnormal MAMs and IP_3-receptor dysfunction on the process. Further model parameters variation will allow to get insight into other aspects of the process. Though our single cell model is not able to take into account the effect of cell connectivity via gap junctions [14] we consider this as an important next step in our modeling research on the problem of metabolic hepatic zonation.

References

1. Agius, L.: Glucokinase and molecular aspects of liver glycogen metabolism. Biochem. J. **414**, 1–18 (2008). https://doi.org/10.1042/BJ20080595
2. Amaya, M.J., Nathanson, M.H.: Calcium signaling in the liver. Compr. Physiol. **3**, 515 (2013). https://doi.org/10.1002/cphy.c120013
3. Arruda, A.P., Hotamisligil, G.S.: Calcium homeostasis and organelle function in the pathogenesis of obesity and diabetes. Cell Metabol. **22**, 381–397 (2015). https://doi.org/10.1016/j.cmet.2015.06.010
4. Bayens, J.W., Dominiczak, M.: Medical Biochemistry, 5th edn. Elsevier, Amsterdam (2018)
5. Ben-Moshe, S., Itzkovitz, S.: Spatial heterogeneity in the mammalian liver. Nat. Rev. Gastroenterol. Hepatol. **16**(7), 395–410 (2019). https://doi.org/10.1038/s41575-019-0134-x
6. Berndt, N., Horger, M.S., Bulik, S., Holzhütter, H.G.: A multiscale modelling approach to assess the impact of metabolic zonation and microperfusion on the hepatic carbohydrate metabolism. PLoS Comput. Biol. **14**(2), e1006005 (2018). https://doi.org/10.1371/journal.pcbi.1006005
7. Berndt, N., et al.: Functional consequences of metabolic zonation in murine livers: insights for an old story. Hepatology **73**(2), 795–810 (2021). https://doi.org/10.1002/hep.31274

8. Berridge, M.J., Bootman, M.D., Roderick, H.L.: Calcium signalling: dynamics, homeostasis and remodelling. Nat. Rev. Mol. Cell Biol. **4**, 517–529 (2003). https://doi.org/10.1038/nrm1155

9. Bhagavan, N.: Medical Biochemistry, 4th edn. Elsevier, Amsterdam (2021). https://doi.org/10.1016/B978-012095440-7/50017-2

10. Chamlian, A., Benkoel, L., Minko, D., Njee, T., Gulian, J.M.: Ultrastructural heterogeneity of glycogen in human liver. Liver **9**(6), 346–350 (1989). https://doi.org/10.1111/j.1600-0676.1989.tb00422.x

11. Cunningham, R.P., Porat-Shliom, N.: Liver zonation - revisiting old questions with new technologies. Front. Physiol. **12**, 1433 (2021). https://doi.org/10.3389/fphys.2021.732929

12. Day, C.P., James, O.F.: Steatohepatitis: a tale of two 'hits'? Gastroenterology **114**, 842–845 (1998). https://doi.org/10.1016/S0016-5085(98)70599-2

13. Dokukina, I.V., Yamashev, M.V., Samarina, E.A., Tilinova, O.M., Grachev, E.A.: Calcium-dependent insulin resistance in hepatocytes: mathematical model. J. Theor. Biol. **522**, 110684 (2021). https://doi.org/10.1016/j.jtbi.2021.110684

14. Dokukina, I., Tsukanov, A., Gracheva, M., Grachev, E.: Effect of the tissue architecture on cell-to-cell calcium signaling. Biofizika **53**(2), 305–314 (2008)

15. Foguet, C., et al.: HepatoDyn: a dynamic model of hepatocyte metabolism that integrates 13C isotopomer data. PLoS Computat. Biol. **12**, e1004899 (2016). https://doi.org/10.1371/journal.pcbi.1004899

16. Freckmann, G., et al.: Continuous glucose profiles in healthy subjects under everyday life conditions and after different meals. J. Diab. Sci. Technol. **1**(5), 695–703 (2007). https://doi.org/10.1177/193229680700100513

17. Grzegorczyk, E., et al.: Effect of sleeve gastrectomy on proprotein convertase subtilisin/kexin type 9 (Pcsk9) content and lipid metabolism in the blood plasma and liver of obese wistar rats. Nutrients **11**(9), 2174 (2019). https://doi.org/10.3390/nu11092174

18. Hall, J.E., Hall, M.: Guyton and Hall Textbook of Medical Physiology, 14th edn. Elsevier, Amsterdam (2021)

19. Heacock, A.M., Agranoff, B.W.: CDP-diacylglycerol synthase from mammalian tissues. Biochim. Biophys. Acta - Lipids Lipid Metab. **1348**, 166–172 (1997). https://doi.org/10.1016/S0005-2760(97)00096-9

20. Hirata, K., et al.: Regulation of CA^{2+} signaling in rat bile duct epithelia by inositol 1,4,5-trisphosphate receptor isoforms. Hepatology **36**, 284–296 (2002). https://doi.org/10.1053/jhep.2002.34432

21. Jelic, K., Hallgreen, C.E., Colding-Jørgensen, M.: A model of NEFA dynamics with focus on the postprandial state. Ann. Biomed. Eng. **37**(9), 1897–1909 (2009). https://doi.org/10.1007/s10439-009-9738-6

22. Jungermann, K., Kietzmann, T.: Zonation of parenchymal and nonparenchymal metabolism in liver. Ann. Rev. Nutr. **16**, 179–203 (1996). https://doi.org/10.1146/annurev.nu.16.070196.001143

23. Khan, M.T., Wagner, L., Yule, D.I., Bhanumathy, C., Joseph, S.K.: AKT kinase phosphorylation of inositol 1, 4, 5-trisphosphate receptors. J. Biol. Chem. **281**, 3731–3737 (2006). https://doi.org/10.1074/jbc.M509262200

24. Kietzmann, T.: Metabolic zonation of the liver: the oxygen gradient revisited. Redox Biol. **11**, 622–630 (2017). https://doi.org/10.1016/j.redox.2017.01.012

25. Kim, J.Y., Hickner, R.C., Cortright, R.L., Dohm, G.L., Houmard, J.A.: Lipid oxidation is reduced in obese human skeletal muscle. Am. J. Physiol. Endocrinol. Metab. **279**, E1039–E1044 (2000). https://doi.org/10.1152/ajpendo.2000.279.5.e1039

26. König, M., Bulik, S., Holzhütter, H.G.: Quantifying the contribution of the liver to glucose homeostasis: a detailed kinetic model of human hepatic glucose metabolism. PLoS Computat. Biol. **8**, e1002577 (2012). https://doi.org/10.1371/journal.pcbi. 1002577

27. Lee, K., Berthiaume, F., Stephanopoulos, G.N., Yarmush, M.L.: Profiling of dynamic changes in hypermetabolic livers. Biotechnol. Bioeng. **83**, 400–415 (2003). https://doi.org/10.1002/bit.10682

28. de Mas, I.M., et al.: Compartmentation of glycogen metabolism revealed from 13C isotopologue distributions. BMC Syst. Biol. **5**, 1–14 (2011). https://doi.org/10. 1186/1752-0509-5-175

29. Mine, T., Kojima, I., Ogata, E.: Role of calcium fluxes in the action of glucagon on glucose metabolism in rat hepatocytes. Am. J. Physiol. - Gastrointest. Liver Physiol. **265**, G35–G42 (1993). https://doi.org/10.1152/ajpgi.1993.265.1.g35

30. Okar, D.A., Lange, A.J., Manzano, À., Navarro-Sabatè, A., Riera, L., Bartrons, R.: PFK-2/FBPase-2: maker and breaker of the essential biofactor fructose-2,6-bisphosphate. Trends Biochem. Sci. **26**, 30–35 (2001). https://doi.org/10.1016/ S0968-0004(00)01699-6

31. Orman, M.A., Arai, K., Yarmush, M.L., Androulakis, I.P., Berthiaume, F., Ierapetritou, M.G.: Metabolic flux determination in perfused livers by mass balance analysis: effect of fasting. Biotechno. Bioeng. **107**, 825–835 (2010). https://doi. org/10.1002/bit.22878

32. Perry, R.J., et al.: Leptin mediates a glucose-fatty acid cycle to maintain glucose homeostasis in starvation. Cell **172**(1–2), 234–248 (2018). https://doi.org/10.1016/ j.cell.2017.12.001

33. Phung, T.L., Roncone, A., Jensen, K.L.D.M., Sparks, C.E., Sparks, J.D.: Phosphoinositide 3-kinase activity is necessary for insulin-dependent inhibition of apolipoprotein b secretion by rat hepatocytes and localizes to the endoplasmic reticulum. J. Biol. Chem. **272**, 30693–30702 (1997). https://doi.org/10.1074/jbc. 272.49.30693

34. Pratt, A.C., Wattis, J.A., Salter, A.M.: Mathematical modelling of hepatic lipid metabolism. Math. Biosci. **262**, 167–181 (2015). https://doi.org/10.1016/j.mbs. 2014.12.012

35. Previs, S.F., Cline, G.W., Shulman, G.I.: A critical evaluation of mass isotopomer distribution analysis of gluconeogenesis in vivo. Am. J. Physiol. - Endocrinol. Metab. **277**, E154–E160 (1999). https://doi.org/10.1152/ajpendo.1999.277.1.e154

36. Roach, P.: Glycogen and its metabolism. Curr. Mol. Med. **2**, 101–120 (2005). https://doi.org/10.2174/1566524024605761

37. Samuel, V.T., Shulman, G.I.: The pathogenesis of insulin resistance: integrating signaling pathways and substrate flux. J. Clin. Invest. **126**, 12–22 (2016). https:// doi.org/10.1172/JCI77812

38. Schleicher, J., et al.: Zonation of hepatic fatty acid metabolism - the diversity of its regulation and the benefit of modeling. Biochim. Biophys. Acta Mol. Cell Biol. Lipids **1851**(5), 641–656 (2015). https://doi.org/10.1016/j.bbalip.2015.02.004

39. Schleicher, J., Dahmen, U., Guthke, R., Schuster, S.: Zonation of hepatic fat accumulation: insights from mathematical modelling of nutrient gradients and fatty acid uptake. J. Roy. Soc. Interface **14**(133), 20170443 (2017). https://doi.org/10. 1098/rsif.2017.0443

40. Sparks, J.D., Sparks, C.E., Adeli, K.: Selective hepatic insulin resistance, VLDL overproduction, and hypertriglyceridemia. Arterioscler. Thromb. Vasc. Biol. **32**, 2104–2112 (2012). https://doi.org/10.1161/ATVBAHA.111.241463

41. Stephens, F.B., Constantin-teodosiu, D., Greenhaff, P.L.: New insights concerning the role of carnitine in the regulation of fuel metabolism in skeletal muscle. J. Physiol. **581**, 431–444 (2007). https://doi.org/10.1113/jphysiol.2006.125799
42. Stümpel, F., Ott, T., Willecke, K., Jungermann, K.: Connexin 32 gap junctions enhance stimulation of glucose output by glucagon and noradrenaline in mouse liver. Hepatology **28**(6), 1616–1620 (1998). https://doi.org/10.1002/hep.510280622
43. Sydne J. Carlson-Newberry, R.B.C.: Emerging Technologies for Nutrition Research: Potential for Assessing Military Performance Capability. National Academies Press (1997). https://doi.org/10.17226/5827
44. Theurey, P., et al.: Mitochondria-associated endoplasmic reticulum membranes allow adaptation of mitochondrial metabolism to glucose availability in the liver. J. Mol. Cell Biol. **8**, 129–143 (2016). https://doi.org/10.1093/jmcb/mjw004
45. Thorens, B.: GLUT2, glucose sensing and glucose homeostasis. Diabetologia **58**(2), 221–232 (2014). https://doi.org/10.1007/s00125-014-3451-1
46. Willebrords, J., et al.: Structure, regulation and function of gap junctions in liver. Cell Commun. Adhesion **22**(2–6), 29–37 (2015). https://doi.org/10.3109/15419061.2016.1151875

On the Period Length Modulo p of the Numerators of Convergents for the Square Root of a Prime Number p

S. V. Sidorov[(⊠)] and P. A. Shcherbakov

Lobachevsky State University of Nizhni Novgorod, Nizhni Novgorod, Russia
sergey.sidorov@itmm.unn.ru, woolfer707@gmail.com

Abstract. In this paper, we investigate the properties of the sequence of numerators of convergents for the square root of a prime number. It is proved that the period length L of this sequence modulo p is equal to l, $2l$ or $4l$, where l is the period length of the continued fraction for the square root of the prime p. Namely, if the remainder of dividing p by 8 is equal to 7, then $L = l$; if the remainder of dividing p by 8 is equal to 3, then $L = 2l$; if the remainder of dividing p by 4 is equal to 1, then $L = 4l$.

Keywords: Continued fraction · Convergent · Prime number · Numerator · Square root

1 Introduction

A continued fraction is a classical concept of number theory, which is the subject of extensive literature (see [3,8–10,16,17,19]). Continued fractions have been used since ancient times to approximate real numbers with rational numbers (Diophantine approximation, see [12]). In particular, the continued fractions can be used in calendars to create intercalation cycles. In addition, continued fractions can be used to solve Pell's equation $x^2 - Dy^2 = 1$ (also called the Pell-Fermat equation) and generalized Pell's equation $x^2 - Dy^2 = N$, where $D > 1$ and is not a perfect square, $N \in \mathbb{Z}$ (see [13]). One application of Pell's equation is creating a public key cryptosystem based on it (see [21]).

2 Preliminaries

It is known that any real number x can be represented as a continued fraction

$$x = [q_0, q_1, q_2, \ldots, q_n, \ldots] = q_0 + \cfrac{1}{q_1 + \cfrac{1}{q_2 + \ldots}},$$

where $q_0 \in \mathbb{Z}$, and $q_i \in \mathbb{N}$ for $i = 1, 2, \ldots$. A continued fraction is finite if and only if $x \in \mathbb{Q}$. The integers $q_0, q_1, q_2, \ldots$ are called the coefficients, the

D. Balandin et al. (Eds.): MMST 2022, CCIS 1750, pp. 136–147, 2022.
https://doi.org/10.1007/978-3-031-24145-1_11

terms or partial quotients of the continued fraction. The continued fraction $\alpha_i = [q_i, q_{i+1}, q_{i+2}, \ldots]$ is the i-th complete quotient of x; and the continued fraction $[q_0, q_1, \ldots, q_i]$ is the i-th **convergent** to x. A convergent, being defined as a finite continued fraction, is always a rational number, so $[q_0, q_1, \ldots, q_i] = \dfrac{A_i}{B_i}$, where A_i, B_i are positive integers (except possibly $A_0 = q_0$, that can be negative). A continued fraction is called a periodic continued fraction if its terms eventually repeat from some point onwards. The minimal number of repeating terms is called the period length of the continued fraction. In general, a periodic continued fraction has the form $[q_0, q_1, \ldots, q_m, \overline{q_{m+1}, \ldots, q_{m+l}}]$, where the bar indicates the periodic part, which is repeated indefinitely, l is the period length. Lagrange proved in 1770 that a continued fraction is periodic if and only if x is a quadratic irrationality, that is, x arises as the solution of quadratic equation with integral coefficients.

There are the following recursive formulas for the numerator and the denominator of the ith convergent $\dfrac{A_i}{B_i}$:

$$\begin{cases} A_{-1} = 1, \\ A_0 = q_0, \\ A_{i+1} = q_{i+1}A_i + A_{i-1}, i \geq 0, \end{cases} \qquad \begin{cases} B_{-1} = 0, \\ B_0 = 1, \\ B_{i+1} = q_{i+1}B_i + B_{i-1}, i \geq 0. \end{cases}$$

The algorithm for calculating the terms q_i of a continued fraction for an irrational number x is as follows. Initial settings: $\alpha_0 = x, q_0 = \lfloor \alpha_0 \rfloor$. Recursive formulas are

$$\alpha_{i+1} = \frac{1}{\alpha_i - q_i}, q_{i+1} = \lfloor \alpha_{i+1} \rfloor, i \geq 0.$$

If $D > 0$ is an integer which is not a perfect square, the continued fraction for $\sqrt{D}$ is not only periodic, but also has remarkable properties. Namely (see [3, 17]),

$$\sqrt{D} = [q_0, \overline{q_1, q_2, \ldots, q_{l-2}, q_{l-1}, q_l}],$$

where l is the period length, $q_l = 2q_0$ and $q_i = q_{l-i}$ for $i = 1, \ldots, l-1$, i.e. the sequence $q_1, q_2, \ldots, q_{l-2}, q_{l-1}$ is a palindrome.

Note that the classical method of extracting the integer part and «reversing» the remainder is not suitable for calculating the period of the continued fraction for $\sqrt{D}$ using a computer. At some point, we will not have enough accuracy. But there is an algorithm that allows you to calculate the terms q_i of the continued fraction for $\sqrt{D}$ without losing accuracy. This algorithm is given below and it is sometimes called PQ-algorithm (see [15, 17, 20]).

Initial settings:

$$P_0 = 0, \quad Q_0 = 1, \quad \alpha_0 = \frac{P_0 + \sqrt{D}}{Q_0} = \sqrt{D}, \quad q_0 = \left\lfloor \frac{P_0 + \sqrt{D}}{Q_0} \right\rfloor.$$

For $i \geq 0$ set

$$P_{i+1} = q_i Q_i - P_i, \quad Q_{i+1} = \frac{D - P_{i+1}^2}{Q_i},$$

$$\alpha_{i+1} = \frac{P_{i+1} + \sqrt{D}}{Q_{i+1}}, \quad q_{i+1} = \left\lfloor \frac{P_{i+1} + \sqrt{D}}{Q_{i+1}} \right\rfloor.$$

It is easy to verify that this algorithm reproduces the continued fraction expansion of $\sqrt{D}$.

To find of the period length of the continued fraction for $\sqrt{D}$, it need to finish the calculation when the condition $q_i = 2q_0$ is hold. Under this condition, we get the period length $l = i$.

The following properties of sequences $\{A_i\}_{i=-1}^{\infty}$, $\{B_i\}_{i=-1}^{\infty}$, $\{P_i\}_{i=1}^{\infty}$, $\{Q_i\}_{i=0}^{\infty}$ are well known.

Theorem 1 ([17]). *If l is the period length of the continued fraction for $\sqrt{D}$, where D is not a perfect square, then sequences $\{P_i\}_{i=1}^{\infty}$, $\{Q_i\}_{i=0}^{\infty}$ are purely periodic with the period length l, that is*

$$P_{l+i+1} = P_{i+1}, \quad Q_{l+i} = Q_i \quad \text{for all} \quad i \geq 0,$$

and

$$P_{i+1} = P_{l-i}, \quad 0 \leq i \leq l-1,$$
$$Q_i = Q_{l-i}, \quad 0 \leq i \leq l.$$

In addition,

$$1 \leq Q_i \leq 2q_0 = 2\lfloor \sqrt{D} \rfloor \quad \text{for all} \quad i \geq 0,$$

and

$$Q_i = 1 \quad \text{if and only if} \quad l \mid i.$$

Theorem 2 ([17], p. 92). *If D is not a perfect square, then for all $i \geq -1$ the following equality holds*

$$A_i^2 - DB_i^2 = (-1)^{i+1} Q_{i+1}.$$

Theorem 3 ([17], pp. 84–85). *If l is the period length of the continued fraction for $\sqrt{D}$, where D is not a perfect square, and $\alpha_i = \dfrac{P_i + \sqrt{D}}{Q_i}$, $i = 1, 2, \ldots, l-1$ are the complete quotients of $\sqrt{D}$, then $Q_i \geq 3$, $i = 1, 2, \ldots, l-1$ except for even $l = 2k$, while $Q_k = 2$.*

Theorem 4 ([3], p. 108). *Let l be the length of the period of the continued fraction for $\sqrt{p}$, where p is an odd prime number. Then*

1. *l is odd if and only if $p \equiv 1 \pmod 4$.*
2. *l is even if and only if $p \equiv 3 \pmod 4$.*

The following theorem is proved in [7], but we give a new proof using some results from [17].

Theorem 5. *Let p be a prime number, $p \equiv 3 \pmod 4$, and let l be the length of the period of the continued fraction for $\sqrt{p}$. Then*

1. $l \equiv 0 \pmod 4$ *if and only if $p \equiv 7 \pmod 8$.*
2. $l \equiv 2 \pmod 4$ *if and only if $p \equiv 3 \pmod 8$.*

Proof. By Theorem 4, the period length l is even, so $l = 2k$. By Theorem 2 for $i = k - 1$, we get

$$A_{k-1}^2 \equiv (-1)^k Q_k \pmod p.$$

By Theorem 3, it follows that if $l = 2k$, then $Q_k = 2$, so

$$A_{k-1}^2 \equiv (-1)^k \cdot 2 \pmod p.$$

If $k = 2t$ is even, then $l = 4t \equiv 0 \pmod 4$ and

$$A_{k-1}^2 \equiv 2 \pmod p.$$

As is well known (see [8,14]), 2 is a quadratic residue modulo $p \equiv 3 \pmod 4$ if and only if $p \equiv 7 \pmod 8$. Therefore, $l \equiv 0 \pmod 4$ iff $p \equiv 7 \pmod 8$.

If $k = 2t + 1$ is odd, then $l = 4t + 2 \equiv 2 \pmod 4$ and

$$A_{k-1}^2 \equiv -2 \pmod p.$$

As is well known (see [8,14]), -2 is a quadratic residue modulo $p \equiv 3 \pmod 4$ if and only if $p \equiv 3 \pmod 8$. Therefore, $l \equiv 2 \pmod 4$ iff $p \equiv 3 \pmod 8$.   □

We will consider the case when $D = p$ is a prime number. To the end of this paper, we will be interested in the properties of the sequence $\{A_i\}_{i=-1}^{\infty}$ for $\sqrt{p}$ modulo p.

3   The Main Result

Lemma 1. *Let $k \in \mathbb{N}$, and let l be the length of the period of the continued fraction for $\sqrt{p}$, where p is an odd prime number. Then the following congruences modulo p hold:*

$$A_{kl-1}^2 \equiv (-1)^{kl} \pmod p,$$
$$A_{kl-2}^2 \equiv (-1)^{kl} q_0^2 \pmod p,$$
$$A_{kl-1} A_{kl-2} \equiv (-1)^{kl-1} q_0 \pmod p.$$

Proof. We have

$$\sqrt{p} = [q_0, \overline{q_1, \ldots, q_{l-1}, 2q_0}],$$

where $q_i = q_{l-i}$ for $i = 1, \ldots, l - 1$.

Let A_n and B_n, $n \geq -1$ be the numerator and the denominator of the n-th convergent for $\sqrt{p}$, respectively. According to Theorem 2 we have the equality

$$A_n^2 - pB_n^2 = (-1)^{n+1} Q_{n+1}. \tag{1}$$

Therefore, reducing the equality (1) modulo p, we obtain

$$A_n^2 \equiv (-1)^{n+1} Q_{n+1} \pmod{p} \tag{2}$$

for $n \geq -1$. Theorem 1 and the PQ-algorithm imply $P_1 = q_0 Q_0 - P_0 = q_0$, $Q_{kl} = 1$, $Q_{kl-1} = Q_{kl+1} = Q_1 = \dfrac{p - P_1^2}{Q_0} = p - q_0^2$ for all $k \in \mathbb{N}$. Substituting $n = kl - 1$ into (2), we get

$$A_{kl-1}^2 \equiv (-1)^{kl} \pmod{p}.$$

Substituting $n = kl - 2$ into (2), we get

$$A_{kl-2}^2 \equiv (-1)^{kl-1} Q_{kl-1} \equiv (-1)^{kl-1}(p - q_0^2) \equiv (-1)^{kl} q_0^2 \pmod{p}.$$

Hence, we have

$$A_{kl-2}^2 \equiv (-1)^{kl} q_0^2 \pmod{p}.$$

It remains to prove only the congruence $A_{kl-1} A_{kl-2} \equiv (-1)^{kl-1} q_0 \pmod{p}$. To do this, squaring the equality

$$A_n = q_n A_{n-1} + A_{n-2},$$

we get

$$A_n^2 = q_n^2 A_{n-1}^2 + 2 q_n A_{n-1} A_{n-2} + A_{n-2}^2,$$

so

$$2 q_n A_{n-1} A_{n-2} = A_n^2 - q_n^2 A_{n-1}^2 - A_{n-2}^2. \tag{3}$$

Considering the equality (3) modulo p and substituting the congruences (2) into its right side, we get

$$2 q_n A_{n-1} A_{n-2} \equiv (-1)^{n+1} Q_{n+1} - q_n^2 (-1)^n Q_n - (-1)^{n-1} Q_{n-1} \equiv$$
$$\equiv (-1)^{n-1} Q_{n+1} + q_n^2 (-1)^{n-1} Q_n - (-1)^{n-1} Q_{n-1} \equiv$$
$$\equiv (-1)^{n-1}(Q_{n+1} + q_n^2 Q_n - Q_{n-1}) \pmod{p},$$

so for all $n \geq 1$

$$2 q_n A_{n-1} A_{n-2} \equiv (-1)^{n-1}(Q_{n+1} + q_n^2 Q_n - Q_{n-1}) \pmod{p}. \tag{4}$$

Since Theorem 1 implies that $Q_{kl} = 1$, $Q_{kl+1} = Q_{kl-1}$ for all $k \in \mathbb{N}$, then from (4) for $n = kl$ we obtain

$$2 q_{kl} A_{kl-1} A_{kl-2} \equiv (-1)^{kl-1} q_{kl}^2 \pmod{p}.$$

Since $q_{kl} = 2 q_0 = 2 \lfloor \sqrt{p} \rfloor < p$ for all odd primes p, then q_{kl} is not divisible by p, so, reducing by $2 q_{kl}$, we get

$$A_{kl-1} A_{kl-2} \equiv (-1)^{kl-1} q_0 \pmod{p}.$$

$\square$

Lemma 2. *Let l be the length of the period of the continued fraction for $\sqrt{p}$, where p is a prime number. Then*

$$A_{l-k} \equiv (-1)^{k-1} A_{l-1} A_{k-2} \pmod{p}$$

for all $k = 1, \ldots, l + 1$.

Proof. By Lemma 1 for $k = 1$ we have:

$$A_{l-1} A_{l-2} \equiv (-1)^{l-1} q_0 \pmod{p}, \tag{5}$$
$$A_{l-1}^2 \equiv (-1)^l \pmod{p}. \tag{6}$$

It follows from (6) that $A_{l-1} \not\equiv 0 \pmod{p}$, so dividing (5) by (6), we get $A_{l-2} A_{l-1}^{-1} \equiv -q_0 \pmod{p}$. Hence $A_{l-2} \equiv -A_{l-1} q_0 \pmod{p}$.

Let's prove by induction on k that $A_{l-k} \equiv (-1)^{k-1} A_{l-1} A_{k-2} \pmod{p}$, $k = 1, \ldots, l + 1$. For $k = 1$, it's obvious. For $k = 2$, it's true, as we have just proved $A_{l-2} \equiv -A_{l-1} q_0 \equiv -A_{l-1} A_0 \pmod{p}$. By the induction hypothesis, $A_{l-(k-1)} \equiv (-1)^{k-2} A_{l-1} A_{k-3} \pmod{p}$ and $A_{l-(k-2)} \equiv (-1)^{k-3} A_{l-1} A_{k-4} \pmod{p}$. Hence

$$A_{l-k} = A_{l-k+2} - q_{l-k+2} A_{l-k+1} \equiv (-1)^{k-3} A_{l-1} A_{k-4} - q_{k-2}(-1)^{k-2} A_{l-1} A_{k-3}$$
$$= (-1)^{k-3} A_{l-1}(q_{k-2} A_{k-3} + A_{k-4}) = (-1)^{k-1} A_{l-1} A_{k-2} \pmod{p}.$$

$\square$

Corollary 1. *Let l be the length of the period of the continued fraction for $\sqrt{p}$, where p is an odd prime number.*

1. *If $p \equiv 7 \pmod 8$, then $l \equiv 0 \pmod 4$, $A_{l-1} \equiv 1 \pmod p$, and*

$$A_{l-k} \equiv (-1)^{k-1} A_{k-2} \pmod{p}, \ k = 1, \ldots, l + 1.$$

2. *If $p \equiv 3 \pmod 8$, then $l \equiv 2 \pmod 4$, $A_{l-1} \equiv -1 \pmod p$, and*

$$A_{l-k} \equiv (-1)^k A_{k-2} \pmod{p}, \ k = 1, \ldots, l + 1.$$

3. *If $p \equiv 1 \pmod 4$, then l is odd, $A_{l-1}^2 \equiv -1 \pmod p$, and*

$$A_{l-k} \equiv (-1)^{k-1} A_{l-1} A_{k-2} \pmod{p}, \ k = 1, \ldots, l + 1.$$

Proof. 1, 2. If l is even, then, by Theorem 4, $p \equiv 3 \pmod 4$, and, by Lemma 1, for $k = 1$ we have:

$$A_{l-1} A_{l-2} \equiv -q_0 \pmod{p}, \quad A_{l-1}^2 \equiv 1 \pmod{p}, \quad A_{l-2}^2 \equiv q_0^2 \pmod{p}.$$

Therefore, there are 2 cases.

Case 1: $A_{l-1} \equiv 1 \pmod p$, $A_{l-2} \equiv -q_0 \pmod p$.
Case 2: $A_{l-1} \equiv -1 \pmod p$, $A_{l-2} \equiv q_0 \pmod p$.

In the first case, by Corollary 1, we have $A_{l-k} \equiv (-1)^{k-1} A_{k-2} \pmod{p}$, $k = 1, \ldots, l + 1$.

In the second case, by Corollary 1, we have $A_{l-k} \equiv (-1)^k A_{k-2} \pmod{p}$, $k = 1, \ldots, l+1$.

Note that $p \nmid A_i$ for all $i \geq 0$. Indeed, by Theorem 2, we have $A_i^2 \equiv (-1)^{i+1} Q_{i+1} \pmod{p}$, so $p \mid A_i$ if and only if $p \mid Q_{i+1}$. Since, by Theorem 1, we have $1 \leq Q_{i+1} \leq 2\lfloor \sqrt{p} \rfloor < p$ for all $i \geq 0$, so $1 \leq Q_{i+1} < p$. Thus, $p \nmid Q_{i+1}$, so $p \nmid A_i$.

If $A_{l-1} \equiv 1 \pmod{p}$, then for $k = \frac{l}{2}+1$, $M = l-k = k-2$ we have $A_M \equiv (-1)^{\frac{l}{2}} A_M \pmod{p}$. Hence $l \equiv 0 \pmod 4$ (since otherwise $l \equiv 2 \pmod 4$ we get a contradiction $2A_M \equiv 0 \pmod p$). Thus, by Theorem 5, a prime p has the form $p \equiv 7 \pmod 8$.

If $A_{l-1} \equiv -1 \pmod{p}$, then for $k = \frac{l}{2}+1$, $M = l-k = k-2$ we have $A_M \equiv (-1)^{\frac{l}{2}+1} A_M \pmod{p}$. Hence $l \equiv 2 \pmod 4$ (since otherwise $l \equiv 0 \pmod 4$ we get a contradiction $2A_M \equiv 0 \pmod p$). Thus, by Theorem 5, a prime p has the form $p \equiv 3 \pmod 8$.

3. If l is odd, then, by Theorem 4, $p \equiv 1 \pmod 4$, and, by Corollary 1, we have $A_{l-k} \equiv (-1)^{k-1} A_{l-1} A_{k-2} \pmod{p}$, $k = 1, \ldots, l+1$. Moreover, it follows from (6) that $A_{l-1}^2 \equiv -1 \pmod{p}$. □

The proof of Corollary 1 does not answer the question which of two solutions of the congruence $x^2 \equiv -1 \pmod{p}$ congruent to A_{l-1} modulo p, where $p \equiv 1 \pmod 4$. Now denote by r exactly the one of two solutions that satisfies the condition $1 \leq r \leq \frac{p-1}{2}$. Then we can't know in advance whether the congruences $A_{l-1} \equiv r \pmod{p}$ or $A_{l-1} \equiv -r \pmod{p}$ holds. The results of calculations for primes $p \equiv 1 \pmod 4$, $p \leq 181$ are summarized below.

p	5	13	17	29	37	41	53	61	73	89	97	101	109	113	137	149	157	173	181
$p \bmod 8$	5	5	1	5	5	1	5	5	1	1	1	5	5	1	1	5	5	5	5
$A_{l-1} \bmod p$	2	5	4	12	6	32	23	11	46	55	75	10	33	98	100	44	129	80	162
r	2	5	4	12	6	9	23	11	27	34	22	10	33	15	37	44	28	80	19
l	1	5	1	5	1	3	5	11	7	5	11	1	15	9	9	9	17	5	21

We propose the following conjecture.

Conjecture 1. Denote by $\mathbb{P}_{4,1} = \{p_1, p_2, \ldots, p_N, \ldots\}$ the set of all primes $p \equiv 1 \pmod 4$. Let X_N be the number of primes in the set $\{p_1, p_2, \ldots, p_N\}$ such that $A_{l-1} \equiv r \pmod{p}$. Let Y_N be the number of primes in the set $\{p_1, p_2, \ldots, p_N\}$ such that $A_{l-1} \equiv -r \pmod{p}$. It's obvious that $X_N + Y_N = N$ for all $N \geq 1$. Then

1. $X_N > Y_N$ for all $N \geq 1$.
2. $\lim\limits_{N \to \infty} \dfrac{X_N}{Y_N} = 1$.

The experimental results give support for this conjecture (see Fig. 1).

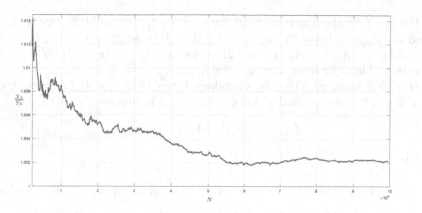

Fig. 1. The graph of the ratio of the number of primes $p \equiv 1 \pmod 4$ with $A_{l-1} \equiv r$ (mod p) to the number of primes $p \equiv 1 \pmod 4$ with $A_{l-1} \equiv -r \pmod p$, $N < 10^7$.

The following theorem is the main result of this paper.

Theorem 6. *Suppose p is a prime number, l is the length of the period of the continued fraction for $\sqrt{p}$, and L is the length of the period of the numerators modulo p of the corresponding convergents.*

1. *If $p = 2$, then $L = l = 1$.*
2. *If $p \equiv 7 \pmod 8$, then $L = l$, $l = 4t$ for some $t \in \mathbb{N}$, and the periodic part is*

$$\underbrace{1, A_0, A_1, \ldots, A_{2t-3}, A_{2t-2}, A_{2t-1}, -A_{2t-2}, A_{2t-3}, \ldots, A_1, -A_0}_{l}.$$

3. *If $p \equiv 3 \pmod 8$, then $L = 2l$, $l = 4t + 2$ for some $t \in \mathbb{N} \cup \{0\}$, and the periodic part is*

$$1, A_0, A_1, \ldots, A_{2t-1}, A_{2t}, -A_{2t-1}, \ldots, -A_1, A_0,$$
$$-1, -A_0, -A_1, \ldots, -A_{2t-1}, -A_{2t}, A_{2t-1}, \ldots, A_1, -A_0.$$

4. *If $p \equiv 1 \pmod 4$, then $L = 4l$, $l = 2t + 1$ for some $t \in \mathbb{N} \cup \{0\}$, and the periodic part is*

$$1, A_0, A_1, \ldots, A_{t-2}, A_{t-1}, (-1)^t r A_{t-1}, (-1)^{t-1} r A_{t-2}, \ldots, r A_1, -r A_0,$$
$$r, r A_0, r A_1, \ldots, A_{t-2}, r A_{t-1}, (-1)^{t-1} A_{t-1}, (-1)^{t-2} A_{t-2} \ldots, -A_1, A_0,$$
$$-1, -A_0, -A_1, \ldots, -A_{t-2}, -A_{t-1}, (-1)^{t-1} r A_{t-1}, (-1)^{t-2} r A_{t-2}, \ldots, -r A_1, r A_0,$$
$$-r, -r A_0, -r A_1, \ldots, -r A_{t-2}, r A_{t-1}, (-1)^t A_{t-1}, (-1)^{t-1} A_{t-2}, \ldots, A_1, -A_0,$$

where r satisfies the congruence $r^2 \equiv -1 \pmod p$.

Proof. Let's prove that the sequence $\{A_n \bmod p\}_{n=-1}^{\infty}$ is periodic, starting from $A_{-1} = 1$.

For $p = 2$ the theorem is obvious. Indeed, $\sqrt{2} = [1,\overline{2}]$, so $A_{-1} \equiv 1$ (mod 2), $A_0 \equiv 1$ (mod 2), $A_1 = 2 \cdot 1 + 1 \equiv 1$ (mod 2) and so on. In general, $A_k = 2A_{k-1} + A_{k-2} \equiv A_{k-2}$ (mod 2), so we get $A_k \equiv 1$ (mod 2) for $k \geq -1$ by induction. Thus, we have $L = l = 1$ for $p = 2$.

If $p \equiv 7$ (mod 8), then, by Corollary 1, we have $l = 4t$ for some $t \in \mathbb{N}$, $A_{l-k} \equiv (-1)^{k-1}A_{k-2}$ (mod p) for $k = 1, \ldots, l+1$. Hence

$$A_{4t-k} \equiv (-1)^{k-1}A_{k-2} \quad (\text{mod } p),$$

that is

$$A_{2t} \equiv -A_{2t-2} \quad (\text{mod } p), \quad A_{2t+1} \equiv A_{2t-3} \quad (\text{mod } p), \ldots, A_{4t-2} \equiv -A_0 \quad (\text{mod } p).$$

Let's prove by induction on j that $A_{l+j} \equiv A_j$ for $j \geq -1$. For $j = -1, j = 0$ it's true, since

$$A_{l-1} \equiv 1 \quad (\text{mod } p),$$
$$A_l = q_l A_{l-1} + A_{l-2} \equiv 2q_0 \cdot 1 - q_0 = q_0 \equiv A_0 \quad (\text{mod } p).$$

By the induction hypothesis, $A_{l+j-1} \equiv A_{j-1}$ (mod p) and $A_{l+j-2} \equiv A_{j-2}$ (mod p). Therefore,

$$A_{l+j} = q_{l+j}A_{l+j-1} + A_{l+j-2} \equiv q_j A_{j-1} + A_{j-2} \equiv A_j \quad (\text{mod } p).$$

Thus, $L = l$.

If $p \equiv 3$ (mod 8), then, by Corollary 1, we have $l = 4t + 2$ for some $t \geq 0$, $A_{l-k} \equiv (-1)^k A_{k-2}$ (mod p) for $k = 1, \ldots, l+1$. Hence

$$A_{4t+2-k} \equiv (-1)^k A_{k-2} \quad (\text{mod } p),$$

that is

$$A_{2t+1} \equiv -A_{2t-1} \quad (\text{mod } p), \quad A_{2t+2} \equiv A_{2t-2} \quad (\text{mod } p), \ldots, A_{4t} \equiv A_0 \quad (\text{mod } p).$$

Let's prove by induction on j that $A_{l+j} \equiv -A_j$ (mod p) for $j \geq -1$. For $j = -1, j = 0$ it's true, since

$$A_{l-1} \equiv -1 \quad (\text{mod } p),$$
$$A_l = q_l A_{l-1} + A_{l-2} \equiv 2q_0 \cdot (-1) + A_0 = -2q_0 + q_0 = -q_0 = -A_0 \quad (\text{mod } p).$$

By the induction hypothesis, $A_{l+j-1} \equiv -A_{j-1}$ (mod p) and $A_{l+j-2} \equiv -A_{j-2}$ (mod p). Therefore,

$$A_{l+j} = q_{l+j}A_{l+j-1} + A_{l+j-2} \equiv q_j(-A_{j-1}) - A_{j-2} = -A_j \quad (\text{mod } p).$$

It remains to prove that for $p \equiv 3$ (mod 8) the period of the sequence $\{A_i$ (mod $p)\}_{i=-1}^{\infty}$ is equal to $2l$, that is $A_{2l+j} \equiv A_j$ for $j \geq -1$. In fact,

$$A_{2l+j} \equiv -A_{l+j} \equiv -(-A_j) = A_j \quad (\text{mod } p).$$

Hence $L = 2l$.

If $p \equiv 1 \pmod 4$, then, by Corollary 1, we have

$$A_{l-k} \equiv (-1)^{k-1} r A_{k-2} \pmod p, \quad k = 1, \ldots, l+1,$$

where $A_{l-1} \equiv r \pmod p$, $r^2 \equiv -1 \pmod p$. Let's prove by induction on j that $A_{l+j} \equiv r A_j \pmod p$ for $j \geq -1$. For $j = -1, j = 0$ it's true, since

$$A_{l-1} \equiv r A_{-1} = r \pmod p,$$
$$A_l = q_l A_{l-1} + A_{l-2} \equiv 2 q_0 r A_{-1} - r A_0 \equiv 2 q_0 r - r q_0 = r q_0 = r A_0 \pmod p.$$

By the induction hypothesis, $A_{l+j-1} \equiv r A_{j-1} \pmod p$ and $A_{l+j-2} \equiv r A_{j-2} \pmod p$. Therefore,

$$A_{l+j} = q_{l+j} A_{l+j-1} + A_{l+j-2} \equiv q_j r A_{j-1} + r A_{j-2} = r(q_j A_{j-1} + A_{j-2}) = r A_j \pmod p.$$

Hence

$$A_{2l+j} \equiv r A_{l+j} \equiv r^2 A_j \equiv -A_j \pmod p,$$
$$A_{3l+j} \equiv r A_{2l+j} \equiv -r A_j \pmod p,$$
$$A_{4l+j} \equiv r A_{3l+j} \equiv -r^2 A_j \equiv A_j \pmod p.$$

Thus, $L = 4l$. □

We attach some examples of the periodic parts. The periods are marked in bold in the tables.

1. $\sqrt{31} = [5, \overline{1, 1, 3, 5, 3, 1, 1, 10}]$. Here $L = l = 8$.

n	-1	0	1	2	3	4	5	6	7	8	9	10
q_n		5	**1**	**1**	**3**	**5**	**3**	**1**	**1**	**10**	1	1
$A_n \bmod 31$	1	5	**6**	**11**	**8**	**20**	**6**	**26**	**1**	**5**	6	11

2. $\sqrt{19} = [4, \overline{2, 1, 3, 1, 2, 8}]$. Here $L = 2l = 12$.

n	-1	0	1	2	3	4	5	6	7	8	9	10	11	12
q_n		4	**2**	**1**	**3**	**1**	**2**	**8**	**2**	**1**	**3**	**1**	**2**	8
$A_n \bmod 19$	1	4	**9**	**13**	**10**	**4**	**18**	**15**	**10**	**6**	**9**	**15**	**1**	4

3. $\sqrt{13} = [3, \overline{1, 1, 1, 1, 6}]$. Here $L = 4l = 20$.

n	-1	0	1	2	3	4	5	6	7	8	9	10	11	12	13	14	15	16	17	18	19	20
q_n		3	**1**	**1**	**1**	**1**	**6**	**1**	**1**	**1**	**1**	**6**	**1**	**1**	**1**	**1**	**6**	**1**	**1**	**1**	**1**	6
$A_n \bmod 13$	1	3	**4**	**7**	**11**	**5**	**2**	**7**	**9**	**3**	**12**	**10**	**9**	**6**	**2**	**8**	**11**	**6**	**4**	**10**	**1**	3

4   Related Works

The sequences of numerators and denominators of convergents of the continued fraction for $\sqrt{p}$ satisfy a second order linear recurrence relation with non-constant coefficients q_i (which are periodic). There are many works that are devoted to the study of periods of sequences, that satisfy linear recurrences with constant coefficients (see [1,2,4–6,18,22,23]). But we do not know the existence of works in which the periods of linear recurrent sequences with non-constant coefficients (for example, with periodic coefficients) were studied. Most of them consider perhaps the most famous recurrent sequence, namely, the Fibonacci sequence $\{F_i\}_{i=0}^{\infty}$, i.e. $F_0 = 0, F_1 = 1, F_n = F_{n-1} + F_{n-2}$ for $n \geq 2$. It was known to Lagrange [11] that the sequence of Fibonacci numbers is periodic modulo m for any integer $m > 1$. The period of this sequence modulo m is called the Pisano period. Let's call the rank of F_i modulo m the minimal integer $r > 0$ such that $p \mid F_r$. Let's denote the rank and the Pisano period by $\alpha(p)$ and $\pi(p)$, respectively. The order F_i modulo m is $\beta(p) = \frac{\pi(p)}{\alpha(p)}$. Then $\beta(p)$ takes only three values: 1, 2 and 4 (see [18], p. 37, Corollary 2.39). We would like to note that these are the same coefficients that appear in Theorem 6 ($L = 1 \cdot l$, $L = 2 \cdot l$ and $L = 4 \cdot l$ depending on a prime p).

5   Conclusion

The next natural problem we would like to consider is the study of the sequence of denominators of convergents modulo p for $\sqrt{p}$ and its relation to the sequence of numerators. It would be interesting to investigate how the period of the sequence of denominators depends on p. In addition, we would like to generalize the obtained results for $\sqrt{D}$, where D is not a prime number.

We put forward the following conjectures. Let's denote the length of the period of the sequences of the numerators and the denominators of convergents for $\sqrt{D}$ modulo D by $L_A(\sqrt{D})$ and $L_B(\sqrt{D})$, respectively (D is not a perfect square).

Conjecture 2. For any prime p, the following equality holds

$$L_B(\sqrt{p}) = p \cdot L_A(\sqrt{p}).$$

Conjecture 3. Let D be a positive square-free integer. Let $x + y\sqrt{D}$ be the fundamental unit of the field $\mathbb{Q}(\sqrt{D})$. If y is divisible by D then

$$L_B(\sqrt{D}) = L_A(\sqrt{D}).$$

Acknowledgements. The authors are grateful to anonymous reviewers for valuable comments.

References

1. Bloom, D.M.: Periodicity in generalized Fibonacci sequences. Amer. Math. Monthly **72**(8), 856–861 (1965)
2. Brent, R.P.: On the periods of generalized Fibonacci sequences. Math. Comput. **63**(207), 389–401 (1994)
3. Davenport, H.: The Higher Arithmetic: An Introduction to the Theory of Numbers. Cambridge University Press, Cambridge (1999)
4. Engstrom, H.T.: On sequences defined by linear recurrence relations. Trans. Am. Math. Soc. **33**(1), 210–218 (1931)
5. Falcon, S., Plaza, A.: k-Fibonacci sequence modulo m. Chaos, Solitons Fractals **41** (1), 497–504 (2009)
6. Fulton, J.D., Morris, W.L.: On arithmetical functions related to the Fibonacci numbers. Acta Arith **16**, 105–110 (1969)
7. Golubeva, E.P.: Quadratic irrationalities with a fixed length of the expansion period into a continued fraction. Zap. Nauch. Sem. LOMI **196**, 5–30 (1991). [in Russian]
8. Hardy, G.H., Wright, E.M.: An Introduction to the Theory of Numbers. Oxford University Press, Oxford (1962)
9. Khinchin, A.Y.: Continued fractions (1978). In Russian
10. Khovansky, A.N.: Application of continued fractions and their generalizations to questions of approximate analysis, 206 (1956). In Russian
11. Lagrange, J.L.: Oeuvres de Lagrange. Gauthier Villars, Paris (1877)
12. Lang, S.: Introduction to Diophantine Approximations. Addison-Wesley, Boston (1966)
13. Matthews K.R.: The Diophantine equation $x^2 - Dy^2 = N, D > 0$. Expositiones Math. **18**, 323–331 (2000)
14. Nagell, T.: Introduction to Number Theory. Wiley, New York (1981)
15. Niven, I., Zuckerman, H.S., Montgomery, H.L.: An Introduction to the Theory of Numbers, 5th edn. Wiley, New York (1991)
16. Olds, C.D.: Continued Fractions. Random House, New York (1963)
17. Perron, O.: Die Lehre von den Kettenbrüchen. - Stuttgart: Teubner. Bd. 1. Elementare Kettenbrüche. - Reprograph. Nachdr. d. 3, verb. u. erw. Aufl. Stuttgart, p. 202 (1977)
18. Renault, M.: The Fibonacci Sequence Under Various Moduli. Master's Thesis. Wake Forest University, p. 82 (1996)
19. Rockett, A.M., Szüsz, P.: Continued Fractions. World Scientific, New York (1992)
20. Rosen, K.H.: Elementary Number Theory And Its Applications. Addison-Wesley, Boston (1984)
21. Sarma K.V.S.S.R.S.S., Avadhani P.S.: Public Key Cryptosystem based on Pell's Equation Using The Gnu Mp Library. Int. J. Comput. Sci. Eng. **4**, 739–743 (2012)
22. Wall, D.D.: Fibonacci series modulo m. Amer. Math. Monthly **67**(6), 525–532 (1960)
23. Ward, M.: The characteristic number of a sequence of integers satisfying a linear recursion relation. Trans. Am. Math. Soc. **33**(1), 153–165 (1931)

Investigation of a Queueing System with Two Classes of Jobs, Bernoulli Feedback, and a Threshold Switching Algorithm

Andrei V. Zorine[✉]

Lobachevsky State University of Nizhni Novgorod,
Nuzhni Novgorod 603022, Russian Federation
andrei.zorine@itmm.unn.ru

Abstract. A server with jobs of two classes, Bernoulli feedback, and setup times is considered. Jobs arrive in batches according to a Poisson process. Different job classes have different arrival intensities and different batch size distributions. Service times and setup times are random with exponential probability distributions. A control algorithm is parametrized by a threshold L: first-class jobs are taken for service only if the number of the second-class jobs in the system doesn't exceed L. This kind of queueing models is often used to model computer systems and other information-processing systems. On the other hand, threshold-type controls for multiclass jobs have not been widely studied yet. A time-homogeneous Markov process describing the server state and numbers in the queues is introduced, its infinitesimal intensities are identified, a necessary and sufficient condition for the existence of the stationary probability distribution is found. Steady-state probabilities for the server states are found explicitly, and independence of the threshold parameter L is established. An algorithm for the numerical solution of a system of functional equations for the steady-state probability generating functions, the main objective of the talk, is presented for several particular values of the threshold L. This algorithm is a result of the problem investigation by means of computer algebra software since the size of intermediate formulas exceeds human capabilities to manage them by hand.

Keywords: Multiclass jobs · Queueing system with dynamic priorities · Threshold control algorithm · Steady-state probabilities · Solution in terms of probability generating functions

1 Introduction

Multi-class job queueing systems with dynamic priorities (e.g. priority indices for queues, dependent on numbers in the queues at decision epochs) and feedback were studied in papers [1–3]. Such systems are adequate models for information processing processes in computer systems. The main practical result of these

D. Balandin et al. (Eds.): MMST 2022, CCIS 1750, pp. 148–159, 2022.
https://doi.org/10.1007/978-3-031-24145-1_12

papers was as follows: it was shown that an algorithm for optimal priority indices assignment such that the mathematical expectation of sojourn cost of all jobs in the system per unit of time or per a working tact is the least is a classic rule with time-independent priority indices. Such indices can be pre-computed based on data about expected service durations and sojourn costs per time unit for each job class.

At the same time, another optimization problems are also of certain interest. In paper [4], a minimization problem for the expected time of the first hit of a given subset of states in presence of a subset of taboo states by a Markov random process was studied. It turned out that under a particular choice of the subsets of permitted states, target states, and taboo states, better results (in comparison with a classic priority algorithm and serve-the-longest-queue algorithm) were observed from a threshold-type algorithm. Now it is natural to conduct a deeper study of threshold control in a multi-class jobs queueing system with feedback. In the present paper we focus on building a numerical algorithm to obtain a stationary probability distribution for the queueng system state in terms of probability generating functions. In the future it will enable obtaining theoretical mean and variance and other numerical characteristics for the numbers in the queues and stationary probabilities for the server states by mean of the theory of a complex variable.

2   The Problem Statement

Two Poisson batch flows of jobs Π_1, Π_2 enter the queueing system, one flow per job class. Arrival intensity of batches from the flow Π_j, $j = 1, 2$, is a constant $\lambda_j > 0$, and a batch size equals n with a known probability $f(b, j) \geqslant 0$, $b = 1$, $2, \ldots$; $\sum_{b=1}^{\infty} f(b, j) = 1$. Jobs from the flow Π_j join a dedicated buffer O_j with unlimited capacity. Service time for a job from the buffer O_j has the exponential probability distribution with parameter $\beta_j > 0$. A served job from the queue O_j either becomes class r job with probability $p_{j,r}$ and joins O_r for another service quantum, or leaves the queueing system with probability $p_{j,0} = 1 - p_{j,1} - p_{j,2} \geqslant 0$ and joins an output flow. After every service operation for a job from O_j, the server needs a setup time which is also random with an exponential probability distribution with parameter $\bar{\beta}_j$. After a setup period, if the number in the queues are described by a non-zero vector (x_1, x_2), the next service begins for a job from a queue $s = h(x_1, x_2)$ where $h(\cdot, \cdot)$ естьis a given mapping of a non-negative integer lattice $X = \{0, 1, \ldots\} \times \{0, 1, \ldots\}$ onto $\{0, 1, 2\}$. This mapping should obey the following: $s = h(x_1, x_2)$ implies $x_s > 0$ and the only point which is mapped to 0 is the zero vector $\bar{0} = (0, 0)$ in X. If, upon a setup period termination, the queues are empty then the server goes into an idle state and waits for the first batch of jobs to arrive from any of the two flows. When a new batch of jobs arrives into an empty queueing system, a service begins for one job from the batch and the rest join its dedicated queue.

Let $\kappa_j(t)$ be the number in the queue O_j at the time instant $t \geqslant 0$, $\kappa(t) = (\kappa_1(t), \kappa_2(t))$. Let $\Gamma = \{\Gamma^{(0)}, \Gamma^{(1)}, \ldots, \Gamma^{(4)}\}$ be the set of the server states; here

$\Gamma^{(0)}$ means the idle state when waiting for new jobs, $\Gamma^{(j)}$ for $j = 1$, 2 means processing a job in the queue O_j, while for $j = 3$, 4 it means a setup period after processing a job in the queue O_{j-2}. Let the random element $\Gamma(t) \in \Gamma$ denote the server state at the time instant $t \geqslant 0$.

The random process $\{(\Gamma(t), \kappa(t)); t \geqslant 0\}$ is a time-homogeneous continuous-time Markov process. Its state space can be taken as

$$\{(\Gamma^{(0)}, 0, 0)\} \cup \{\Gamma^{(1)}, \Gamma^{(2)}, \Gamma^{(3)}, \Gamma^{(4)}\} \times X.$$

Set

$$Q(r, x_1, x_2; t) = \mathbf{P}(\{\Gamma(t) = \Gamma^{(r)}, \kappa(t) = (x_1, x_2)\});$$

$$f_j(z) = \sum_{b=1}^{\infty} z^b f(b, j), \qquad |z| \leqslant 1;$$

$$\Psi(z_1, z_2, r; t) = \mathbf{E}(z_1^{\kappa_1(t)} z_2^{\kappa_2(t)} I(\Gamma(t) = \Gamma^{(r)}))$$

$$= \sum_{x_1=0}^{\infty} \sum_{x_2=0}^{\infty} z_1^{x_1} z_2^{x_2} Q(r, x_1, x_2; t), \qquad |z_1| < 1. |z_2| < 1.$$

Theorem 1. *For $r = 1$, 2 the following differential equations hold:*

$$\frac{\partial}{\partial t} \Psi(z_1, z_2, r; t) = \Psi(z_1, z_2, r; t)(\lambda_1(f_1(z_1) - 1) + \lambda_2(f_2(z_2) - 1) - \beta_r)$$

$$+ \sum_{j=1}^{2} \bar{\beta}_j \mathbf{E}(z_1^{\kappa_1(t)} z_2^{\kappa_2(t)} I(\{\Gamma(t) = \Gamma^{(2+j)}, h(\kappa_1(t), \kappa_2(t)) = r\}))$$

$$+ \lambda_r f_r(z_r) Q(0, 0, 0; t),$$

$$\frac{\partial}{\partial t} \Psi(z_1, z_2, 2 + r; t) = \Psi(z_1, z_2, 2 + r; t)(\lambda_1(f_1(z_1) - 1) + \lambda_2(f_2(z_2) - 1) - \bar{\beta}_r)$$

$$+ \beta_r z_r^{-1}(1 + p_{r,1}(z_1 - 1) + p_{r,2}(z_2 - 1))\Psi(z_1, z_2, r; t),$$

$$\frac{d}{dt} Q(0, 0, 0, t) = -(\lambda_1 + \lambda_2)Q(0, 0, 0; t) + \bar{\beta}_1 Q(3, 0, 0, t) + \bar{\beta}_2 Q(4, 0, 0, t).$$

3   Essays on Solving the Stationary Equations

In the rest of the paper assume that the Markov process $\{(\Gamma(t), \kappa(t)); t \geqslant 0\}$ is stationary and will focus on its stationary probabilities. If we set the initial probability distribution we guarantee the existence of limits equal to the stationary probability distribution

$$Q(r, x_1, x_2) = \lim_{t \to \infty} Q(r, x_1, x_2; t), \qquad \Psi(z_1, z_2, r) = \lim_{t \to \infty} \Psi(z_1, z_2, r; t),$$

and the differential equations of Theorem 1 turn info functional algebraic equations

$$\Psi(z_1, z_2, r)(\lambda_1(1 - f_1(z_1)) + \lambda_2(1 - f_2(z_2)) + \beta_r)$$

$$= \sum_{j=1}^{2} \bar{\beta}_j \mathbf{E}\big(z_1^{\kappa_1(t)} z_2^{\kappa_2(t)} I(\{\Gamma(t) = \Gamma^{(2+j)}, \kappa(t) \in X_r\})\big)$$

$$+ \lambda_r f_r(z_r) Q(0,0,0), \qquad r = 1,2; \tag{1}$$

$$\Psi(z_1, z_2, 2 + r)(\lambda_1(1 - f_1(z_1)) + \lambda_2(1 - f_2(z_2)) + \bar{\beta}_r)$$
$$= \beta_r z_r^{-1}(1 + p_{r,1}(z_1 - 1) + p_{r,2}(z_2 - 1))\Psi(z_1, z_2, r), \qquad r = 1,2 \tag{2}$$
$$(\lambda_1 + \lambda_2)Q(0,0,0) = \bar{\beta}_1 Q(3,0,0) + \bar{\beta}_2 Q(4,0,0). \tag{3}$$

Let us denote by $\mu_j = f_j'(1)$ the mean batch size for the flow Π_j. Let us introduce vectors and a matrix

$$\beta = (\beta_1^{-1}, \beta_2^{-1}), \ \bar{\beta} = (\bar{\beta}_1^{-1}, \bar{\beta}_2^{-1}), \ \bar{\lambda} = \begin{pmatrix} \lambda_1 \mu_1 \\ \lambda_2 \mu_2 \end{pmatrix}, \ \mathbf{Q} = \begin{pmatrix} p_{1,1} \ p_{1,2} \\ p_{2,1} \ p_{2,2} \end{pmatrix}, \ \mathbf{I} = \begin{pmatrix} 1 \ 0 \\ 0 \ 1 \end{pmatrix}.$$

We assume that the matrix $(\mathbf{I} - \mathbf{Q})$ is invertible. From the definition of the generating functions it follows that $\Psi(1,1,r)$ is the stationary probability of the server state $\Gamma^{(r)}$ no matter what the queues are, $r = 1, 2, 3, 4$.

Theorem 2. *Some of the stationary probabilities are*

$$Q(0,0,0) = 1 - (\beta + \bar{\beta})(\mathbf{I} - \mathbf{Q}^T)^{-1}\bar{\lambda},$$

$$\Psi(1,1,1) = \frac{\beta_2 \beta(\mathbf{I} - \mathbf{Q}^T)^{-1}\bar{\lambda} - (1,1)(\mathbf{I} - \mathbf{Q}^T)^{-1}\bar{\lambda}}{\beta_2 - \beta_1},$$

$$\Psi(1,1,2) = \frac{(1,1)(\mathbf{I} - \mathbf{Q}^T)^{-1}\bar{\lambda} - \beta_1 \beta(\mathbf{I} - \mathbf{Q}^T)^{-1}\bar{\lambda}}{\beta_2 - \beta_1},$$

$$\Psi(1,1,3) = \frac{\bar{\beta}_2 \bar{\beta}(\mathbf{I} - \mathbf{Q}^T)^{-1}\bar{\lambda} - (1,1)(\mathbf{I} - \mathbf{Q}^T)^{-1}\bar{\lambda}}{\bar{\beta}_2 - \bar{\beta}_1},$$

$$\Psi(1,1,4) = \frac{(1,1)(\mathbf{I} - \mathbf{Q}^T)^{-1}\bar{\lambda} - \bar{\beta}_1 \bar{\beta}(\mathbf{I} - \mathbf{Q}^T)^{-1}\bar{\lambda}}{\bar{\beta}_2 - \bar{\beta}_1}.$$

These formulas do not depend on the switching function $h(\cdot)$.

Proof. Set $z_j = 1 - \theta_j u$ in a neighborhood of the zero, the quantities $\theta_1 > 0$ and $\theta_2 > 0$ to be defined later. Set $\theta = (\theta_1, \theta_2)$. Then expansions take place:

$$\lambda_r f_r(z_r) = \lambda_r + \lambda_r \mu_r u + o(u),$$
$$(\lambda_1(1 - f_1(z_1)) + \lambda_2(1 - f_2(z_2)) + \beta_r) = \beta_r - (\lambda_1 \mu_1 \theta_1 + \lambda_2 \mu_2 \theta_2)u + o(u),$$
$$(\lambda_1(1 - f_1(z_1)) + \lambda_2(1 - f_2(z_2)) + \bar{\beta}_r) = \bar{\beta}_r - (\lambda_1 \mu_1 \theta_1 + \lambda_2 \mu_2 \theta_2)u + o(u),$$
$$\beta_r z_r^{-1}(1 + p_{r,1}(z_1 - 1) + p_{r,2}(z_2 - 1)) = \beta_r(1 + (p_{r,1}\theta_1 + p_{r,2}\theta_2 - \theta_r)u + o(u))$$

Summing up Eqs. (1) and (2) for $r = 1$, 2 we get:

$$\sum_{r=1}^{2} (\Psi(z_1, z_2, r)(\beta_r - \theta\bar{\lambda}u + o(u)) + \Psi(z_1, z_2, 2 + r)(\bar{\beta}_r - \theta\bar{\lambda}u + o(u)))$$

$$= \sum_{j=1}^{2} \bar{\beta}_r(\Psi(z_1, z_2, 2 + j) - \Psi(0, 0, 2 + j)) + Q(0, 0, 0)(\lambda_1 + \lambda_2 + \theta\bar{\lambda}u + o(u))$$

$$+ \sum_{r=1}^{2} \Psi(z_1, z_2, r)(\beta_r + \beta_r(p_{r,1}\theta_1 + p_{r,2}\theta_2 - \theta_r)u + o(u)).$$

After collecting terms, divide by u and send $u \to 0$, so that $z_1 \to 1$, $z_2 \to 1$. In result, we obtain:

$$- \theta\bar{\lambda}(\Psi(1, 1, 1) + \Psi(1, 1, 2) + \Psi(1, 1, 3) + \Psi(1, 1, 4))$$

$$= \theta\bar{\lambda}Q(0, 0, 0) + \sum_{r=1}^{2} \Psi(1, 1, r)(p_{r,1}\theta_1 + p_{r,2}\theta_2 - \theta_r).$$

The normalization condition here is

$$\Psi(1, 1, 1) + \Psi(1, 1, 2) + \Psi(1, 1, 3) + \Psi(1, 1, 4) + Q(0, 0, 0) = 1.$$

So, we necessarily get:

$$\theta\bar{\lambda} + \sum_{r=1}^{2} \Psi(1, 1, r)\beta_r(p_{r,1}\theta_1 + p_{r,2}\theta_2 - \theta_r) = 0.$$

If we let θ_1 and θ_2 to be the solution of a system of equations

$$\theta_r = p_{r,1}\theta_1 + p_{r,2}\theta_2 + \beta_r^{-1}, \quad r = 1, 2,$$

i.e. $\theta = \beta(\mathbf{I} - \mathbf{Q}^T)^{-1}$, then

$$\beta(\mathbf{I} - \mathbf{Q}^T)^{-1}\bar{\lambda} - \Psi(1, 1, 1) - \Psi(1, 1, 2) = 0.$$

Next, substituting $z_1 = z_2 = 1$ into Eq. (2) we get:

$$\Psi(1, 1, 2 + r)\bar{\beta}_r = \beta_r\Psi(1, 1, r).$$

Now consider a system of equations

$$\theta_r = p_{r,1}\theta_1 + p_{r,2}\theta_2 + \bar{\beta}_r^{-1}, \quad r = 1, 2,$$

i.e. let $\theta = \beta(\mathbf{I} - \mathbf{Q}^T)^{-1}$. Then

$$\Psi(1, 1, 3) + \Psi(1, 1, 4) = \frac{\beta_1}{\bar{\beta}_1}\Psi(1, 1, 1) + \frac{\beta_2}{\bar{\beta}_2}\Psi(1, 1, 2) = \bar{\beta}(\mathbf{I} - \mathbf{Q}^T)^{-1}\bar{\lambda}.$$

By virtue of the normalization condition we get

$$Q(0,0,0) = 1 - (\beta + \bar{\beta})(\mathbf{I} - \mathbf{Q}^T)^{-1}\bar{\lambda}.$$

Now, let us consider a system of equations

$$\theta_r = p_{r,1}\theta_1 + p_{r,2}\theta_2 + 1, \quad r = 1, 2,$$

then we get

$$\beta_1\Psi(1,1,1) + \beta_2\Psi(1,1,2) = (1,1)(\mathbf{I} - \mathbf{Q}^T)^{-1}\bar{\lambda}.$$

So,

$$\Psi(1,1,1) = \frac{\beta_2\beta(\mathbf{I} - \mathbf{Q}^T)^{-1}\bar{\lambda} - (1,1)(\mathbf{I} - \mathbf{Q}^T)^{-1}\bar{\lambda}}{\beta_2 - \beta_1},$$

$$\Psi(1,1,2) = \frac{(1,1)(\mathbf{I} - \mathbf{Q}^T)^{-1}\bar{\lambda} - \beta_1\beta(\mathbf{I} - \mathbf{Q}^T)^{-1}\bar{\lambda}}{\beta_2 - \beta_1}.$$

then,

$$\Psi(1,1,3) + \Psi(1,1,4) = 1 - Q(0,0,0) - (\Psi(1,1,1) + \Psi(1,1,2))$$
$$= \bar{\beta}(\mathbf{I} - \mathbf{Q}^T)^{-1}\bar{\lambda},$$
$$\bar{\beta}_1\Psi(1,1,3) + \bar{\beta}_2\Psi(1,1,4) = \beta_1\Psi(1,1,1) + \beta_2\Psi(1,1,2)$$
$$= (1,1)(\mathbf{I} - \mathbf{Q}^T)^{-1}\bar{\lambda}.$$

Solving a system of equations

$$\Psi(1,1,3) = \frac{\bar{\beta}_2\bar{\beta}(\mathbf{I} - \mathbf{Q}^T)^{-1}\bar{\lambda} - (1,1)(\mathbf{I} - \mathbf{Q}^T)^{-1}\bar{\lambda}}{\bar{\beta}_2 - \bar{\beta}_1},$$

$$\Psi(1,1,4) = \frac{(1,1)(\mathbf{I} - \mathbf{Q}^T)^{-1}\bar{\lambda} - \bar{\beta}_1\bar{\beta}(\mathbf{I} - \mathbf{Q}^T)^{-1}\bar{\lambda}}{\bar{\beta}_2 - \bar{\beta}_1}.$$

we prove the claim.

Moreover, using methods from paper [4], we can prove that the inequality

$$(\beta + \bar{\beta})(\mathbf{I} - \mathbf{Q}^T)^{-1}\bar{\lambda} < 1 \tag{4}$$

is a necessary and sufficient for the the existence of the stationary probability distribution.

A switching function for a threshold control algorithm with a threshold parameter $L \geqslant 0$ (an integer) has the form

$$h(x_1, x_2) = \begin{cases} 1, & \text{if either } x_1 > L \text{ or } x_2 = 0, x_1 > 0; \\ 2, & \text{if } x_1 \leqslant L, x_2 > 0; \\ 0, & \text{if } x_1 = x_2 = 0. \end{cases}$$

In this particular class of controls we have also

$$\mathbf{E}\big(z_1^{\kappa_1(t)} z_2^{\kappa_2(t)} I(\{\Gamma(t) = \Gamma^{(2+j)}, h(\kappa_1(t), \kappa_2(t)) = 1\})\big) = \Psi(z_1, z_2, 2+j)$$

$$-\Psi(0, z_2, 2+j) - \sum_{k=1}^{L} \frac{z_1^k}{k!} \frac{\partial^k}{\partial z_1^k} (\Psi(z_1, z_2, 2+j) - \Psi(z_1, 0, 2+j))\Big|_{z_1=0},$$

$$\mathbf{E}\big(z_1^{\kappa_1(t)} z_2^{\kappa_2(t)} I(\{\Gamma(t) = \Gamma^{(2+j)}, \kappa(t) \in X_2\})\big) = \Psi(0, z_2, 2+j) - \Psi(0, 0, 2+j)$$

$$+ \sum_{k=1}^{L} \frac{z_1^k}{k!} \frac{\partial^k}{\partial z_1^k} (\Psi(z_1, z_2, 2+j) - \Psi(z_1, 0, 2+j))\Big|_{z_1=0},$$

$$\Psi(0, 0, 2+j) = Q(2+j, 0, 0).$$

We will need the following new notations

$$q_1(z_1.z_2) = z_1 \frac{(\lambda_1(1 - f_1(z_1)) + \lambda_2(1 - f_2(z_2)) + \bar{\beta}_1)}{\beta_1(1 + p_{1,1}(z_1 - 1) + p_{1,2}(z_2 - 1))}$$

$$\times (\lambda_1(1 - f_1(z_1)) + \lambda_2(1 - f_2(z_2)) + \beta_1) - \bar{\beta}_1,$$

$$q_2(z_1, z_2) = \frac{(\lambda_1(1 - f_1(z_1)) + \lambda_2(1 - f_2(z_2)) + \bar{\beta}_2)}{\beta_2(1 + p_{2,1}(z_1 - 1) + p_{2,2}(z_2 - 1))}$$

$$\times (\lambda_1(1 - f_1(z_1)) + \lambda_2(1 - f_2(z_2)) + \beta_2),$$

$$\tilde{\Psi}_r(z_2, k) = z_2^{-1} \frac{\partial^k (\Psi(z_1, z_2, r) - \Psi(z_1, 0, r))}{\partial z_1^k}\Big|_{z_1=0}, \quad k = 0, 1, \ldots, L; \ r = 3, 4;$$

$$\tilde{\Psi}_{r,k} = \frac{\partial^k \Psi(z_1, 0, r))}{\partial z_1^k}\Big|_{z_1=0} \quad \text{(at that, } \tilde{\Psi}_{r,0} = \Psi(0, 0, r)).$$

Then the stationary equations for the unknown probability degeneration functions can be reduced to the following ones:

$$q_1(z_1, z_2)\Psi(z_1, z_2, 3) - \bar{\beta}_2\Psi(z_1, z_2, 4) = -\bar{\beta}_1 z_2 \sum_{k=0}^{L} \frac{z_1^k}{k!} \tilde{\Psi}_3(z_2, k)$$

$$- \bar{\beta}_2 z_2 \sum_{k=0}^{L} \frac{z_1^k}{k!} \tilde{\Psi}_4(z_2, k) + \lambda_1 f_1(z_1) Q(0, 0, 0), \quad (5)$$

$$q_2(z_1, z_2)\Psi(z_1, z_2, 4) = \bar{\beta}_1 \sum_{k=0}^{L} \frac{z_1^k}{k!} \tilde{\Psi}_3(z_2, k) + \bar{\beta}_2 \sum_{k=0}^{L} \frac{z_1^k}{k!} \tilde{\Psi}_4(z_2, k)$$

$$+ \lambda_2 z_2^{-1} f_2(z_2) Q(0, 0, 0). \quad (6)$$

At the first step of the solution, one can eliminate $\Psi(z_1, z_2, 3)$ and $\Psi(z_1, z_2, 4)$. For this purpose we will need the following lemma.

Lemma 1. *Let β be a positive constants, then the quantity $\beta + \lambda_1(1 - w_1) + \lambda(1 - w_2)$ is nonzero everywhere in the polydisk $|w_1| \leqslant 1$, $|w_2| \leqslant 1$.*

Proof. From a geometric viewpoint, the real parts $\Re(1-w_1) \geqslant 0$, $\Re(1-w_2) \geqslant 0$, because the complex numbers $(1-w_1)$ and $(1-w_2)$ lie in a unit disk centered at 1. So,

$$\Re(\lambda_1(1-w_1) + \lambda_2(1-w_2)) \geqslant 0.$$

Moreover,

$$\Re(\lambda_1(1-w_1) + \lambda_2(1-w_2) + \beta) > 0,$$

hence the claim follows.

As a corollary of the Lemma, the equation $q_2(z_1, z_2) = 0$ has no solutions in the polydisk $|z_1| \leqslant 1$, $|z_2| \leqslant 1$. Finding out the number of solutions $z_1 = z_1(z_2)$ $q_1(z_1, z_2) = 0$ such that $|z_1| < 1$ for $|z_2| < 1$ can be done using the well-known formula from complex analysis

$$\frac{1}{2\pi\sqrt{-1}} \int\limits_{|z_1|=1.01} \frac{\frac{\partial}{\partial z_1} q_1(z_1, z_2)}{q_1(z_1, z_2)} dz_1$$

and using numerical quadratures. Let $z_1 = z_1(w)$ be the (unique) solution of the equation $q_1(z_1, w) = 0$ such that it takes on value $|z_1| \leqslant 1$ when $|w| \leqslant 1$. Let us substitute $z_1 = z_1^*$ into Eq. (5) and use (6). After rearranging the terms we get:

$$\bar{\beta}_1(\bar{\beta}_2 - z_2 q_2(z_1^*, z_2)) \sum_{k=0}^{L} \frac{(z_1^*)^k}{k!} \tilde{\Psi}_3(z_2, k) + \bar{\beta}_2(\bar{\beta}_2 - z_2 q_2(z_1^*, z_2))$$

$$\times \sum_{k=0}^{L} \frac{(z_1^*)^k}{k!} \tilde{\Psi}_4(z_2, k) = Q(0,0,0)\big(q_2(z_1^*, z_2)(\lambda_1(1 - f_1(z_1^*)) + \lambda_2)$$

$$- \bar{\beta}_2 \lambda_2 z_2^{-1} f_2(z_2)\big). \quad (7)$$

Substituting $z_1 = 0$ into (6) we get:

$$\tilde{\Psi}_4(z_2, 0)(z_2 q_2(0, z_2) - \bar{\beta}_2) = \bar{\beta}_1 \tilde{\Psi}_3(z_2, 0) - q_2(0, z_2)\Psi(0, 0, 4)$$

$$+ \lambda_2 z_2^{-1} f_2(z_2) Q(0,0,0). \quad (8)$$

3.1 Priority Algorithm (Case $L = 0$)

For $L = 0$ we have the classic priority control algorithm. In this case equations (7) and (8) take the form

$$\bar{\beta}_1(z_2 q_2(z_1^*, z_2) - \bar{\beta}_2)\tilde{\Psi}_3(z_2, 0) + \bar{\beta}_2(z_2 q_2(z_1^*, z_2) - \bar{\beta}_2)\tilde{\Psi}_4(z_2, 0)$$

$$= \big(\bar{\beta}_2 \lambda_2 z_2^{-1} f_2(z_2) + q_2(z_1^*, z_2)(\lambda_1(f_1(z_1^*) - 1) - \lambda_2)\big)Q(0,0,0), \quad (9)$$

$$- \bar{\beta}_1 \tilde{\Psi}_3(z_2, 0) + (z_2 q_2(0, z_2) - \bar{\beta}_2)\tilde{\Psi}_4(z_2, 0)$$

$$= \lambda_2 z_2^{-1} f_2(z_2) Q(0,0,0) - \Psi(0, 0, 4)q_2(0, z_2). \quad (10)$$

When $z_2 = 0$ the determinant of the system of Eqs. (9), (10) with respect to unknowns $\tilde{\Psi}_3(z_2)$, $\tilde{\Psi}_4(z_2)$ is

$$\begin{vmatrix} -\bar{\beta}_1\bar{\beta}_2 - (\bar{\beta}_2)^2 \\ -\bar{\beta}_1 \quad -\bar{\beta}_2 \end{vmatrix} = \bar{\beta}_2 \cdot \begin{vmatrix} -\bar{\beta}_1 & -\bar{\beta}_2 \\ -\bar{\beta}_1 & -\bar{\beta}_2 \end{vmatrix} = 0.$$

So, we can multiply the system of Eqs. (9), (10) by an adjoint matrix and set $z_2 = 0$ in order to get linear equations for unknown constant $\Psi(0,0,3)$ and $\Psi(0,0,4)$. By combining it with Eq. (6) we finally obtain a system of linear equations for $\Psi(0,0,3)$ and $\Psi(0,0,4)$. Now that we have these constants, we can find both $\tilde{\Psi}_3(z_2)$, $\tilde{\Psi}_4(z_2)$, and $\Psi(z_1,z_2,3)$, $\Psi(z_1,z_2,4)$, by backtracking.

3.2   Case $L = 1$

Let us take the derivative of Eqs. (5), (6) with respect to z_1 and then set $z_1 = 0$. Altogether with Eqs. (7) and (8) we get the following system of equations

$$\bar{\beta}_1(\bar{\beta}_2 - z_2 q_2(z_1^*, z_2))(\tilde{\Psi}_3(z_2,0) + z_1^*\tilde{\Psi}_3(z_2,1)) + \bar{\beta}_2(\bar{\beta}_2 - z_2 q_2(z_1^*, z_2))$$
$$\times (\tilde{\Psi}_4(z_2,0) + z_1^*\tilde{\Psi}_4(z_2,1)) = Q(0,0,0)$$
$$\times (q_2(z_1^*, z_2)(\lambda_1(1 - f_1(z_1^*)) + \lambda_2) - \bar{\beta}_2\lambda_2 z_2^{-1} f_2(z_2)),$$
$$-\bar{\beta}_1\tilde{\Psi}_3(z_2,0) + (z_2 q_2(0,z_2) - \bar{\beta}_2)\tilde{\Psi}_4(z_2,0)$$
$$= -q_2(0,z_2)\tilde{\Psi}_{4,0} + \lambda_2 z_2^{-1} f_2(z_2)Q(0,0,0),$$
$$\bar{\beta}_1\tilde{\Psi}_3(z_2,1) - \frac{\partial q_2}{\partial z_1}(0,z_2)z_2\tilde{\Psi}_4(z_2,0) + (\bar{\beta}_2 - z_2 q_2(0,z_2))\tilde{\Psi}_4(z_2,1)$$
$$= \frac{\partial q_2}{\partial z_1}(0,z_2)\tilde{\Psi}_{4,0} + q_2(0,z_2)\tilde{\Psi}_{4,1},$$
$$\frac{\partial q_1}{\partial z_1}(0,z_2)z_2\tilde{\Psi}_3(z_1,0) = -\frac{\partial q_1}{\partial z_1}(0,z_2)\tilde{\Psi}_{3,0} + \bar{\beta}_1\tilde{\Psi}_{3,1} + \bar{\beta}_2\tilde{\Psi}_{4,1} + \lambda_1 p(1,1)Q(0,0,0).$$

Consider it as a system of equations for unknowns $\tilde{\Psi}_3(z_2,0)$, $\tilde{\Psi}_3(z_2,1)$, $\tilde{\Psi}_4(z_2,0)$, $\tilde{\Psi}_4(z_2,1)$. The system's main matrix $B(z_2)$ is (two first columns first, then the last two columns in the next line)

$$\begin{pmatrix} \bar{\beta}_1(\bar{\beta}_2 - z_2 q_2(z_1^*, z_2)) & z_1^*\bar{\beta}_1(\bar{\beta}_2 - z_2 q_2(z_1^*, z_2)) \\ -\bar{\beta}_1 & 0 \\ 0 & \bar{\beta}_1 \\ z_2\dfrac{\partial q_1}{\partial z_1}(0,z_2) & 0 \end{pmatrix}$$

$$\begin{pmatrix} \bar{\beta}_2(\bar{\beta}_2 - z_2 q_2(z_1^*, z_2)) & z_1^*\bar{\beta}_2(\bar{\beta}_2 - z_2 q_2(z_1^*, z_2)) \\ (z_2 q_2(0,z_2) - \bar{\beta}_2) & 0 \\ -\dfrac{\partial q_2}{\partial z_1}(0,z_2)z_2 & (\bar{\beta}_2 - z_2 q_2(0,z_2)) \\ 0 & 0 \end{pmatrix}$$

In the matrix form, with notations easy to guess, the system of equations looks like

$$B(z_2)\tilde{\Psi}(z_2) = C_1(z_2)\tilde{\Psi}_0 + Q(0,0,0) \cdot C_2(z_2).$$

Multiplying it by the adjoint matrix we get

$$(\mathrm{adj}B(z_2))B(z_2)\tilde{\Psi}(z_2) = \big(\mathrm{adj}B(z_2)C_1(z_2)\big)\tilde{\Psi}_0 + Q(0,0,0)\mathrm{adj}B(z_2)\cdot C_2(z_2). \quad (11)$$

To proceed, we need to answer to the questions: how many values z_2 exist, with $|z_2| < 1$, such that $\det B(z_2) = 0$? And what are they exactly? Computing this determinant manually is a tedious task, with little hope to answer to the two questions above in general, for arbitrary parameters satisfying the stationarity condition. Instead, we can numerically test for the number of zeros of the determinant at any particular values of the parameters, and then find approximate values for the zeros. Here we need symbolic computations software to generate code for an integration routine (we used MAXIMA in this research [5]), taking into account that z_1^* is an implicitly defined function.

In out experiments we saw that the zeros of the equation $\det B(z_2) = 0$ are $z_2 = 0$ with multiplicity 2 and another simple zero $z_2^* \neq 0$. Furthermore, the rank of the matrix $B(z_2)$ at $z_2 = 0$ equals 2 and not 3 so that the adjoint matrix $\mathrm{adj}B(0)$ is degenerate. The rank of $B(z_2)$ at $z_2 = z_2^*$ equals 3, so the rank of $\mathrm{adj}B(z_2^*)$ turns out to be 1 and we can pick any nonzero element of the matrix identity (11).

The product matrix $(\mathrm{adj}B(z_2))B(z_2)$ is diagonal with $\det B(z_2)$ on the main diagonal.

Since the rank of the product of tw matrices is no greater that the ranks of the factors, setting $z_2 = 0$ into (11) gives a zero vector on the left, but the rank of the matrix $\mathrm{adj}B(0)C_1(0)$ is also zero, i.e. the coefficients in front of $\tilde{\Psi}_{3,0}$, $\tilde{\Psi}_{3,1}$, $\tilde{\Psi}_{4,0}$, $\tilde{\Psi}_{4,1}$, vanish too.

Taking the derivative of (11) with respect to z_2 and then set $z_2 = 0$, we get an identity

$$0 = \frac{d}{dz_2}\Big((\mathrm{adj}B(z_2))C_1(z_2)\Big)\Big|_{z_2=0} \cdot \tilde{\Psi}_0$$

$$+ Q(0,0,0)\frac{d}{dz_2}\Big((\mathrm{adj}B(z_2))C_2(z_2)\Big)\Big|_{z_2=0}. \quad (12)$$

Taking (12) together with $(\lambda_1 + \lambda_2)Q(0,0,0) = \bar{\beta}_1\tilde{\Psi}_{3,0} + \bar{\beta}_2\tilde{\Psi}_{4,0}$ we get a system of linear equations for the four unkowns $\tilde{\Psi}_{3,0}$, $\tilde{\Psi}_{4,0}$, $\tilde{\Psi}_{4,0}$, and $\tilde{\Psi}_{4,1}$ which can be solved numerically.

3.3   Some Observations Concerning the Case $L= 2$

Our experiments using the developed program in the MAXIMA programming language allow to make a few observations. Let us take the derivatives of Eqs. (5),

(6) with respect to z_1 up to order L and then set $z_1 = 0$. Thus we'll get $2(L+1)$ equations which in the matrix form have the same structure as above:

$$B(z_2)\tilde{\Psi}(z_2) = C_1(z_2)\tilde{\Psi}_0 + Q(0,0,0) \cdot C_2(z_2).$$

- The equation $\det B(z_2) = 0$ has a root $z_2 = 0$ of multiplicity $(L+1)$ and the root $z_2 = z_2^*$ of multiplicity L;
- when $L = 2$, the rank of the matrix $\mathrm{adj}B(z_2^*)C_1(z_2^*)$ equals 1 and the rank of matrix $\frac{d}{dz_2}(\mathrm{adj}B(z_2^*)C_1(z_2^*))\big|_{z_2=z_2^*}$ equals 2;
- when $L = 2$, $\mathrm{adj}B(0)C_1(0)$ and $\frac{d}{dz_2}(\mathrm{adj}B(z_2)C_1(z_2))\big|_{z_2=0}$ are zero matrices, while the rank of $\frac{d^2}{dz_2^2}(\mathrm{adj}B(z_2)C_1(z_2^*))\big|_{z_2=0}$ is 3.

From these one might conclude that we can pick five equations to determine unknown constants, but it is hard to give recommendations how to organize an automatic selection of such equations. Recall that these are empirical hypotheses. For now it is unclear how they can be proved analytically.

3.4   Numerical Example

Let us consider a set of parameters:

$$Q = \begin{pmatrix} 0.1 & 0.3 \\ 0.2 & 0.1 \end{pmatrix}, \qquad \lambda_1 = 0.025, \quad \lambda_2 = 0.05,$$

$$\beta_1 = 1, \quad \beta_2 = 2, \quad \bar{\beta}_1 = 0.5, \quad \bar{\beta}_2 = 0.8,$$

$$f_1(z) = 0.9z_1 + 0.1z_1^2, \quad f_2(z) = 0.5z + 0.5z^2.$$

Here, the batches have either one of two jobs. This flow was studied in [6] and thus is sometimes called the Gnedenko–Kovalenko flow. Mean batch sizes are $\mu^{(1)} = 1.1$, $\mu^{(2)} = 1.5$. Since the left-hand side of Eq. (4) equals 0.33575, the stationary probability distribution exists.

The stationary probability of idle server is $Q(0,0,0) = 0.66425$ for all L. The nonzero root of $\det B(z_2)$ here equals $z_2^* = 0.7151503296576\ldots$. Values of the constants $\tilde{\Psi}_{3,0}$, $\tilde{\Psi}_{4,0}$, $\tilde{\Psi}_{3,1}$, $\tilde{\Psi}_{4,1}$ for $L = 1$ and values of $\tilde{\Psi}_{3,0}$, $\tilde{\Psi}_{4,0}$, $\tilde{\Psi}_{3,1}$, $\tilde{\Psi}_{4,1}$, $\tilde{\Psi}_{3,1}$, and $\tilde{\Psi}_{4,1}$ for $L = 2$ and for the parameters as above are presented in Table 1.

Table 1. Comparison of thresholds $L = 1$ and $L = 2$ for one-or-two batch sizes

L	$\tilde{\Psi}_{3,0}$	$\tilde{\Psi}_{4,0}$	$\tilde{\Psi}_{3,1}$	$\tilde{\Psi}_{4,1}$	$\tilde{\Psi}_{3,2}$	$\tilde{\Psi}_{4,2}$
$L = 1$	0.03990032	0.03733571	0.01512898	0.02324446	—	—
$L = 2$	0.03990036	0.03733571	0.01771008	0.02163128	0.00789297	0.01322855

If we take the batch size distribution generating functions of the form

$$f_1(z_1) = \frac{(10/11)z_1}{1 - (1/11)z_1}, \qquad f_2(z_2) = \frac{(2/3)z_2}{1 - (1/3)z_2},$$

Table 2. Comparison of thresholds $L = 1$ and $L = 2$ for batch sizes with geometric distribution

L	$\tilde{\Psi}_{3,0}$	$\tilde{\Psi}_{4,0}$	$\tilde{\Psi}_{3,1}$	$\tilde{\Psi}_{4,1}$	$\tilde{\Psi}_{3,2}$	$\tilde{\Psi}_{4,2}$
$L = 1$	0.03919354	0.03777748	0.01447431	0.02255470	—	—
$L = 2$	0.03919354	0.03777748	0.01740392	0.02072370	0.00821225	0.01331250

with same mean batch sizes, then $z_2^* = 0.716057562914$. The results are presented in Table 2.

From these two examples we see that the batch size distribution affects the stationary probabilities. On the other hand, the threshold level L seems to keep empty queue probabilities $\tilde{\Psi}_{3,0}$, $\tilde{\Psi}_{4,0}$ untouched.

Acknowledgements. This work is dedicated to the memory of my father, Vladimir Aleksandrovich Zorin, who passed away on 10.09.2022, being a docent of the department of probability theory and data analysis of the N.I. Lobachevsky State University of Nizhni Novgorod.

References

1. Klimov, G.P.: Time-sharing service systems. I. Theory Probab. Appl. **19**(3), 532–551 (1975)
2. Kitaev, M.Y., Rykov, V.V.: On a queuing system with the branching flow of secondary demands. Autom. Remote. Control. **9**, 52–61 (1980)
3. Fedotkin, M.A.: Optimal control for conflict flows and marked point processes with selected discrete component. I. Liet. Mat. Rinkinys. **28**(4), 783–794 (1988)
4. Zorine, A.V.: On ergodicity conditions in a polling model with Markov modulated input and state-dependent routing. Queueing Syst. **76**(2), 223–241 (2014). https://doi.org/10.1007/s11134-013-9385-3
5. The Maxima Project Homepage. https://maxima.sourceforge.io
6. Gnedenko, B.V., Kovalenko, I.N.: Introduction to Queueing Theory, 2nd edn. Birkhäuser, Boston (1989)

Coexistence of Dissipative and Conservative Regimes in Unidirectionally Coupled Maps

Dmitry Lubchenko$^{(\boxtimes)}$ (ID) and Alexey Savin (ID)

Institute of Physics, Saratov State University, 83 Astrakhanskaya str., 410012 Saratov, Russia
dima4398lub@mail.ru

Abstract. We consider the billiard system consisting of a particle moving between two walls one of which is plane and fixed and the other one is harmonically corrugated (like in Tennyson-Lieberman-Lichtenberg system) and oscillates harmonically. The collisions of the particle and the wall are suggested to be elastic. We assume that the oscillation and the corrugation amplitudes are weak so some significant simplifications of the system are justified which results in the system of two unidirectionally coupled 2D maps. The master system is the original Tennison-Lieberman-Lichtenberg system with fixed walls and the slave system is Ulam map parametrically driven by the master system. The variables of the slave system are the velocity of a particle before the collision and the time between the collisions. We calculate numerically the Jacobian of various trajectories of the system and reveal that the regions of conservative (with the Jacobian very close to zero) and dissipative dynamics coexist in the phase space of the system.

Keywords: Time-dependent billiards · Numerical research · Mixed dynamics

1 Introduction

Billiard-like dynamical systems are of great interest, both from applied and fundamental points, since a wide variety of nonlinear phenomena is observed in them and they are easy to research [1–4]. Usually billiards are assumed to be conservative (i.e. without loss of energy) but also they can be dissipative [5] (e.g. with energy loss by friction or inelastic collisions). Dissipative dynamics is characterized by the existence of attracting invariant sets [6]. The conservative behavior of billiards is well described by Hamiltonian systems [7]. For example, the phase space of Tennyson-Lieberman-Lichtenberg system [8] is typical for non-integrable two-dimensional Hamiltonian systems. It has regular trajectories that are quasi-periodic (KAM) tori and chaotic ones that are destructed tori as a result of the perturbations of an integrable Hamiltonian system [9].

© The Author(s), under exclusive license to Springer Nature Switzerland AG 2022
D. Balandin et al. (Eds.): MMST 2022, CCIS 1750, pp. 160–166, 2022.
https://doi.org/10.1007/978-3-031-24145-1_13

In this paper we consider a system that consists of a particle moving between two walls one of which is plane and fixed and the other one is harmonically corrugated (like in Tennyson-Lieberman-Lichtenberg system) and oscillates harmonically. The collisions of the particle and the wall are suggested to be elastic. We assume that the oscillation and the corrugation amplitudes are weak so some significant simplifications of the system are justified, which results in the system of two unidirectionally coupled 2D maps. The master system is original Tennison-Lieberman-Lichtenberg system with fixed walls and the slave system is Ulam map [10] parametrically driven by the master system. We calculate numerically the Jacobian of various trajectories of the system and reveal that the regions of conservative (with the Jacobian very close to zero) and dissipative dynamics coexist in the phase space of the system.

2 Model Description

The original system consists of a particle which moves between two boundaries and elastically collides with them. One boundary is fixed and set by the equation:

$$y_1 = 0. \tag{1}$$

The other boundary is corrugated and can oscillate harmonically. Then its equation is:

$$y_2 = F(x, t) = b \cos kx + a \cos wt + h \tag{2}$$

In (2) a – the oscillation amplitude, b – the corrugation amplitude, h – the average distance between the boundaries.

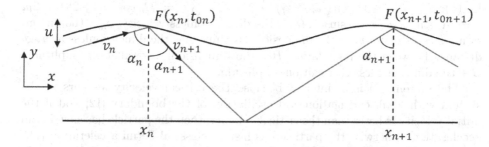

Fig. 1. Illustration of the particle movement between two boundaries. x_n – the coordinate of the n-th collision with the upper boundary; α_n – the angle between the normal to the bottom boundary and the velocity vector at the moment of the n-th collision v_n; v_n – the particle velocity at the moment of the n-th collision with the upper boundary; t_{0_n} – the elapsed time from the moment the particle begins to move until the moment of the n-th collision with the upper boundary.

The model is mechanical and it is not difficult to obtain expressions for $v_{n+1}, \alpha_{n+1}, x_{n+1}, t_{0_{n+1}}$ (Fig. 1) in the case of the weak amplitudes of the corrugation and the oscillation. So, these expressions form the following 4D map:

$$\begin{cases} v_{n+1} = \sqrt{v_{n+1_x}^2 + v_{n+1_y}^2}; \\ \alpha_{n+1} = \arctan \frac{v_{n+1_x}}{v_{n+1_y}}; \\ x_{n+1} = x_n + 2h \frac{v_{n+1_x}}{v_{n+1_y}}; \\ t_{0_{n+1}} = t_{0_n} + \frac{2h}{v_{n+1_y}}. \end{cases} \qquad (3)$$

In (3): $v_{n+1_x} = v_n \sin(\alpha_n + 2\gamma) - 2\gamma u$, $v_{n+1_y} = v_n \cos(\alpha_n + 2\gamma) - 2u$, $u = -aw \sin wt_{0_n}$, $\gamma = -kb \sin kx_n$. The number of the parameters can be reduced via the following replacement:

$$\begin{cases} \phi_n = kx_n; \\ \psi = wt_{0_n}; \\ \Omega = \frac{v_{n_{x,y}}}{2hw}; \\ A = 2hk; \\ B = \frac{a}{h}; \\ C = bk. \end{cases} \qquad (4)$$

It results in the 4D map with four parameters:

$$\begin{cases} \Omega_{n+1} = \sqrt{\Omega_{n+1_x}^2 + \Omega_{n+1_y}^2}; \\ \alpha_{n+1} = \arctan \left[\frac{\Omega_{n+1_x}}{\Omega_{n+1_y}} \right]; \\ \phi_{n+1} = \phi_n + A \frac{\Omega_{n+1_x}}{\Omega_{n+1_y}}; \\ \psi_{n+1} = \psi_n + \frac{1}{\Omega_{n+1_y}}. \end{cases} \qquad (5)$$

In (5): $\Omega_{n+1_x} = \Omega_n \sin(\alpha_n + 2\gamma) - 2\gamma u$, $\Omega_{n+1_y} = \Omega_n \cos(\alpha_n + 2\gamma) - 2u$, $\gamma = -C \sin \phi_n$, $u = -B \sin \psi_n$, Ω_n – the dimensionless velocity, ϕ_n – the dimensionless coordinate, ψ_n – the dimensionless time, A – the dimensionless average distance between the boundaries, B – the dimensionless oscillation amplitude, C – the dimensionless corrugation amplitude.

The system (5) is of interest because the critical velocity appears in the system with weak corrugation and oscillation of the boundary [12] and if the initial velocity is lower than the critical velocity, then the particle has slow Fermi acceleration, otherwise the particle has fast or classical Fermi acceleration [11]. If the particle is moving with slow Fermi acceleration, then $u << \Omega_n$ and (5) can be simplified as follows:

$$\begin{cases} \alpha_{n+1} = \alpha_n - 2C \sin \phi_n \\ \phi_{n+1} = \phi_n + A \tan \alpha_{n+1} \end{cases} \qquad (6)$$

$$\begin{cases} \Omega_{n+1} = \Omega_n + 2B \sin \psi_n \cos \alpha_{n+1} \\ \psi_{n+1} = \psi_n + \frac{1}{\Omega_n + 2B \sin \psi_n} \end{cases} \qquad (7)$$

The four-dimensional map is split into two-dimensional ones, where (6) is the Tennison-Lieberman-Lichtenberg map that affects the system (7) which is similar to Ulam map. The first map is the master system and the second one is the slave system.

3   Jacobian of System

The Jacobian of the system (6) is identically equal to one, which means that the system is conservative. However the Jacobian of the system (7) depends on the variables:

$$J = 1 - 2B \left(\frac{\sin\left(\alpha_n - 2C\sin\phi_n\right)}{\Omega_n \cos\left(\alpha_n - 2C\sin\phi_n\right) + 2B\sin\psi_n} \right)^2 \cos\psi_n. \tag{8}$$

The Jacobian of the full system consisting of these two maps is the same. Since the Jacobian depends on the state of the system, we should calculate the iteration-averaged (average over the number of iterations) Jacobian along the trajectory to find out if the regime is conservative or dissipative.

Let us fix the parameters $A = 2, B = 0.03, C = 0.05$ and plot the map of the Jacobian values, on which: vertical α_0 – the initial angle, horizontal ϕ_0 – the initial coordinate. The color indicates the absolute value of the Jacobian averaged along the trajectory: orange - impossible to determine (we will discuss why below), red - less than one, blue - equal to one, green - larger than one. The initial velocity Ω_0 is different for each figure. Initial time $\psi_0 = 0$ is selected for all figures.

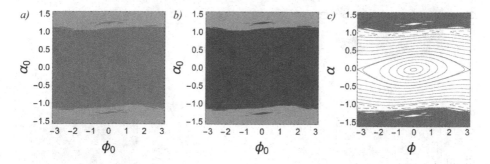

Fig. 2. Map of the Jacobian of the system (7) with the parameters: $A = 2, B = 0.03, C = 0.05$ and $\psi_0 = 0$, (a) $\Omega_0 = 0.1$; (b) $\Omega_0 = 1.1$; α_0 and ϕ_0 are coordinates of the map and colors marked the value of Jacobian. (c) Phase portrait of the Tennyson-Lieberman-Lichtenberg system with the same parameters.

The map of the Jacobian at the initial velocity $\Omega_0 = 0.1$ is shown in Fig. 2a. The red parts are the areas of dissipative dynamics and the orange ones are the areas of the initial conditions in which the numerical calculation of the Jacobian gives significant errors, since the Jacobian changes significantly when the number of iterations for its averaging is changed. If we compare the map of the Jacobian with the phase portrait of the system (6) we can see that the orange regions are situated in the chaotic layer of the phase space. The chaotic trajectories are located close to the angle $\frac{\pi}{2}$. It means that the collisions of a particle with the wall occur almost tangentially to it. Due to this reason the time between hits becomes extremely large. This situation seems to be similar to Levy's flight [13] i.e. the extremely long flight without collisions. It causes that the values of the time derivatives become large and they significantly affect the value of the local Jacobian, thereby leading to a significant change in averaged Jacobian.

If we increase the initial velocity to $\Omega_0 = 1.1$ then most of the dissipative regions become the conservative ones, the numerical value of the Jacobian is equal to one with adequate accuracy (Fig. 2b). We assumed that the region is conservative when $|\overline{J}| < 1 - |\epsilon|$. If ϵ is gradually reduced, then at one moment all values of the average Jacobians that fell into this region stop doing this. Also, this region is resistant to weak variation of the parameters or the initial conditions. We calculated the largest Lyapunov exponent, which is zero with adequate accuracy for this region, to validate that these areas are not the regions of unstable initial conditions. All this makes it possible to confirm that the blue parts are the areas of conservative dynamics. Note that there are small green areas where attractors exist, but their Jacobian is also calculated incorrectly due to Levy flights.

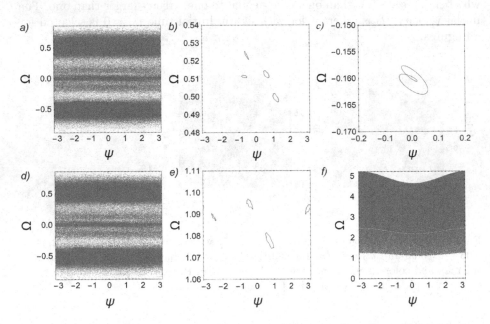

Fig. 3. Some trajectories of the system (7) with the parameters: $A = 2, B = 0.03, C = 0.05$ and $\psi_0 = 0$; (a, b, c) $\Omega_0 = 0.1$; (d, e, f) $\Omega_0 = 1.1$; (a, d) $\alpha_0 = 0.41\pi, \phi_0 = -0.57\pi$; (b, e) $\alpha_0 = 0.382\pi, \phi_0 = -0.19\pi$; (c, f) $\alpha_0 = 0.05\pi, \phi_0 = 0$.

It should be noted that we also calculated values of the largest Lyapunov exponent for such initial conditions and we found that it is positive in the orange areas and is equal to zero with adequate accuracy in the other areas. Examples of numerical values of the largest Lyapunov exponent are provided below.

Let us consider some of the phase trajectories of the system (7). For all trajectories in Fig. 3 the initial time $\psi_0 = 0$, in Fig. 3a, b, c $\Omega_0 = 0.1$, and in Fig. 3d, e, f $\Omega_0 = 1.1$. The trajectory in Fig. 3a with $\phi_0 = -0.57\pi$ and $\alpha_0 = 0.41\pi$ (chaotic region) has the largest Lyapunov exponent $\Lambda = 1.49$ (the Jacobian is incorrect). This is a chaotic trajectory and it has not undergone significant changes while the initial velocity increases. (Fig. 3e). The trajectory in Fig. 3b with $\phi_0 = -0.19\pi$ and $\alpha_0 = 0.382\pi$ (the stability island) has the dissipative Jacobian $J = -0.61$ and the largest Lyapunov exponent $\Lambda = 6.76 \times 10^{-6}$ which can be assumed equal to zero. Since there are only two Lyapunov exponents of the system (7), the second one has to be negative because for dissipative Jacobians the sum of the Lyapunov exponents has to be negative. In 2D maps one zero and one negative Lyapunov exponents indicate that there is a two-frequency torus in the phase space. The attractor changes with the increase of the initial velocity (Fig. 3e) and this trajectory has $J = 0.95$ and $\Lambda = 6.85 \times 10^{-6}$, which indicates the existence of the multistability in the system. The trajectory in Fig. 3c with $\phi_0 = 0$ and $\alpha_0 = 0.05\pi$ (close to the elliptic point) has the $J = 0.97$ and $\Lambda = 5.29 \times 10^{-6}$, which indicates the existence of the attractor. The attractor disappears as the initial velocity increases (Fig. 3f), $J = 1 - 0.2 \times 10^{-6}$ and $\Lambda = 1.10 \times 10^{-6}$. We believe that the Jacobian is equal to one with adequate accuracy, which indicates either conservative dynamics or instability. Since the largest Lyapunov exponent is equal to zero with adequate accuracy, it means that blue parts are regions of the conservative dynamics. Note that a part of the trajectory is shown in Fig. 3f. In fact, it continues to move upwards in the same way and its velocity increases without limit.

4   Conclusion

The research shows numerically that in the Ulam-like map parametrically driven by Tennison-Lieberman-Lichtenberg map the dissipative and the conservative regimes coexist in the phase space. For both maps the Jacobian was found. The Tennison-Lieberman-Lichtenberg map is conservative while the other one has the Jacobian which depends on the system's state and has to be either larger than one or less than one. When we calculated the Jacobian averaged along the trajectory for different initial conditions, it turned out that if the initial velocity is small then the average Jacobian is less than one and if the initial velocity is large enough then the system has to behave conservatively. This consideration is true for trajectories that are not located in the area of chaos of the Tennison-Lieberman-Lichtenberg system. Another problem is that for chaotic trajectories it is impossible to calculate the Jacobian numerically.

The similar phenomenon of conservative regimes and attractors coexisting in the phase space was studied in [14] and was called mixed dynamics there. It was shown that such dynamics occur if the system is time-reversible with involution. That means that some homeomorphism in the phase space exists when there is some transformation of the orbits into themself with the reversion of time. However there is no such homeomorphism (at least, the evident one) in our system, so we think that the phenomenon in our system is not exactly the same.

References

1. Bunimovich, L.A.: Mushrooms and other billiards with divided phase space. Chaos **11**, 802 (2001)
2. Chernov, N., Markarian, R.: Chaotic Billiards. Mathematical Surveys and Monographs, USA (2006)
3. Loskutov, A.: Dynamical chaos: systems of classical mechanics. Phys. Usp. **50**, 939–964 (2007)
4. Sinai, Y.G.: Dynamical systems with elastic reflections. Russ. Math. Surv. **25**, 137 (1970)
5. Oliveira, D.F.M., Leonel, E.D.: Evolution to the equilibrium in a dissipative and time dependent billiard. Phys. A **465**, 043122 (2011)
6. Kuznetsov, Y.A.: Elements of Applied Bifurcation Theory, 2nd edn. Applied Mathematical Sciences, New York (1998). https://doi.org/10.1007/978-1-4757-3978-7
7. Zaslasvsky, G.M.: Hamiltonian Chaos and Fractional Dynamics. Oxford University Press, Oxford (2005)
8. Tennyson, J.L., Lieberman, M.A., Lichtenberg, A.J.: Diffusion in near-integrable Hamiltonian systems with three degrees of freedom. AIP Conf. Proc. **57**, 272 (1980)
9. Lichtenberg, A.J., Lieberman, M.A.: Regular and Chaotic Dynamics, 1st edn. Applied Mathematical Sciences, New York (1983)
10. Ulam, S.M.: On Some Statistical Properties of Dynamical Systems. Berkeley Sympos. Math. Stat. Prob. **4**(3), 315–320 (1961)
11. Fermi, E.: On the origin of the cosmic radiation. Phys. Rev. **75**, 1169 (1949)
12. Lubchenko, D.O., Savin, A.V.: Critical velocity for the onset of fast fermi acceleration. IJBC **32**(12), 2250177 (2022)
13. Lévy, P.: Téorie de l'addition des variables aléatoires. Gauthier-Villars, Paris (1937)
14. Gonchenko, S.V., Turaev, D.V.: On three types of dynamics and the notion of attractor. Trudy Mat. Inst. Steklova **297**, 133–157 (2017)

Regulation of Neural Network Activity by Extracellular Matrix Molecules

Sergey Stasenko[1,2]($\boxtimes$) and Victor Kazantsev[1,3]

[1] Lobachevsky University, Nizhny Novgorod 603950, Russia
stasenko@neuro.nnov.ru
[2] Institute of Applied Physics RAS, Nizhny Novgorod 603022, Russia
[3] Immanuel Kant Baltic Federal University, Kaliningrad 236016, Russia

Abstract. The study of the mechanisms of synchronization of neural ensembles is an important task of modern neurodynamics. The synchronous dynamics of neuron activity underlies many cognitive functions, and its violation leads to socially significant diseases of the brain. In this paper, the mechanism of regulation of the neural activity of the spike neural network by the extracellular matrix of the brain is considered. It is assumed that such regulation can lead to the emergence of quasi-synchronous activity.

Keywords: Spiking neural network · Synchronization · Extracellular matrix molecules

1 Introduction

Understanding the mechanisms and principles of transmission, storage and processing of information in the brain is an important problem in modern neurodynamics and requires the construction of biologically relevant models of these processes. Traditionally, it was believed that only neurons, which form the concept of a two-partite synapse, are predominantly involved in the transmission of information in the brain. With the development of experimental techniques, new data were obtained indicating the participation in the process of transmission and processing of information between neurons of structural formations.

Burst activity is a dynamic state in which sets of spikes are repeatedly formed on a neuron, alternating with a state of rest. The neurons that form burst dynamics are important for generating and synchronizing motor patterns. They can be found in many areas of the brain, such as: the neocortex (intrinsically ruptured neurons [1], vibrating neurons [2]), the hippocampus [3], the thalamus and cerebellum [4]. Burst activity is widely represented in the brain and is often associated with the ensemble dynamics of neurons in the brain. It should be noted that burst activity is usually associated with various brain states, both representing normal and pathological states of the brain (for example, in epilepsy). Burst activity has been shown in a series of experiments to be involved in various cognitive functions of the brain, in particular with the development of the

D. Balandin et al. (Eds.): MMST 2022, CCIS 1750, pp. 167–175, 2022.
https://doi.org/10.1007/978-3-031-24145-1_14

visual system [5], sensory processing [6], neural transmission [7], learning and memory [8]. Burst activity was studied on cultures of neurons placed and developed on multielectrode arrays (MEA) [9]. At the same time, most often in such experiments, he observes changes in the structural or functional connections of neurons [10–15]. The dynamic properties of burst activity, such as self-adjusting complexity [16] or dynamic attractors [17], have also been studied in detail.In the work of Ben-Jacob, the role of burst activity in information encoding [18] and memory size [16] was studied. Understanding the principles and mechanisms of burst activity can lead to the creation of a living neural chip [19] with predefined functions.

The structure of the brain that affects synaptic transmission is the extracellular matrix of the brain (ECM) [20]. It has been shown that brain structure (ECM) has a significant role in the homeostatic regulation of neuronal activity at times of the order of hours, days and months [20,21]. At a functional level, ECM-maintained homeostatic plasticity can regulate neuronal hypo- and hyper-excitation, leading to a pathological brain condition that can lead to neuronal death. For example, the synaptic scaling mechanism can keep neurons active for various sensory inputs [22,23]. Changes in expression of ECM receptors (integrins) on the postsynaptic side cause changes in AMPA receptor expression, thus modulating the effective synaptic weight [20]. Another ECM-mediated pathway of synaptic modulation in through the action of hyaluronic acid and heparan sulfate proteoglycans on voltage-dependent L-type calcium channels (L-VDCC) [24]. ECM concentration is determined by the activity of various proteases (such as tissue plasminogen activator, plasmin, matrix metalloproteinases 2 and 9, agrecanases 1 and 2, neuropsin and neurotrypsin), which are released pre- and postsynaptically and cleave ECM molecules. Thereby ECM can act as excitator/inhibitor of neural activity.

In this paper, we consider the effect of such feedback on synaptic transmission and the formation of quasi-synchronous dynamics. Based on a mean-field approach we propose a multiscale mathematical model accounting neuronal activity modulation by the extracellular matrix molecular. The model predicts several dynamic effects of homeostatic neural activity at the time scale of seconds up to days, including coordination between excitation and inhibition and very slow "homeostatic" oscillations of neuronal activity, which are spontaneously generated due to coherent activation of the extracellular matrix of the brain.

2   Materials

2.1   Mathematical Model of Single Neuron

To describe the dynamics of a single neuron, the Hodgkin-Huxley model [26,27] was used, which is determined by the dynamics of sodium, I_{Na}, potassium, I_K, and leakage current, I_L, with the corresponding gate variables (n, m, h) (Eq. (1)):

$$C\frac{dV}{dt} = I_{inj} - \sum_{i=K,Na,L} g_i p_i (V - V_i) \tag{1}$$

$$\frac{dp}{dt} = \alpha_p(V)(1-p) - \beta_p(V)p, \tag{2}$$

where:

$$I_{inj} = I_{th} + I_{noise} + I_{syn} \tag{3}$$

$$\alpha_p(V) = \frac{p_\infty(V)}{\tau_p} \tag{4}$$

$$\beta_p(V) = \frac{1 - p_\infty(V)}{\tau_p}, \tag{5}$$

for $p = (n, m, h)$.

The shifted Nernst equilibrium potentials for I_{Na}, I_K and I_L are $V_{Na} = 50\,\mathrm{mV}$, $V_K = -77\,\mathrm{mV}$ and $V_L = -54.4\,\mathrm{mV}$, respectively. Typical values of maximal conductances for I_{Na}, I_K and I_L are $\bar{g}_{Na} = 36\,\mathrm{mS/cm^2}$, $\bar{g}_K = 120\,\mathrm{mS/cm^2}$ and $\bar{g}_L = 0.3\,\mathrm{mS/cm^2}$, respectively. Equations from 4 to 5 describe the transition rates between the open and closed states of the respective channels. $C = 1\mu\mathrm{F/cm^2}$ is the membrane capacitance and I_{inj} (Eq.(3)) is the applied current which consists from two parts: I_{th}, I_{noise} and I_{syn}. I_{th} describes the activation threshold of a neuron, which is regulated by the ECM. The thalamic input (I_{noise}) in our model is represented by a noise signal coming in addition to the synaptic input to the neuron. The noisy thalamic input is set in a random way for all neurons in the range from 0 to A_{noise}. The synaptic current, I_{syn}, was described by the following equations:

$$I_{syn} = g_e(Ee - V) + g_i(Ei - V) \tag{6}$$

$$\dot{g}_e = -g_e \frac{1}{\tau_e} \tag{7}$$

$$\dot{g}_i = -g_i \frac{1}{\tau_i}, \tag{8}$$

where g_e, g_i - the excitatory (inhibitory) conductance in the postsynaptic neuron, respectively.

2.2 Mathematical Model of ECM

To describe the dynamics of the extracellular matrix of the brain, a mean-field approach was used, presented by the authors in the works [21,25]. The ECM dynamics equations can be written as follows

$$\dot{Z} = -(\alpha_Z + \gamma_P P)Z + \beta_Z H_Z(E) \tag{9}$$

$$\dot{P} = -\alpha_P P + \beta_P H_P(E), \tag{10}$$

Z and P are concentration of extracellular matrix molecular and proteasis, respectively. In equations from 9 to 10, all activation functions $H_{Z,P}$ have a sigmoid form that is canonical for biological processes [21, 25].

3   Results

To demonstrate the effect of the formation of ECM burst dynamics, a spike neural septum was constructed. described in the next Subsect. 3.1.

3.1   Neural Network

The neural network consists of 100 neurons. In accordance with the experimental data of the mammalian cortex, the ratio of excitatory and inhibitory neurons was chosen as 4 to 1. The probability of connection of excitatory neurons is 1%, the probability of connection of inhibitory neurons is 20%.

3.2   Quasi-synchronous Neural Network Activity

Consider the regulation of neuronal activity by the extracellular matrix of the brain. In the absence of influence on the excitability threshold of a neuron, spontaneous activity is observed in the network (Fig. 1).

The following figure (Fig. 2) shows the neural network activity in the case of regulation of the neuron excitability threshold by the extracellular matrix of the brain. As you can see, a quasi-synchronous dynamics is formed in the network with a period of repetition of the activity of the extracellular matrix of the brain. It can also be seen that in this mode, the ECM goes into the oscillatory mode, which modulates the formation of the quasi-synchronous activity of the neural network on the order of the ECM time.

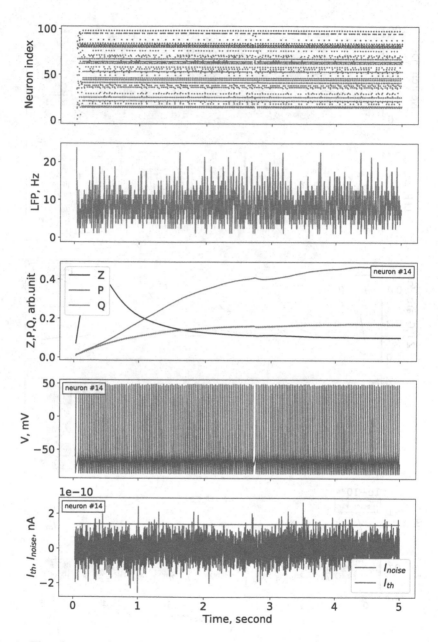

Fig. 1. The absence of regulation of neuronal activity by the extracellular matrix of the brain: rastr neural activity (top panel), local field potential and time series for concentration of extracellular matrix molecular (Z), proteases (P), average activity of neuron (Q), membrane potential of neuron (V) and applied currents (I_{th}, I_{noise}), respectively.

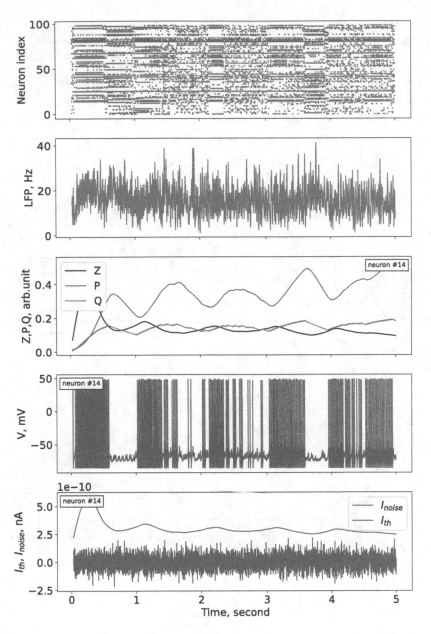

Fig. 2. The presence of regulation of neuron activity by the extracellular matrix of the brain: rastr neural activity (top panel), local field potential and time series for concentration of exctracellular matrix molecular (Z), proteases (P), average activity of neuron (Q), membrane potential of neuron (V) and applied currents (I_{th}, I_{noise}), respectively.

4   Conclusion

Compared to Hebbian synaptic plasticity, which may be initiated relatively fast in response to afferent stimulation, homeostatic plasticity associated with ECM molecules are implemented relatively slowly. Homeostatic processes help to regulate neuronal hypo- and hyperexcitation, leading to a pathological brain condition that can lead to neuronal death.

In this study, based on the simplified phenomenological description of the available experimental data, we proposed a mathematical model of the spiking neural network that considers - for the first time - interactions between neurons and ECM.

Acknowledgements. The work in terms of developing a mathematical model was supported by RSCF grant # 19-72-10128, in term of data analysis within the framework of the Development Program of the Regional Scientific and Educational Mathematical Center "Mathematics of Future Technologies", project No. 075-02-2020-1483/1 and in terms of numerical calculations was supported by the grant of the President of the Russian Federation NSh-2256.2022.1.2.

References

1. Connors, B., Gutnick, M.: Intrinsic firing patterns of diverse neocortical neurons. Trends Neurosci. **13**(3), 99–104 (1990). https://www.sciencedirect.com/science/article/abs/pii/016622369090185D
2. Gray, C., McCormick, D.: Chattering cells: superficial pyramidal neurons contributing to the generation of synchronous oscillations in the visual cortex. Science. **274**, 109–113 (1996). https://www.science.org/doi/10.1126/science.274.5284.109
3. Su, H., Alroy, G., Kirson, E. Yaari, Y.: Extracellular calcium modulates persistent sodium current-dependent burst-firing in hippocampal pyramidal neurons. J. Neurosci. **21**, 4173–4182 (2001). https://www.jneurosci.org/lookup/doi/10.1523/JNEUROSCI.21-12-04173.2001
4. Womack, M., Khodakhah, K.: Active contribution of dendrites to the tonic and trimodal patterns of activity in cerebellar purkinje neurons. J. Neurosci. **22**, 10603–10612 (2002). https://www.jneurosci.org/lookup/doi/10.1523/JNEUROSCI.22-24-10603.2002
5. Meister, M., Wong, R., Baylor, D., Shatz, C.: Synchronous bursts of action potentials in ganglion cells of the developing mammalian retina. Science. **252**, 939–943 (1991,5), https://www.science.org/doi/10.1126/science.2035024
6. Krahe, R., Gabbiani, F.: Burst firing in sensory systems. Nat. Rev. Neurosci. **5**(1), 13–23 (2004)
7. Salinas, E., Sejnowski, T.: Correlated neuronal activity and the flow of neural information. Nat. Rev. Neurosci. **2**, 539–550 (2001). http://www.nature.com/articles/35086012
8. Axmacher, N., Mormann, F., Fernández, G., Elger, C., Fell, J.: Memory formation by neuronal synchronization. Brain Res. Rev. **52**, 170–182 (2006). https://linkinghub.elsevier.com/retrieve/pii/S0165017306000099
9. Wagenaar, D., Pine, J., Potter, S.: An extremely rich repertoire of bursting patterns during the development of cortical cultures. BMC Neurosci. **7**, 11 (2006). https://bmcneurosci.biomedcentral.com/articles/10.1186/1471-2202-7-11

10. Ito, D., Tamate, H., Nagayama, M., Uchida, T., Kudoh, S., Gohara, K.: Minimum neuron density for synchronized bursts in a rat cortical culture on multi-electrode arrays. Neuroscience. **171**, 50–61 (2010). https://linkinghub.elsevier.com/retrieve/pii/S0306452210011668
11. Ivenshitz, M., Segal, M.: Neuronal density determines network connectivity and spontaneous activity in cultured hippocampus. J. Neurophysiol, **104**, 1052–1060 (2010). https://www.physiology.org/doi/10.1152/jn.00914.2009
12. Wilson, N., Ty, M., Ingber, D., Sur, M., Liu, G.: Synaptic Reorganization in Scaled Networks of Controlled Size. J. Neurosci. **27**, 13581–13589 (2007). https://www.jneurosci.org/lookup/doi/10.1523/JNEUROSCI.3863-07.2007
13. Biffi, E., Regalia, G., Menegon, A., Ferrigno, G., Pedrocchi, A.: The influence of neuronal density and maturation on network activity of hippocampal cell cultures: a methodological study. PLoS ONE. **8**, e83899 (2013). https://dx.plos.org/10.1371/journal.pone.0083899
14. VanPelt, J., Wolters, P., Corner, M., Rutten, W., Ramakers, G.: Long-term characterization of firing dynamics of spontaneous bursts in cultured neural networks. IEEE Trans. Biomed. Eng. **51**, 2051–2062 (2004). http://ieeexplore.ieee.org/document/1344208/
15. Penn, Y., Segal, M., Moses, E.: Network synchronization in hippocampal neurons. Proc. Natl Acad. Sci. **113**, 3341–3346 (2016). https://pnas.org/doi/full/10.1073/pnas.1515105113
16. Hulata, E., Baruchi, I., Segev, R., Shapira, Y., Ben-Jacob, E.: Self-regulated complexity in cultured neuronal networks. Phys. Rev. Lett. **92**, 198105 (2004). http://www.ncbi.nlm.nih.gov/pubmed/15169451
17. Wagenaar, D., Nadasdy, Z., Potter, S.: Persistent dynamic attractors in activity patterns of cultured neuronal networks. Physical Review E - Statistical, Nonlinear, And Soft Matter Physics (2006)
18. Segev, R., Baruchi, I., Hulata, E., Ben-Jacob, E.: Hidden neuronal correlations in cultured networks. Phys. Rev. Lett. **92**, 118102 (2004). https://link.aps.org/doi/10.1103/PhysRevLett.92.118102
19. Baruchi, I., Ben-Jacob, E.: Towards neuro-memory-chip: Imprinting multiple memories in cultured neural networks. Phys. Rev **E. 75**, 050901 (2007). https://link.aps.org/doi/10.1103/PhysRevE.75.050901
20. Dityatev, A., Rusakov, D.: Molecular signals of plasticity at the tetrapartite synapse. Current Opin. Neurobiol. **21**, 353–359 (2011). https://linkinghub.elsevier.com/retrieve/pii/S0959438810002084
21. Kazantsev, V., Gordleeva, S., Stasenko, S., Dityatev, A.: A Homeostatic Model of Neuronal Firing Governed by Feedback Signals from the Extracellular Matrix. Public Library of Science San Francisco, USA (2012)
22. Turrigiano, G.: Homeostatic signaling: the positive side of negative feedback. Curr. Opin. Neurobiol. **17**, 318–324 (2007)
23. Rich, M., Wenner, P.: Sensing and expressing homeostatic synaptic plasticity. Trends Neurosci. **30**, 119–125 (2007)
24. Kochlamazashvili, G., et al.: The extracellular matrix molecule hyaluronic acid regulates hippocampal synaptic plasticity by modulating postsynaptic L-type $Ca2+$ channels. Neuron. **67**, 116–128 (2010)
25. Lazarevich, I., Stasenko, S., Rozhnova, M., Pankratova, E., Dityatev, A., Kazantsev, V.: Activity-dependent switches between dynamic regimes of extracellular matrix expression. PLoS ONE **15**, e0227917 (2020)

26. Izhikevich, E.: Dynamical systems in neuroscience: the geometry of excitability and bursting. Dynamical Syst. First, 441 (2007). http://www.amazon.com/Dynamical-Systems-Neuroscience-Excitability-Computational/dp/0262090430

27. Hodgkin, A., Huxley, A.: A quantitative description of membrane current and its application to conduction and excitation in nerve. J. Physiol. **117**, 500–544 (1952)

Investigation of Ice Rheology Based on Computer Simulation of Low-Speed Impact

Evgeniya K. Guseva[1](✉) (ID), Katerina A. Beklemysheva[1] (ID),
Vasily I. Golubev[1](✉) (ID), Viktor P. Epifanov[2] (ID), and Igor B. Petrov[1](✉) (ID)

[1] Moscow Institute of Physics and Technology (National Research University),
Dolgoprudny, Russia
guseva.ek@phystech.edu, {golubev.vi,petrov}@mipt.ru
[2] Ishlinsky Institute for Problems in Mechanics of the Russian Academy of Sciences,
Moscow, Russia

Abstract. Ice is a complex heterogeneous medium that can be described using different mathematical models, for instance, elasticity, viscosity, plasticity, and viscoplasticity models. This work is aimed at the ice properties investigation based on the data of laboratory experiments. The dependencies between instantaneous force on the ball in the impact point and the depth of ball immersion into ice for different striking velocities were obtained experimentally by other scientists. In this work, linear elasticity, elastoplasticity, and Kukudzhanov elastoviscoplasticity models with different parameters were applied to the collision process simulation. The governing system of equations was solved using grid-characteristic method on structured moving meshes. The results of numerical experiments were compared with the dependencies from the laboratory experiments. Qualitative evaluation of the relation between the chosen model parameters and the calculated dependencies was performed.

Keywords: Numerical simulation · Ice rheology · Plasticity model · Viscoplasticity model · Grid-characteristic method

1 Introduction

As more oil deposits are found in the Arctic region, it is becoming more popular for investigation [1]. The numerical simulation is a prominent technology for planning effectively seismic surveys. It should be noticed that the considered geological models contain ice cover area. Ice is a complex heterogeneous medium that can be described using different mechanical models. Furthermore, full-scale experiments show significant differences in ice behavior and a broad range of elastic parameters depending on ice type, its temperature, salinity, etc. [2,3]. This

The reported study was funded by the Russian Foundation for Basic Research, project no. 20-01-00649.

work aims to investigate ice proprieties based on the laboratory experiments conducted by Prof. Epifanov V.P. and his colleagues in the Ishlinsky Institute for
Problems in Mechanics RAS [4]. The problem of the low-speed impact [5] on the
ice specimen is considered. The dependencies between instantaneous force on the
ball in the impact point and the depth of the ball immersion into the ice for different striking velocities were obtained. In this work, linear elasticity, elastoplasticity, and Kukudzhanov elastoviscoplasticity models with different parameters
were used for the numerical simulation. The experiment was conducted using
a software package written in C++ by the Computational Physics Department
and the Informatics and Computational Mathematics Department of the Moscow
Institute of Physics and Technology, RECT (also used, for instance, in [6]), on
a personal computer Intel Core i7.

2   Problem Formulation

In our research, the 2D mechanical problem is considered. The computational
domain (see Fig. 1 on the left) is divided into several parts. Subdomains 1–2
form the ball, 3–4 form the ice. In the ball, curvilinear grids are used, in the ice,
rectangular grids are created initially. In order to fit the used software, which
works with the structured grids, the grids were generated in a form presented
in Fig. 1 on the right with a script written on Python. All edge coordinates
of the curvilinear cells were calculated analytically. Between the ball's parts,
a full adhesion contact condition was applied. At the ball surface and on the
right, left and top sides of the ice, a free boundary condition is used. At the ice
bottom, a zero velocity boundary condition is set. After the beginning of the
collision in contact nodes between the ice and the ball, a full adhesion contact
condition is applied. The computational mesh is rebuilt at each time step based
on the Lagrange corrector. The scheme of the node interaction is identical to
the published ones in [7]. When the contact grids do not coincide, interpolation
procedure is conducted. Previously, the same approach was successfully used on
the unstructured grids [8]. The velocity of the striking ball is the same as in the
published experiment [4] and equals to $0.484\frac{m}{s}$, $0.594\frac{m}{s}$, $0.831\frac{m}{s}$, $0.99\frac{m}{s}$, $1.4\frac{m}{s}$,
$1.87\frac{m}{s}$, $2.23\frac{m}{s}$ in different numerical experiments.

For subdomain 1 in Fig. 1, the number of the grid cells along the horizontal
axis is equal to $N_x = 30$ and along the vertical axis – $N_y = 30$, for region 2*:
$N_x = 15$, $N_y = 30$. Grids for the other regions 2 are formed by an appropriate
grid rotation. The final grid for the ball is presented in Fig. 1 on the right. In
area 3 in Fig. 1, the numbers of the grid cells are $N_x = 500$, $N_y = 275$, in areas
4: $N_x = 100$, $N_y = 55$. The simulation was conducted using a constant time step
$\tau = 0.2\,\mu s$, the calculation process continued until the ball bounced off or the
number of the time steps exceeded 30000.

According to the experimental data [4], ice shows behavior common to viscoelastic materials at the collision beginning. In the later stages, it demonstrates
more complex behavior. In our work, three models (linear isotropic elasticity,
elastoplasticity, and elastoviscoplasticity) are used.

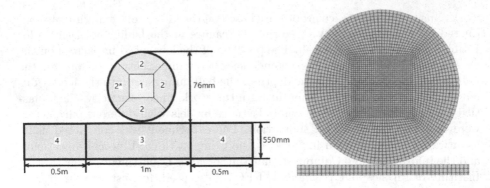

Fig. 1. Computational domain - on the left, the ball's grid - on the right.

Let's first consider at the isotropic linear elasticity model. It can be formulated in terms of the velocity **v** and the stress tensor σ using the following hyperbolic system of equations [9]:

$$\rho\dot{\mathbf{v}} = \nabla\cdot\sigma + \mathbf{f}, \tag{1}$$

$$\dot{\sigma} = \lambda(\nabla\cdot\mathbf{v})I + \mu(\nabla\otimes\mathbf{v} + (\nabla\otimes\mathbf{v})^T) + F. \tag{2}$$

Here, λ and μ are the Lame parameters, which relate to the Young's modulus E and the Poisson coefficient ν with formulas: $\lambda = \frac{E\nu}{(1+\nu)(1-2\nu)}$, $\mu = \frac{E}{2(1+\nu)}$. Here, ρ is the medium density, **f** is the external force, and the additional term F is equal to zero. This system of equations produces two types of waves. The velocity of the pressure wave $c_p = \sqrt{\frac{\lambda+2\mu}{\rho}}$, the velocity of the shear wave $c_s = \sqrt{\frac{\mu}{\rho}}$. For this model, several wave velocities are considered. The velocities $c_p = 3940 \frac{m}{s}$, $c_s = 2493 \frac{m}{s}$ are chosen as an upper boundary, which is common to fresh ice. Another velocities $c_p = 3600 \frac{m}{s}$, $c_s = 1942 \frac{m}{s}$ are calculated using the Berdennicov's formula ($E = (87,6-0,21T-0,0017T^2)\cdot10^8 Pa$, [10]) for ice with a temperature of $-10\,°C$, which is the same as in the laboratory experiment and the constant Poisson coefficient $\nu = 0.295$. Finally, the velocities $c_p = 2450 \frac{m}{s}$, $c_s = 1450 \frac{m}{s}$ are taken as a lower boundary, which is typical of pack-ice. The density of the ice is set to $\rho = 917 \frac{kg}{m^3}$ for each experiment. For the ball, the wave velocities are equal to $c_p = 5700 \frac{m}{s}$, $c_s = 3100 \frac{m}{s}$ and the density is $\rho = 7800 \frac{kg}{m^3}$.

Next, let's discuss the elastoplasticity model. In this work, a simplified version of the Prandtl-Reuss flow rule [11] based on the von Mises yield criterion is applied. If $\frac{1}{2}s_{ij}s_{ij} - k^2 < 0$, where $s_{ij} = \sigma_{ij} - \frac{\sigma_{ll}}{3}\delta_{ij}$ is the stress deviator and k is the maximum sheer stress, then the elasticity model is used. If the von Mises criterion is not satisfied, then the stress deviator is normalized $s_{ij} = s_{ij}^0\frac{\sqrt{2k}}{\sqrt{s_{el}s_{el}}}$. In this work, the maximum sheer stress equals to 10, 10^3, 10^5, $2.5\cdot10^5$, $3\cdot10^5$, $5\cdot10^5$, 10^6, $2.2\cdot10^6$, 10^9 in different tests.

Now let's proceed to the Kukudzhanov elastoviscoplasticity model [12]. Its equations are similar to the elastic model, but the additional term F is non zero and $F_{ij} = -s_{ij}\dfrac{2\mu}{\tau_0}\dfrac{\sqrt{\frac{\sqrt{s_{el}s_{el}}}{k}-\sqrt{2}}}{\sqrt{s_{el}s_{el}}}$, where τ_0 is the relaxation time. In this work, it is set to $0.5\,\mathrm{s}$, $1\,\mathrm{s}$ and $2\,\mathrm{s}$ in different experiments. For the Kukudzhanov model, the splitting along the physical processes is used. So, after each elastic step, the differential equation is integrated as: $s_{ij} = s_{ij}^0\exp(-2\mu\dfrac{\tau}{\tau_0}\dfrac{\sqrt{\frac{\sqrt{s_{el}s_{el}}}{k}-\sqrt{2}}}{\sqrt{s_{el}s_{el}}})$, where τ is the time step.

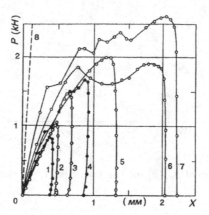

Fig. 2. Dependencies $P(x)$ from the laboratory experiment [4] for different striking velocities, 1 state for the minimum velocity, 7 state for the maximum velocity in the experiment.

3   Simulation Results

The presented above hyperbolic system of equations is solved using the grid-characteristic method [13–15]. Each of 1D transport equations are solved using the third approximation order Rusanov scheme [16] modified by the grid-characteristic monotonicity criterion [17]. In this work the simulation results are evaluated qualitatively. The dependence of projection σ_{yy} which is orthogonal to the ice surface on depth x is checked to resemble the real experiments' graph at Fig. 2. In Fig. 2 the dependencies from the laboratory experiment between instantaneous force P on the ball in the impact point and the depth of ball immersion into the ice x for different striking velocities which are presented.

Let's first take a look at the simulation results with the linear isotropic elasticity model used for the ice (see Fig. 3). They show that the elasticity model with physically correct wave velocities c_p, c_s produce less depths of the ball immersion x and not similar later stages of the collision in regards to the full-scale experiment in Fig. 2. According to the obtained dependence, the following trends can be seen: less wave velocities and less striking velocities v result in less amplitude of σ_{yy}. Nevertheless, less wave velocities lead to greater depths of the

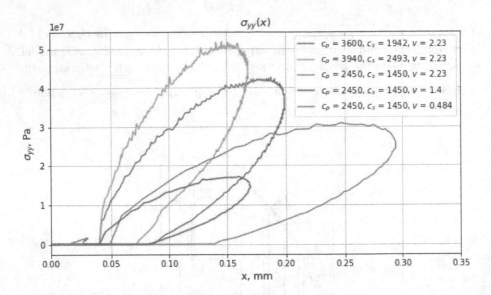

Fig. 3. Dependency of $\sigma_{yy}(x)$ for the linear isotropic elasticity ice model.

ball immersion, which is opposite to the cases of less striking velocities that also move the point when the ice with the immersed ball start to have stress along the vertical axes to the earlier stages of the collision.

Next, let's proceed to the case of elastoplasticity used for the ice rheology simulation. According to Fig. 4, lessening of the maximum sheer stress k leads to lessening of the amplitude of σ_{yy}, while the ball goes into the ice deeper. The point when the ice with the immersed ball starts to have stress along the vertical axes moves to the later stages of the collision. The situation changes when the shear stress is around $k = 10^3 Pa$, and the behavior is similar to $k = 10 Pa$, where the ball continues to go into the ice until the end of the simulation. The case with $k = 10^9 Pa$ is not included into the figures because the calculated dependence is fully consistent with the results of the linear isotropic elasticity model. The reason for it is the fact that the von Mises criterion is not fulfilled during the simulation.

Lessening of the wave velocities c_p, c_s does not any produce explicit effect: the amplitude of σ_{yy} is similar, the maximum depth of the ball immersion and the point when the ice with the immersed ball starts to have stress along the vertical axes change. Overall, the behavior of the $\sigma_{yy}(x)$ in the late collision stage is similar to the experimental data in Fig. 2. Among the used sheer stresses, the value $k = 3 \cdot 10^5 Pa$ produced the maximum depth close to the laboratory experimental maximum depth. Thus, this value was chosen for further numerical tests with different striking velocities (see Fig. 5). According to Fig. 5, lessening of the striking velocities results in less maximum depth of the ball's immersion in the ice, which is compatible with the full-scale experiment.

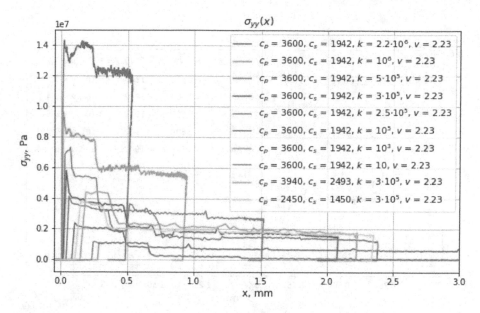

Fig. 4. Dependency of $\sigma_{yy}(x)$ for the elastoplasticity ice model with different maximum shear stresses k and different wave velocities c_p, c_s.

The results of the simulations using the elastoviscoplasticity with different parameters are presented in Fig. 6. The overall profile of the dependence looks like two hills that merge to one a in different test, which does not correspond the dependence from the real experiment. Changing of the parameters τ and k does not have any explicit effect on the amplitude of σ_{yy}, nonetheless, their decrease leads to greater maximum depths of the ball immersion into the ice. The lessening of the striking velocities lessens the amplitude of σ_{yy} and the immersion depth. Notwithstanding, the decrease of the wave velocities c_p, c_s results in an increase of the σ_{yy} amplitude and deeper ball immersion.

Finally, let's compare all considered models in case of similar parameters. Figure 7 shows that the elasticity model produces the highest amplitude of σ_{yy} but the smallest maximum depth of ball immersion. The usage of the elastoviscoplasticity model results in less amplitude of σ_{yy}, the division of the dependence into two hills and the deeper ball immersion. Finally, the elastoplasticity model keeps the lowest amplitude of σ_{yy} and the greatest depth of the ball immersion, which is close to the value obtained in the laboratory experiment.

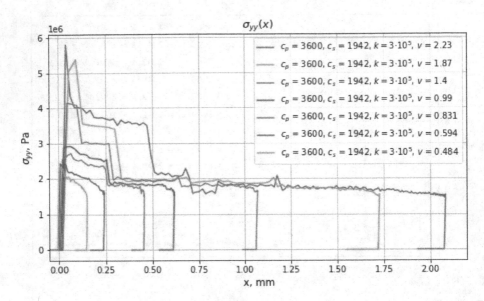

Fig. 5. Dependency of $\sigma_{yy}(x)$ for the elastoplasticity ice model with different striking velocities v.

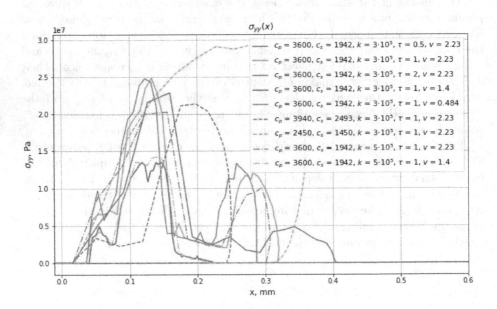

Fig. 6. Dependency of $\sigma_{yy}(x)$ for the elastoviscoplasticity ice model with different parameters.

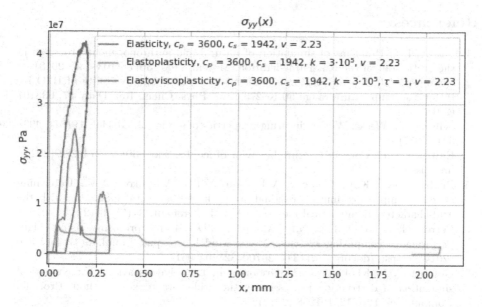

Fig. 7. Comparison of dependencies $\sigma_{yy}(x)$ for all models used for the ice description with similar parameters.

4 Conclusion

In this work, the linear isotropic elasticity, elastoplasticity, and elastoviscoplasticity models with different parameters were used for ice simulation. The results showed that the elastoplasticity model produces results compatible with the laboratory experiments. The parameters that produce depths of the ball's immersion in the ice close to the laboratory ones were selected. Further research should be directed to a three-dimensional case simulation and consideration of more complicated non-linear mechanical models [18]. It should be also noted that the Berdennicov's formula was used to take into account the elastic modulus dependency on the ice temperature. However, during the collision simulation process, these modules were treated as a constant. At this stage, we neglected the thermal effect. During the laboratory experiment, the temperature variation was not measured either. This improvement should be a perspective direction of the further investigation too.

Acknowledgements. We thank Dr. E. Bazanova for critical reading of the paper and helpful recommendations. The reported study was funded by the Russian Foundation for Basic Research, project no. 20-01-00649.

References

1. Petrov, I.B.: Problems of simulation of natural and anthropogenous processes in the arctic zone of the Russian federation. Matem. Mod. **30**(7), 103–136 (2018)
2. Neumeier, J.J.: Elastic constants, bulk modulus, and compressibility of H2O ice Ih for the temperature range 50 K–273 K. J. Phys. Chem. Ref. Data **47**, 033101 (2018)
3. Schwarz, J., Weeks, W.: Engineering properties of sea ice. J. Glaciol. **19**(81), 499–531 (1977)
4. Epifanov, V.P.: Contact fracture behavior of ice. Ice Snow **60**(2), 274–284 (2020). [in Russian]
5. Beklemysheva, K.A., Golubev, V.I., Petrov, I.B., Vasyukov, A.V.: Determining effects of impact loading on residual strength of fiber-metal laminates with the grid-characteristic numerical method. Chin. J. Aeronaut. **34**(7), 1–12 (2021)
6. Petrov, I.B., Khokhlov, N.I.: Modeling 3D seismic problems using high-performance computing systems. Math. Models Comput. Simul. **6**(4), 342–350 (2014). https://doi.org/10.1134/S2070048214040061
7. Golubev, V.I., Khokhlov, N.I., Grigorievyh, D.P., Favorskaya, A.V.: Numerical simulation of destruction processes by the grid-characteristic method. Procedia Comput. Sci. **126**, 1281–1288 (2018)
8. Beklemysheva, K.A., Petrov, I.B., Favorskaya, A.V.: Numerical simulation of processes in solid deformable media in the presence of dynamic contacts using the grid-characteristic method. Math. Models Comput. Simul. **6**(3), 294–304 (2014). https://doi.org/10.1134/S207004821403003X
9. Novatskii, V.: Theory of Elasticity. Mir, Moscow (1975). [in Russian]
10. Berdennikov, V.P.: Izuchenie modulya uprugosti lda. Trudi GGI **7**(61), 13–23 (1948). [in Russian]
11. Lee, Yung-Li, Barkey, M.E.: Chapter 7 - Fundamentals of Cyclic Plasticity Theories. Metal Fatigue Analysis Handbook, pp. 253–297 (2012)
12. Kukudzhanov, V.N.: Numerical Solution of Stress Non-One-Dimensional Wave Propagation Problems in Solids. Vychisl. Tsentr Akad, Nauk SSSR, Moscow (1976). [in Russian]
13. Golubev, V.I., Guseva, E.K., Petrov, I.B.: Application of quasi-monotonic schemes in seismic arctic problems. Smart Innovations, Syst. Technol. **274**, 289–307 (2022)
14. Golubev, V.I., Shevchenko, A.V., Khokhlov, N.I., Petrov, I.B., Malovichko, M.S.: Compact grid-characteristic scheme for the acoustic system with the piece-wise constant coefficients. Int. J. Appl. Mech. **14**(2), 2250002 (2022)
15. Golubev, V.I., Shevchenko, A.V., Petrov, I.B.: Simulation of seismic wave propagation in a multicomponent oil deposit mode. Int. J. Appl. Mech. **12**(8), 2050084 (2020)
16. Rusanov, V.: The calculation of the interaction of non-stationary shock waves with barriers. J. Comput. Math. Phys. USSR. **1**, 267–279 (1961)
17. Kholodov, A.S., Kholodov, Y.A.: Monotonicity criteria for difference schemes designed for hyperbolic equations. Comput. Math. Math. Phys. **46**(9), 1560–1588 (2006)
18. Nikitin, I.S., Golubev, V.I.: Higher order schemes for problems of dynamics of layered media with nonlinear contact conditions. Smart Innovations, Syst. Technol. **274**, 273–287 (2022)

Computation in Optimization
and Optimal Control

Global Optimization Method Based on the Survival of the Fittest Algorithm

Oleg Kuzenkov[ID] and Dmitriy Perov[✉][ID]

Lobachevsky State University, Gagarin Av. 23, Nizhny Novgorod 603950, Russia
unn@unn.ru, diper1998@yandex.ru
http://www.unn.ru

Abstract. One of the most important theoretical questions for evolutionary methods of global optimization is their convergence. The majority of evolutionary methods do not guarantee that the generated sequence
of test points converges to a global extremum in any sense. The purpose
of this paper is to construct and prove convergence of a new evolutionary
global optimization algorithm. This algorithm is created on the base of
the Survival of the Fittest algorithm using ideas of Differential Evolution.
It is proved that the sequence of test points of this algorithm converges
to the solution with probability one. The new method is compared with
other evolutionary algorithms. It is shown that the method has higher
efficiency for some classes of relevant multidimensional functions.

Keywords: Global optimization · Convergence proof · Differential
evolution · Survival of the fittest algorithm

1 Introduction

Global optimization problems appear in many areas of theoretical and practical
research, and there are many efficient algorithms for solving them [1–4]. Evolutionary algorithms occupy an important place among the algorithms of global
optimization. They are based on modeling various aspects of biological evolution, such as mutations and selection [5]. Currently, evolutionary computations
are widely used for global optimization in various application areas including
neural networks, machine learning and artificial intelligence [6–9]. New modifications of evolutionary algorithms are permanently created to increase efficiency
of the optimization and take into account specifics of different classes of problems
[10–16]. Of all evolutionary algorithms, we can single out Differential Evolution
(DE) [8], which is efficient in solving global optimization problems. There are a
huge number of modifications that improve this algorithm [17–19]. Despite the
existence of different evolutionary optimization methods, there are still crucial
theoretical problem of their usage [20]. One of the most important questions for
evolutionary methods of global optimization is their convergence. When we use
some method, we need to be sure that the generated sequence of test points

D. Balandin et al. (Eds.): MMST 2022, CCIS 1750, pp. 187–201, 2022.
https://doi.org/10.1007/978-3-031-24145-1_16

converges to the global extremum in some sense. In the work [21], the new stochastic global optimization algorithm was proposed, called the Survival of the Fittest Algorithm (SoFA). It is based on fundamental principle of biological evolution - the survival of the fittest. Numerically, the algorithm uses the probability measure concentration in a vicinity of the global maximum [22–25]. It has been rigorously proven that the sequence of points of this algorithm converges to the solution with probability one. It was shown that SoFA efficiently works in multidimensional spaces and infinite-dimensional Hilbert spaces. This algorithm shows a higher convergence rate than several other evolutionary methods (Evolutionary Strategy with Cauchy distribution [26,27], Controlled Random Search with local mutation [28–30], and Multi Level Single-Linkage [31,32]) for some important classes of optimization problems in the spaces of a high dimension. It was tested in a class of relevant objective functions and was successfully applied for investigations of physical and biological models [33,34].

The purpose of this paper is to construct a new, more efficient convergent global optimization algorithm based on Differential Evolution and the Survival of the Fittest algorithm for high-dimensional spaces. We have proved that the sequence of test points generated by this algorithm converges in probability to the global maximum. We compared this method with other evolutionary algorithms. It is shown that the method has higher efficiency for some classes of relevant multidimensional functions.

2   Materials and Methods

Suppose that the continuous positive function $J(x_1, x_2, \ldots, x_D)$ (the objective function or fitness) is defined in a D-dimensional parallelepiped:

$$\Pi = \{X = \{x_1, \ldots, x_D\} : x_{min} \le x_i \le x_{max}, \, i = \overline{1, D}\}. \tag{1}$$

Here x_{min} and x_{max} are some constants. Let $X^* = \{x_1^*, x_2^*, \ldots, x_D^*\}$ be a unique point of global maximum for $J(x_1, x_2, \ldots, x_D)$ in Π, then the goal of optimization is to find the point X^*.

2.1   Differential Evolution

First of all, let us remind the main ideas of Differential Evolution (DE). At each stage, the DE method works with the finite set of points from Π, this set is called the population, its elements are called agents, the stage is called the generation. A population P_g contains NP agents: $P_g = (X_{1,g}, \ldots, X_{NP,g})$. Here g is the generation index, $g = \overline{1, G_{max}}$. Agent $X_{i,g}$ is called the parent for agent $X_{i,g+1}$. Each agent is the vector of D components $X_{i,g} = \{x_{i,1,g}, x_{i,2,g}, \ldots, x_{i,D,g}\}$.

DE sequentially selects new test points with next steps:

1) *Initialization.* At the beginning, the initial population is randomly created. Each component of each vector receives a random value uniformly distributed between the lower and upper bounds: $x_{i,j,0} = rand(x_{min}, x_{max})$.

Further, the method consists of the sequence of steps (generations), the new current population is created at each step using three operations Mutation, Crossover and Selection.

2) *Mutation.* The new (mutant) vectors $\widetilde{X}_{i,g+1}$, $i = \overline{1,NP}$, are generated by a mutation operator [17]. Some of the most popular operators are described below.

DE rand 1:
Three vectors are randomly selected from the current population; the difference between two first vectors is multiplied by the scaling factor $F_{i,g}$, and the result is added to the third vector:

$$\widetilde{X}_{i,g+1} = X_{r_1,g} + F_{i,g}(X_{r_2,g} - X_{r_3,g}), \qquad (2)$$

where r_1, r_2, r_3 are indexes from the set $\{1,\ldots,NP\}$ and $r_1 \neq r_2 \neq r_3 \neq i$.

DE best 1:
Two vectors are randomly selected from the current population; the difference between them is multiplied by the scaling factor $F_{i,g}$, and the result is added to the vector $X_{best,g}$, which corresponds to the maximum value of the fitness function of the current generation:

$$\widetilde{X}_{i,g+1} = X_{best,g} + F_{i,g}(X_{r_1,g} - X_{r_2,g}). \qquad (3)$$

DE current to best 1:
Two vectors are randomly selected from the current population; the difference between them is multiplied by the scaling factor $(1 - F_{i,g})$, and the result is added to the difference between the vector $X_{best,g}$ and the parent vector from population $X_{i,g}$. This difference is multiplied by the scaling factor $F_{i,g}$ and the result is added to the parent vector $X_{i,g}$:

$$\widetilde{X}_{i,g+1} = X_{i,g} + F_{i,g}(X_{best,g} - X_{i,g}) + (1 - F_{i,g})(X_{r_1,g} - X_{r_2,g}). \qquad (4)$$

DE current to pBest 1:
Two vectors are randomly selected from the current population; the difference between two vectors is multiplied by the scaling factor $(1 - F_{i,g})$, the result is added to the difference between the vector $X_{pBest,g}$, which is uniformly selected from p vectors with the maximum value of the fitness function for the current generation, and the parent vector from population $X_{i,g}$. This difference is multiplied by the scaling factor $F_{i,g}$ and the result is added to the parent vector $X_{i,g}$:

$$\widetilde{X}_{i,g+1} = X_{i,g} + F_{i,g}(X_{pBest,g} - X_{i,g}) + (1 - F_{i,g})(X_{r_1,g} - X_{r_2,g}). \qquad (5)$$

3) *Crossover.* The created mutant vector $\widetilde{X}_{i,g+1}$ participates in the formation of the test vector $\bar{X}_{i,g+1}$ as follows:

$$\bar{x}_{i,j,g+1} = \widetilde{x}_{i,j,g+1}, \text{ if } rand(0,1) \leq CR_{i,g} \text{ or } j = j_r, \text{ else } x_{i,j,g}, \qquad (6)$$

where $i = \overline{1,NP}$ and $j = \overline{1,D}$. Here $CR_{i,g} \in [0,1]$ is the crossover parameter, which represents the probability of selecting components for the test vector from the mutant vector. The randomly selected index $j_r \in \{1, 2, \ldots, D\}$ is responsible for ensuring that the test vector contains at least one component from the mutant vector. If the component was not selected from the mutant vector, then it is taken from the parent vector $X_{i,g}$.

4) *Selection.* After the crossover operation, the test vector is evaluated – the fitness function $J(\bar{X}_{i,g+1})$ is calculated, then its value is compared with the corresponding value from the population $J(X_{i,g})$. The best vector will remain in the next generation:

$$X_{i,g+1} = \bar{X}_{i,g+1}, \text{ if } J(X_{i,g}) \leq J(\bar{X}_{i,g+1}), \text{ else } X_{i,g}. \tag{7}$$

5) *The stopping criteria.* It can be expressed by the maximum number of fitness function calculations, a time limit, or reaching the required accuracy. For example, let's introduce the maximum number of iterations – calculations of the fitness function K_{max} and a variable that tracks the current number of calculations of the fitness function $k = (g-1)NP+i$, $g = \overline{1, G_{max}}$, $i = \overline{1, NP}$. The algorithm finishes its work if $k > K_{max}$.

Classical differential evolution has several configurable hyperparameters: NP, $CR_{i,g}$, $F_{i,g}$, which can significantly affect the optimization process. There are a large number of ways to set parameters and corresponding modifications of Differential Evolution [17–19]. This paper considers a simple approach to choosing parameters $CR_{i,g}$, $F_{i,g}$, called jDE [17]:

1) Initialization of parameters $CR_{i,0} = CR_i = 0.9$, $F_{i,0} = F_i = 0.5$.
2) Updating parameters:

$$F_{i,g} = rand(0.1, 1), \text{ if } rand(0, 1) \leq 0.1, \text{ else } F_i. \tag{8}$$

$$CR_{i,g} = \Big\{ rand(0, 1), \text{ if } rand(0, 1) \leq 0.1, \text{ else } CR_i. \tag{9}$$

3) Saving parameters when successfully replacing the parent vector with the test vector:

$$CR_i = CR_{i,g}, \text{ if } J(X_{i,g}) \leq J(\bar{X}_{i,g+1}), \tag{10}$$

$$F_i = F_{i,g}, \text{ if } J(X_{i,g}) \leq J(\bar{X}_{i,g+1}), \tag{11}$$

where $i = \overline{1, NP}$ and $g = \overline{1, G_{max}}$.

The population size is often taken proportional to the dimension of the problem $NP = 10D$ [17].

2.2   The New Algorithm SoFDE

We introduce the new algorithm called SoFDE based on DE with a new mutation operator inspired by the SoFA and a modified crossover:

1) *Mutation.* Each vector in the population is assigned once per generation for NP iterations the probability of participation for further mutation as a reference vector:

$$\frac{J^{\psi_g}(X_{i,g})}{J^{\psi_g}(X_{1,g}) + \ldots + J^{\psi_g}(X_{NP,g})}. \tag{12}$$

Here ψ_g is a parameter of the method, an infinitely increasing sequence, depending on the generation, regulating the rate of convergence. Given the probabilities found, the reference vector $X_{r,g}$, $r \in \{1, \ldots NP\}$ and its corresponding coordinates $x_{r,j,g}$ are randomly selected from the population. The mutant vector $X_{i,g+1}$ is created randomly, the components of which $x_{i,j,g+1}$ take values at the segment $[x_{min}, x_{max}]$ with probability density:

$$\frac{A_{i,r,j,g}\varepsilon_{i,g}}{\varepsilon_{i,g}^2 + (\widetilde{x}_{i,j,g+1} - x_{r,j,g})^2}. \tag{13}$$

Here $\varepsilon_{i,g}$ is a sequence decreasing to zero and $A_{i,r,j,g}$ is the normalizing probability density constant for the segment $[x_{min}, x_{max}]$:

$$A_{i,r,j,g} = (arctan(\frac{x_{max} - x_{r,j,g}}{\varepsilon_{i,g}}) - arctan(\frac{x_{min} - x_{r,j,g}}{\varepsilon_{i,g}}))^{-1}. \tag{14}$$

In other words, a mutant vector is obtained with the following mutation of its components:

$$\widetilde{x}_{i,j,g+1} = x_{r,j,g} + \varepsilon_{i,g}tan((rand(0,1) - \frac{1}{2})A_{i,r,j,g}^{-1}). \tag{15}$$

2) *Crossover.* The population P_g is divided into two parts; the crossover operator is not applied for the first part $\bar{X}_{q,g+1}$, $q = \overline{1, Q_{max}}$. This means that the test vectors of the first part will receive a mutation for all their components: $\bar{x}_{q,j,g+1} = \widetilde{x}_{q,j,g+1}$, $j = \overline{1, D}$, where and $g = \overline{1, G_{max}}$. For the second part the created mutant vector $\widetilde{X}_{l,g+1}$ participates in the formation of the test vector $\bar{X}_{l,g+1}$ as follows:

$$\bar{x}_{l,j,g+1} = \widetilde{x}_{l,j,g+1}, \text{ if } rand(0,1) \le CR_{l,g} \text{ or } j = j_r, \text{ else } x_{l,j,g}. \tag{16}$$

Here $l = \overline{Q_{max} + 1, NP}$, $j = \overline{1, D}$. Crossover parameter $CR_{l,g} \in [0,1]$ represents the probability of selecting components for the test vector from the mutant vector. The randomly selected index $j_r \in \{1, 2, \ldots, D\}$ is responsible for ensuring that the test vector contains at least one component from the mutant vector. If the component was not selected from the mutant vector, then it is taken from the parent vector $X_{l,g}$.

Thus, the new SoFDE algorithm was introduced, which follows the steps from DE with the new mutation operator and the crossover constraint. The first part of the population converges to the global maximum under the condition:

$$\varepsilon_{q,g} = \frac{1}{((g-1)NP + q)^{\frac{1}{2D}}}, \tag{17}$$

the proof of this is described below. The second part of the population may refine the solution due to the faster tendency of $\varepsilon_{l,g}$ to zero.

Due to the fact that the new algorithm based on DE with the new mutation operator and the modified crossover, it is assumed that it is possible to use the modifications of DE presented in [17–19] without much difficulty. Therefore, the same approach as for DE is used to choosing the hyperparameter $\varepsilon_{i,g}$ and $CR_{i,g}$, called jDE [17], but with some changes:

1) Initialization of parameters $CR_{l,0} = CR_l = 0.9$, $\varepsilon_{q,0} = \varepsilon_{l,0} = \varepsilon_l = 1$.
2) Updating parameters:

$$\varepsilon_{q,g} = \frac{1}{((g-1)NP+q)^{\frac{1}{2D}}}. \tag{18}$$

$$CR_{l,g} = rand(0,1), \text{ if } rand(0,1) \le 0.1, \text{ else } CR_l. \tag{19}$$

$$\varepsilon_{l,g} = \frac{1}{((g-1)NP+l)^{\frac{1}{2}}}, \text{ if } rand(0,1) \le 0.1, \text{ else } \varepsilon_l. \tag{20}$$

3) Saving parameters when successfully replacing the parent vector with the test vector:

$$CR_l = CR_{l,g}, \text{ if } J(X_{l,g}) \le J(\bar{X}_{l,g+1}), \tag{21}$$

$$\varepsilon_l = \varepsilon_{l,g}, \text{ if } J(X_{l,g}) \le J(\bar{X}_{l,g+1}), \tag{22}$$

where $q = \overline{1, Q_{max}}$, $l = \overline{Q_{max}+1, NP}$ and $g = \overline{1, G_{max}}$.

The parameter ψ_g, which regulates the convergence rate, takes the following value:

$$\psi_g = ((g-1)NP+1)^{\frac{1}{\lambda}}, \tag{23}$$

where λ is some positive constant.

3   Results

3.1   Convergence Proof for SoFDE

This section demonstrates the proof of convergence of the proposed SoFDE algorithm in the D-dimensional parallelepiped Π for the continuous positive function J. The proof consists of the following three theorems:

Theorem 1. *Let* $\varepsilon_{q,g} = \frac{1}{((g-1)NP+q)^{\frac{1}{2D}}}$, *then the subset of test vectors* $\bar{X}_{q,g}$, $q = \overline{1, Q_{max}}$, *forms an everywhere dense subsequence in* Π *with probability of one at* $g \to \infty$, *i.e., for any vector* $X \in \Pi$ *and for any positive number* σ *the probability of having a vector* $\bar{X}_{q,g}$ *in the neighborhood* $O_\sigma(X)$ *tends to unity at* $g \to \infty$.

Proof. We introduce a cubic sigma neighborhood $O_\sigma(X)$ for $X = \{x_1, \ldots, x_D\}$ with $\sigma < \frac{1}{2}$. If $|\bar{x}_j - x_j| \le \sigma$, $j = \overline{1, D}$, then the vector $\bar{X} = \{\bar{x}_1, \ldots, \bar{x}_D\}$ belongs to $O_\sigma(X)$.

The test vector $\bar{X}_{q,g}$ is constructed on the basis the randomly selected reference vector $X_{r,g} = \{x_{r,1,g}, \ldots, x_{r,D,g}\}$ from the population P_g. The coordinates of the test vector $\bar{X}_{q,g}$ have the following probability density:

$$\frac{A_{q,r,j,g}\varepsilon_{q,g}}{\varepsilon_{q,g}^2 + (\bar{x}_{q,j,g} - x_{r,j,g})^2}, \tag{24}$$

where $A_{q,r,j,g}$ is the normalizing constant for the segment $[x_{min}, x_{max}]$:

$$A_{q,r,j,g} = (arctan(\frac{x_{max} - x_{r,j,g}}{\varepsilon_{q,g}}) - arctan(\frac{x_{min} - x_{r,j,g}}{\varepsilon_{q,g}}))^{-1}. \tag{25}$$

Obviously, the following inequality holds:

$$A_{q,r,j,g} \ge \frac{1}{\pi}. \tag{26}$$

Probability $P_{q,g}(\forall j \in \{1, 2 \ldots, D\}, |\bar{x}_{q,j,g} - x_j| \le \sigma)$ that the test vector $\bar{X}_{q,g}$ will be included in $O_\sigma(X)$ at the generation g is defined as follows:

$$P_{q,g}(\forall j \in \{1, 2 \ldots, D\}, |\bar{x}_{q,j,g} - x_j| \le \sigma)$$

$$= \prod_{j=1}^{D} \frac{\int_{x_j-\sigma}^{x_j+\sigma}(\frac{A_{q,r,j,g}\varepsilon_{q,g}}{\varepsilon_{q,g}^2 + (\bar{x}_{q,j,g} - x_{r,j,g})^2})d\bar{x}_{q,j,g}}{\int_{x_{min}}^{x_{max}}(\frac{A_{q,r,j,g}\varepsilon_{q,g}}{\varepsilon_{q,g}^2 + (\bar{x}_{q,j,g} - x_{r,j,g})^2})d\bar{x}_{q,j,g}}. \tag{27}$$

Assume $\varepsilon_{q,g} = \frac{1}{((g-1)NP+q)^{\frac{1}{2D}}}$. In this case, we can estimate the probability:

$$\prod_{j=1}^{D} \frac{\int_{x_j-\sigma}^{x_j+\sigma}(\frac{A_{q,r,j,g}\varepsilon_{q,g}}{\varepsilon_{q,g}^2 + (\bar{x}_{q,j,g} - x_{r,j,g})^2})d\bar{x}_{q,j,g}}{\int_{x_{min}}^{x_{max}}(\frac{A_{q,r,j,g}\varepsilon_{q,g}}{\varepsilon_{q,g}^2 + (\bar{x}_{q,j,g} - x_{r,j,g})^2})d\bar{x}_{q,j,g}}$$

$$= \prod_{j=1}^{D} \int_{x_j-\sigma}^{x_j+\sigma} (\frac{A_{q,r,j,g}\varepsilon_{q,g}}{\varepsilon_{q,g}^2 + (\bar{x}_{q,j,g} - x_{r,j,g})^2})d\bar{x}_{q,j,g}$$

$$\ge \prod_{j=1}^{D}(\frac{A_{q,r,j,g}\varepsilon_{q,g}}{\varepsilon_{q,g}^2 + (x_{max} - x_{min})^2} \int_{x_j-\sigma}^{x_j+\sigma} d\bar{x}_{q,j,g})$$

$$= \prod_{j=1}^{D} \frac{2A_{q,r,j,g}\varepsilon_{q,g}\sigma}{\varepsilon_{q,g}^2 + (x_{max} - x_{min})^2}$$

$$= \prod_{j=1}^{D} \frac{2A_{q,r,j,g}((g-1)NP+q)^{\frac{1}{2D}}\sigma}{1 + ((g-1)NP+q)^{\frac{1}{D}}(x_{max} - x_{min})^2}$$

$$\geq \prod_{j=1}^{D} \frac{2A_{q,r,j,g}((g-1)NP+q)^{\frac{1}{2D}}\sigma}{((g-1)NP+q)^{\frac{1}{D}}(1+(x_{max}-x_{min})^2)}$$

$$\geq \prod_{j=1}^{D} \frac{2\sigma}{\pi((g-1)NP+q)^{\frac{1}{2D}}(1+(x_{max}-x_{min})^2)}$$

$$= c((g-1)NP+q)^{-\frac{1}{2}}, \tag{28}$$

where the constant $c = \prod_{j=1}^{D} \dfrac{2\sigma}{\pi(1+(x_{max}-x_{min})^2)}$ does not depend on the number of generations.

The probability of not getting into the neighborhood $O_\sigma(X)$ is estimated as:

$$P_{q,g}(\exists j \in \{1,2,\ldots,D\} : |\bar{x}_{q,j,g} - x_j| > \sigma) \leq 1 - c((g-1)NP+q)^{-\frac{1}{2}}, \tag{29}$$

then

$$\lim_{g \to \infty}(1 - c((g-1)NP+q)^{-\frac{1}{2}})^g = 0. \tag{30}$$

In other words, the probability that the vector $\bar{X}_{q,g}$ will not be selected from $O_\sigma(X)$ for g generations tends to zero at $g \to \infty$. Consequently, the subsequence of the generated test vectors $\bar{X}_{q,g}$ is everywhere dense in Π with probability one.

Theorem 2. *Let $J(X)$ be a continuous positive function defined in Π, $X^* = (x_1^*,\ldots,x_D^*)$ is a single point of its global maximum, then for any positive number σ, the probability of choosing a reference vector $X_{r,g}$, $r \in \overline{1,NP}$, from the neighborhood $O_\sigma(X^*)$ tends to unity at $g \to \infty$.*

Proof. We introduce the following definitions: $J_0 = \sup_{X \in \Pi \setminus O_\sigma(X^*)} J(X)$; I_g is the set of indices of vectors $X_{i,g}$, $i = \overline{1,NP}$, belonging to $O_\sigma(X^*)$; $\overline{I}_g$ is the set of indices of vectors $X_{i,g}$, $i = \overline{1,NP}$, that do not belong to $O_\sigma(X^*)$.

By the Theorem 1 $\bar{X}_{q,g}$ is everywhere a dense sequence (with probability one) in Π. Since J is a continuous function, there will always be a vector $X_{p,n} \in O_\sigma(X^*)$ such that $J(X_{p,n}) > J_0$. Then we can estimate the probability of choosing the vector $X_{r,g}$ as a reference vector with index r from the set $\overline{I}_g$ for $g > n$:

$$\frac{\sum_{i \in \overline{I}_g} J^{\psi_g}(X_{r,g})}{\sum_{i=1}^{NP} J^{\psi_g}(X_{i,g})} \leq \frac{\sum_{i \in \overline{I}_g} J_0^{\psi_g} J^{-\psi_g}(X_{p,n})}{\sum_{i=1}^{NP} J^{\psi_g}(X_{i,g}) J^{-\psi_g}(X_{p,n})} \tag{31}$$

$$< (\frac{J_0}{J(X_{p,n})})^{\psi_g} NP \xrightarrow{g \to \infty} 0.$$

It can be seen that this probability tends to zero as g tends to infinity, therefore the probability of choosing the vector $X_{r,g}$ as a reference vector with the index r from the set I_g tends to unity.

Theorem 3. *Let $\varepsilon_{q,g} = \dfrac{1}{((g-1)NP+q)^{\frac{1}{2D}}}$ and the continuous positive function $J(X)$ defined in Π has a unique global maximum, which is reached at $X^* = (x_1^*, \dots, x_D^*)$, then for any positive number σ, the probability of selecting new test vectors $\bar{X}_{q,g}, q = \overline{1, Q_{max}}$, from the neighborhood $O_\sigma(X^*)$ tends to unity at $g \to \infty$. In other words, a subset of the test vectors $\bar{X}_{q,g}, q = \overline{1, Q_{max}}$, forms a probability-convergent subsequence to the global maximum vector:*

$$\forall \sigma > 0, \lim_{g \to \infty} P(||\bar{X}_{q,g} - X^*|| > \sigma) = 0. \tag{32}$$

Proof. Assuming that $O_\sigma(X^*) = \{X = (x_1, \dots, x_m) : |x_i - x_i^*| \leq \sigma, i = \overline{1; D}\}$, we estimate the probability of choosing test vectors $\bar{X}_{q,g}$ from the neighborhood $O_\sigma(X^*)$ provided that $X_{r,g}$ with the index r from the set I_g is taken as the reference vector, i.e., $X_{r,g} \in O_\sigma(X^*)$:

$$P_{q,g}(\forall j, |\bar{x}_{q,j,g} - x_j^*| \leq \sigma) = \prod_{j=1}^{D} \frac{\int_{x_j^*-\sigma}^{x_j^*+\sigma} \left(\dfrac{A_{q,r,j,g}\varepsilon_{q,g}}{\varepsilon_{q,g}^2 + (\bar{x}_{q,j,g} - x_{r,j,g})^2} \right) d\bar{x}_{q,j,g}}{\int_{x_{min}}^{x_{max}} \left(\dfrac{A_{q,r,j,g}\varepsilon_{q,g}}{\varepsilon_{q,g}^2 + (\bar{x}_{q,j,g} - x_{r,j,g})^2} \right) d\bar{x}_{q,j,g}} \tag{33}$$

$$\geq \prod_{j=1}^{D} F(x_j^* + \sigma - x_{r,j,g}) - \prod_{j=1}^{D} F(x_j^* - \sigma - x_{r,j,g}), \tag{34}$$

where

$$F(x) = \frac{1}{\pi} arctan\left(\frac{x}{\varepsilon_{q,g}}\right) + \frac{1}{2}. \tag{35}$$

Since $\varepsilon_{q,g} = \dfrac{1}{((g-1)NP+q)^{\frac{1}{2D}}}$ and $x_j^* + \sigma - x_{r,j,g} > 0$ we have:

$$arctan(x((g-1)NP+q)^{\frac{1}{2D}}) \xrightarrow{g \to \infty} \frac{\pi}{2},$$
$$F(x_j^* + \sigma - x_{r,j,g}) \xrightarrow{g \to \infty} 1. \tag{36}$$

Since $x_j^* - \sigma - x_{r,j,g} < 0$ we have:

$$arctan(x((g-1)NP+q)^{\frac{1}{2D}}) \xrightarrow{g \to \infty} -\frac{\pi}{2},$$
$$F(x_j^* - \sigma - x_{r,j,g}) \xrightarrow{g \to \infty} 0. \tag{37}$$

Therefore:

$$P_{q,g}(\forall j, |\bar{x}_{q,j,g} - x_j^*| \leq \sigma)$$
$$\geq \prod_{j=1}^{D} F(x_j^* + \sigma - x_{r,j,g}) - \prod_{j=1}^{D} F(x_j^* - \sigma - x_{r,j,g}) \xrightarrow{g \to \infty} 1. \tag{38}$$

This fact completes the proof of convergence for the new algorithm.

3.2   Comparison of Evolutionary Algorithms

The new SoFDE algorithm is compared to DE using various mutation strategies that have been described in Sect. 2.1 on multivariate positive continuous sigmoid functions:

$$J_i(x_1, x_2, \ldots, x_D) = \frac{1}{1 + e^{-\frac{1}{D}G_i(x_1, x_2, \ldots, x_D)}}, \tag{39}$$

where the functions $G_i = G_i(x_1, x_2, \ldots, x_D)$, $i = \overline{1,4}$, are described below, and their global maximum value is 0. In this case, the global maximum value of function J_i is 0.5. For example, you can see in Fig. 1 the comparison of the Rastrigin function and the sigmoid function from the Rastrigin function.

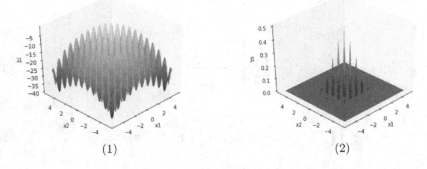

(1)   (2)

Fig. 1. (1) Rastrigin function G_1; (2) Sigmoid function from Rastrigin function J_1.

Rastrigin function

$$G_1 = \sum_{i=1}^{D}(-x_i^2 + 10cos(2\pi x_i) - 10). \tag{40}$$

Global maximum is reached at $x_i^* = 0$, $i = \overline{1,D}$,
with bounds $x_{min} = -5.12$, $x_{max} = 5.12$.

Weierstrass Function

$$G_2 = \sum_{i=1}^{D} -(\sum_{j=0}^{j_{max}}[a^j cos(2\pi b^j(x_i + 0.5))]) + D\sum_{j=0}^{j_{max}} a^j cos(\pi b^j). \tag{41}$$

Here $a = 0.5$, $b = 3$, $j_{max} = 20$. Global maximum is reached at $x_i^* = 0$, $i = \overline{1,D}$, with bounds $x_{min} = -100$, $x_{max} = 100$.

Schwefel Function

$$G_3 = -418.9829D + \sum_{i=1}^{D} x_i sin(\sqrt{|x_i|}). \tag{42}$$

Global maximum is reached at $x_i^* = 420.9687$, $i = \overline{1, D}$,
with bounds $x_{min} = -500$, $x_{max} = 500$.

Griewank Function

$$G_4 = -\sum_{i=1}^{D} \frac{x_i^2}{4000} + 10 \prod_{i=1}^{D} cos(\frac{x_i}{\sqrt{i}}) - 10. \tag{43}$$

Global maximum is reached at $x_i^* = 0$, $i = \overline{1, D}$,
with bounds $x_{min} = -100$, $x_{max} = 100$.

The global maximum is found for functions J_i, $i = \overline{1, 4}$, by setting hyperparameters from Sect. 2.2, with parameters $D = 10$, $NP = 100$, $Q_{max} = 10$, $\lambda = 100$, $p = 10$, $k_{max} = 2 \cdot 10^4$.

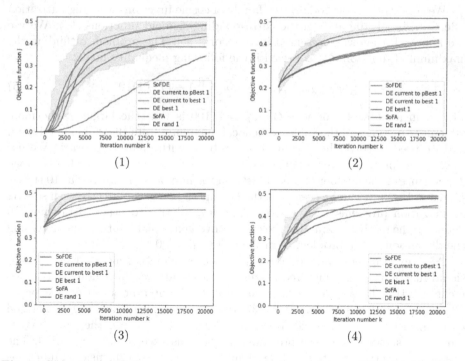

Fig. 2. Comparison of the average convergence rate on the modified (1) Rastrigin function; (2) Weierstrass function; (3) Schwefel function; (4) Griewank function.

The comparison of the convergence rate for the methods is shown in Fig. 2. The average dependence of the found best value of the fitness function on the number of iterations at 100 runs is presented. It can be seen that SoFDE provides a higher convergence rate for the modified Weierstrass and Schwefel functions, shows greater accuracy in finding the maximum for the modified Weierstrass function, competes in the convergence rate and accuracy of DE with the mutation current to pBest for the modified Rastrigin function, shows not the worst convergence and great accuracy for the modified Griewank function. The new method is efficient for functions with a large number of local maxima, due to the fact that the first part of the set is a convergent subsequence to the global maximum, and the second part clarifies the results of the first part.

100 Cosine Functions with Regularization Rule

$$G_{5,j} = \frac{1}{2} \sum_{i=1}^{D} (cos(\frac{2ix_i}{100}) + \frac{i}{D} cos(\frac{2i^2 x_i}{100})) - \frac{1}{4} \sum_{i=1}^{D} \frac{|x_i|}{(i+j)} - \frac{1}{4}(3D+1). \quad (44)$$

Here $j = \overline{1,100}$. Global minimum is reached at $x_i^* = 0$, $i = \overline{1,D}$, with bounds $x_{min} = -100$, $x_{max} = 100$.

We compare the methods on the family of cosine functions with regularization rule $G_{5,j}, j = \overline{1,100}$ and on the corresponding sigmoid functions $J_{5,j}$. We use the approach that was presented in the work [35]. The problem of finding the maximum of function $J_{5,j}$ is solved if the following inequality holds:

$$|J_{5,j}(x_1, x_2, \ldots, x_D) - 0.5| \leq 0.1 - (j-1)10^{-3}. \quad (45)$$

The main feature of functions $G_{5,j}, j = \overline{1,100}$ is that the larger j, the more difficult to distinguish the value of the local maximum from the global maximum, and we need to solve the optimization problem with greater accuracy. We use the same hyperparameter settings as in the previous comparisons. The average dependence of the number of solved tasks on the number of iterations at 10 runs is shown in Fig. 3. It can be seen that the new method copes perfectly with solving optimization problems with a given accuracy, relative to other algorithms. In addition, the method maintains the good convergence scalability when increasing the dimension of the problem from $D = 10$ to $D = 20$.

Also, experiments were performed on 100 randomly generated ten dimensional functions with a hundred randomly chosen attraction regions at the segment $[-1,1]^{10}$. These functions are similar to the hump functions from the paper [36] and paraboloid functions from [37,38]. The example of randomly generated two–dimensional function can be seen in Fig. 4 (1). We use the same hyperparameter settings as in the previous comparisons except $k_{max} = 5 \cdot 10^3$. The averaged convergence rate for all randomly generated ten–dimensional functions on 3 runs is shown in Fig. 4 (2). On these functions the new method shows greater efficiency, relative to other algorithms, and competes with the method DE with the mutation current to pBest.

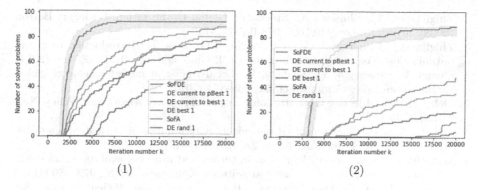

Fig. 3. Comparison of the average number of solved problems on 100 Cosine functions with the regularization rule for (1) $D = 10$; (2) $D = 20$.

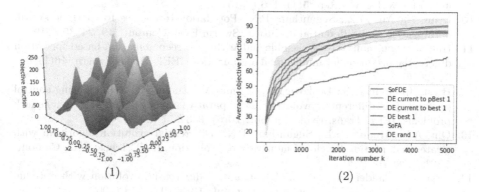

Fig. 4. (1) Example of the randomly generated function; (2) Comparison of the average convergence rate for all 100 randomly generated functions.

4   Summary

In this paper we present the new SoFDE algorithm, which is the modification for Differential Evolution (DE) with the crossover constraint and the new mutation operator that is based on the Survival of the Fittest algorithm (SoFA). We have proved the convergence of SoFDE algorithm to the global maximum with the probability of unite. Then we compared the convergence rate of the new method to the Differential Evolution with various mutation operators and to the original SoFA. We use a well-known multidimensional test functions and demonstrate good performance of the new method.

References

1. Strongin, R., Barkalov, K., Bevzuk, S.: Global optimization method with dual Lipschitz constant estimates for problems with non-convex constraints. Soft. Comput. **24**(16), 11853–11865 (2020)

2. Zhigljavsky, A., Žilinskas, A.: Stochastic Global Optimization. Springer, Boston (2008). https://doi.org/10.1007/978-0-387-74740-8
3. Zhigljavsky, A., Sergeyev, Y.D., Strongin, R.G., Lera, D.: Introduction to global optimization exploiting space-lling curves. J. Glob. Optim. **60**(3), 595–596 (2014)
4. Gergel, V., Grishagin, V., Israfilov, R.: Local tuning in nested scheme of global optimization. Proc. Comput. Sci. **51**, 865–874 (2015)
5. Deb, K.: Multiobjective Optimization Using Evolutionary Algorithms. Wiley, New York (2001)
6. Rai, D., Tyagi, K.: Bio-inspired optimization techniques: a critical comparative study. SIGSOFT Softw. Eng. Notes **38**(4), 1–7 (2013)
7. Galletly, J.: Evolutionary algorithms in theory and practice: evolution strategies, evolutionary programming, genetic algorithms. Kybernetes **27**(8), 979–980 (1998)
8. Storn, R., Price, K.: Differential evolution - a simple and efficient heuristic for global optimization over continuous spaces. J. Glob. Optim. **11**(4), 341–359 (1997)
9. Patti, F.D., Fanelli, D., Piazza, F.: Optimal search strategies on complex multi-linked networks. Sci. Rep. **5**(1), 1–6 (2015)
10. Lynn, N., Ali, M.Z., Suganthan, P.N.: Population topologies for particle swarm optimization and differential evolution. Swarm Evol. Comput. **39**, 24–35 (2018)
11. Hui, S., Suganthan, P.N.: Ensemble and arithmetic recombination-based speciation differential evolution for multimodal optimization. IEEE Trans. Cybern. **46**(1), 64–74 (2016)
12. Brest, J., Greiner, S., Boskovic, B., Mernik, M., Zumer, V.: Self-adapting control parameters in differential evolution: a comparative study on numerical benchmark problems. IEEE Trans. Evol. Comput. **10**(6), 646–657 (2006)
13. Qin, A.K., Huang, V.L., Suganthan, P.N.: Differential evolution algorithm with strategy adaptation for global numerical optimization. IEEE Trans. Evol. Comput. **13**(2), 398–417 (2009)
14. Zhang, J., Sanderson, A.C.: JADE: adaptive differential evolution with optional external archive. IEEE Trans. Evol. Comput. **13**(5), 945–958 (2009)
15. Guo, S.-M., Tsai, J.S.-H., Yang, C.-C., Hsu, P.-H.: A self-optimization approach for l-shade incorporated with eigenvector-based crossover and successful-parent selecting framework on CEC 2015 benchmark set. In: 2015 IEEE Congress on Evolutionary Computation (CEC), pp. 1003–1010 (2015)
16. Das, S., Mullick, S.S., Suganthan, P.N.: Recent advances in differential evolution - an updated survey. Swarm Evol. Comput. **27**, 1–30 (2016)
17. Brest, J., Maučec, M.S., Bošković, B.: The 100-digit challenge: algorithm JDE100. In: 2019 IEEE Congress on Evolutionary Computation (CEC), pp. 19–26 (2019)
18. Viktorin, A., Senkerik, R., Pluhacek, M., Kadavy, T., Zamuda, A.: Dish algorithm solving the CEC 2019 100-digit challenge. In: 2019 IEEE Congress on Evolutionary Computation (CEC), pp. 1–6 (2019)
19. Kumar, A., Misra, R.K., Singh, D., Das, S.: Testing a multi-operator based differential evolution algorithm on the 100-digit challenge for single objective numerical optimization. In: 2019 IEEE Congress on Evolutionary Computation (CEC), pp. 34–40 (2019)
20. Mavrovouniotis, M., Li, C., Yang, S.: A survey of swarm intelligence for dynamic optimization: Algorithms and applications. Swarm Evol. Comput. **33**, 1–17 (2017)
21. Morozov, A.Y., Kuzenkov, O.A., Sandhu, S.K.: Global optimisation in hilbert spaces using the survival of the ttest algorithm. Commun. Nonlinear Sci. Numer. Simul. **103**, 106007 (2021)
22. Gorban, A.N.: Selection theorem for systems with inheritance. Math. Model. Nat. Phenom. **2**(4), 1–45 (2007)

23. Kuzenkov, O.A., Ryabova, E.A.: Limit possibilities of solution of a heredi-tary control system. Differ. Eqn. **51**(4), 523–532 (2015). https://doi.org/10.1134/S0012266115040096

24. Kuzenkov, O.A., Novozhenin, A.V.: Optimal control of measure dynamics. Com-mun. Nonlinear Sci. Numer. Simul. **21**(1), 159–171 (2015)

25. Kuzenkov, O., Morozov, A.: Towards the construction of a mathematically rigorous framework for the modelling of evolutionary fitness. Bull. Math. Biol. **81**(11), 4675–4700 (2019). https://doi.org/10.1007/s11538-019-00602-3

26. Da Silva Santos, C.H., Gonçalves, M.S., Hernández-Figueroa, H.E.: Designing novel photonic devices by bio-inspired computing. IEEE Photonics Technol. Lett. **22**(15), 1177–1179 (2010)

27. Da Silva Santos, C.H.: Parallel and bio-inspired computing applied to analyze microwave and photonic metamaterial strucutures (2010)

28. Kaelo, P., Ali, M.M.: Some variants of the controlled random search algorithm for global optimization. J. Optim. Theory Appl. **130**(2), 253–264 (2006)

29. Price, W.L.: Global optimization by controlled random search. J. Optim. Theory Appl. **40**(3), 333–348 (1983)

30. Price, W.L.: A controlled random search procedure for global optimisation. Com-put. J. **20**(4), 367–370 (1977)

31. Rinnooy Kan, A.H.G., Timmer, G.T.: Stochastic global optimization methods part 1: clustering methods. Math. Programm. **39**(1), 27–56 (1987)

32. Rinnooy Kan, A.H.G., Timmer, G.T.: Stochastic global optimization methods part 2: multi level methods. Math. Program. **39**(1), 57–78 (1987)

33. Kuzenkov, O.A., Grishagin, V.A.: Global optimization in hilbert space. AIP Conf. Proc. **1738**(1), 400007 (2016)

34. Irkhina, A.L., Kuzenkov, O.: Identification of the distribution of deformations in a rod as a problem of optimal control. J. Comput. Syst. Sci. Int. **44**(5), 689–94 (2005)

35. Sergeyev, Y.D., Kvasov, D.E., Mukhametzhanov, M.S.: On the efficiency of natureinspired metaheuristics in expensive global optimization with limited bud-get. Sci. Rep. **8**(1), 453 (2018)

36. Rönkkönen, J., Li, X., Kyrki, V., Lampinen, J.: A framework for generating tunable test functions for multimodal optimization. Soft Computing **15**, 1689–1706 (2011)

37. Gaviano, M., Kvasov, D.E., Lera, D., Sergeyev, Y.D.: Software for generation of classes of test functions with known local and global minima for global optimization (2011)

38. Sergeyev, Y.D., Kvasov, D.E., Mukhametzhanov, M.S.: A generator of multiex-tremal test classes with known solutions for black-box constrained global opti-mization. IEEE Trans. Evol. Comput. (2021)

Solving of the Static Output Feedback Synthesis Problem in a Class of Block-Homogeneous Matrices of Input and Output

A. V. Mukhin[✉] [ID]

Lobachevsky State University, Nizhny Novgorod, Russia
myhin-aleksey@yandex.ru

Abstract. The paper studies the static output feedback synthesis problem solving for unstable linear continuous stationary systems. The problem solution in a special class of input and output matrices is proposed. It is assumed that input and output matrices can be represented as block matrices, one of the block of which is a square matrix of full rank, and the second block is a zero matrix. The necessary condition for stabilization using static output feedback is the existence of a stable diagonal block of arbitrary dimension in the matrix of the system. There are determined some special cases when the problem is solvable. The possible solutions of the synthesis problem with the given input matrix and the given output dimension are considered. It is shown that the problem of static output feedback synthesis can be reduced to solving an auxiliary problem of convex optimization with constraints given in the form of linear matrix inequality.

Keywords: Hurwitz matrix · Static output feedback · Lyapunov theorem · Linear matrix inequality

1 Introduction

The stabilization problem of unstable linear systems with the static output feedback is the one of the central ones in control theory. A large number of works have been proposed for solving the problem. Several of them are mentioned in works [1–6]. Nevertheless the fundamental question of the existence of the stabilizing static output feedback in general case is still open. In the general formulation the problem is NP-hard and is classified as a nonconvex optimization problem [7].

The paper shows that if there is one stable diagonal block of arbitrary dimension in the system matrix, then by means of the special class of input and output matrices the synthesis of the static output controller ensuring the closed loop system matrix stability can be reduced to the linear matrix inequality solving. It is determined all possible cases when the synthesis problem is guarantee solvable. In particular, it is shown that if we organize control and measurement in such a way that the controller matrix allows us to change the unstable diagonal block of the system matrix, then this problem can be reduces to the linear matrix inequality solving. Also the some approaches to the synthesis problem solving with the arbitrary matrices are considered.

The remainder of the paper is organized as follows. The second section presents the formulation of the problem in terms of the linear matrix inequalities. The third section presents the problem solution in the special class of input and output matrices. The Sect. 4 contains the possible synthesis problem solutions for the given input matrix and the given number of outputs. The last one contains the main concluding remarks.

2 Formulation of the Problem

Let us consider a linear unstable continuous stationary control system describes by the equation

$$\dot{x} = Ax + Bu, x(0) = x_0, \tag{1}$$

where $x \in R^n$ is the state of the system, $A \in R^{n \times n}$ is the nonsingular matrix of the system, $B \in R^{n \times m}$ is the input matrix, $u \in R^m$ is the input.

To stabilize the system (1) we apply a static output feedback, which is describes by the equation

$$u = KCx, \tag{2}$$

where $K \in R^{m \times p}$ is the controller matrix, $C \in R^{p \times n}$ is the output matrix.

As a result, the closed loop system takes the next form:

$$\dot{x} = (A + BKC)x = A_c x. \tag{3}$$

The problem is to calculate the controller matrix K, which ensures the asymptotic stability of the closed loop system (3). The most universal and convenient approach to solving this problem is based on the methods developed by A.M. Lyapunov [8]. In accordance with this approach, to solve the static output feedback synthesis problem it is necessary to solve bilinear matrix inequality [9, 10], which is defined as:

$$AY + YA^T + BKCY + YC^T K^T B^T < 0, \tag{4}$$

where $Y = Y^T > 0$.

Solvability (4) means the existence of matrices Y and K, which is equivalent to stability of the closed loop system (3). It is known [11] that inequality (4) is solvable with respect to the matrix K if and only if the matrix Y satisfying the following inequalities:

$$\mathcal{N}_C^T \left(Y^{-1}A + A^T Y^{-1} \right) \mathcal{N}_C < 0, \tag{5}$$

$$\mathcal{N}_{B^T}^T \left(AY + YA^T \right) \mathcal{N}_{B^T} < 0,$$

where $\mathcal{N}_C$ and $\mathcal{N}_{B^T}$ are the kernels of the matrices C and B^T, respectively.

Thus, to solve the static output feedback synthesis problem, it is necessary to solve either inequality (4) with respect to two matrix variables, or system (5) with respect to matrix Y and its inverse matrix. In the both cases, the constraint sets are nonconvex, which does not allow the use of convex optimization methods. To solve (4) and (5), various algorithms have been developed, such as linearization algorithms, or an algorithm for finding reciprocal matrices, and others [11–16]. There are special bilinear matrix inequality solvers such as the commercial software package PENBMI [13].

3 Synthesis of Controllers in a Class of Block-Homogeneous Input and Output Matrices

Let us introduce block-homogeneous input and output matrices, which we define as follows [6]:

$$B = \begin{pmatrix} B_m \\ 0_{(n-m) \times m} \end{pmatrix}, \tag{6}$$

$$B = \begin{pmatrix} 0_{(n-m) \times m} \\ B_m \end{pmatrix}, \tag{7}$$

$$C = \begin{pmatrix} C_p & 0_{p \times (n-p)} \end{pmatrix}, \tag{8}$$

$$C = \begin{pmatrix} 0_{p \times (n-p)} & C_p \end{pmatrix}. \tag{9}$$

where B_m and C_p are nonsingular matrices.

If we set the input matrix in the form (6) and the output matrix in the form (8), then the matrix product BKC can be represented as a block matrix

$$BKC = \begin{pmatrix} \widehat{K}_{m \times p} & 0_{m \times (n-p)} \\ 0_{(n-m) \times p} & 0_{(n-m) \times (n-p)} \end{pmatrix}, \tag{10}$$

where $\widehat{K}_{m \times p} = B_m K C_p$.

Since B_m and C_p are nonsingular matrices, the equation $\widehat{K}_{m \times p} = B_m K C_p$ is always solvable with respect to K. So let us assume that $K = \widehat{K}$. Note that if the conditions (6) and (8) are satisfied, then by means of a linear transformation of the system matrix and renaming of the state variables, the conditions (7) and (9) can be also satisfied.

Let us consider the solving of the static output feedback synthesis problem in such class of input and output matrices. Let us assume that in an arbitrary unstable matrix of the system there is at least one stable subspace in the form of a diagonal block of any nonzero dimension. In accordance with the dimension of the latter, let's divide it into blocks as follows

$$A = \begin{pmatrix} A_{11} & A_{12} \\ A_{21} & A_{22} \end{pmatrix}. \tag{11}$$

Let, for definiteness, $A_{11} \in R^{k \times k}$ is an unstable matrix, and $A_{22} \in R^{(n-k) \times (n-k)}$ is Hurwitz matrix, i.e. the spectral abscissa of this matrix is negative quantity. Firstly we consider the stabilization of the matrix using (10). Let us write the closed loop system matrix:

$$A_c = \begin{pmatrix} A_{11} & A_{12} \\ A_{21} & A_{22} \end{pmatrix} + \begin{pmatrix} K_{m \times p} & 0_{m \times (n-p)} \\ 0_{(n-m) \times p} & 0_{(n-m) \times (n-p)} \end{pmatrix}. \tag{12}$$

We should match the dimensions of the matrix blocks. To do this we size the following condition on the output dimension

$$p = k. \tag{13}$$

Then the matrix (12) takes the form

$$A_c = \begin{pmatrix} A_{11} + B_{11}K & A_{12} \\ A_{21} & A_{22} \end{pmatrix},$$ (14)

where $B_{11} = \begin{pmatrix} I_m \\ 0_{(k-m)\times m} \end{pmatrix} \in R^{k\times m}$.

If $p = m$ then $B_{11} = I$. However, this condition is a redundant. Instead we impose the rank condition on the input dimension and the corresponding matrix [9]:

$$rank\left(B_{11} \ A_{11}B_{11} \ \ldots \ A_{11}^{k-m}B_{11} \right) = k.$$ (15)

Since A_{11} is an arbitrary unstable matrix, then (13) and (15) are necessary and sufficient conditions for its stabilization. If at least one of these conditions is not fulfilled, then the matrix A_{11} cannot be stabilized. Moreover the matrix A_{11} defines an unstable subspace of A_c, since the stabilization of the latter in this case is also impossible. Therefore, if there is not a single stable diagonal block in the original matrix, then stabilization of such a matrix in the class of block-homogeneous input and output matrices by static output feedback is not possible.

Let us show that the static output feedback synthesis problem under conditions (13) and (15) is solvable, and it reduces to solving a linear matrix inequality. Prove the following theorem:

Theorem 1. *For the existence of a static output feedback ensuring the stability of the matrix (14), it is necessary and sufficient that there exists the matrix $Y = Y^T > 0$, satisfying the linear matrix inequality*

$$\mathcal{N}_{B^T}^T\left(AY + YA^T \right)\mathcal{N}_{B^T} < 0.$$ (16)

Proof. Need. Rewrite the Lyapunov inequality (4) as the symmetric matrix

$$A_cY + YA_c^T = \begin{pmatrix} \mathcal{H}_{11} & \mathcal{H}_{12} \\ \mathcal{H}_{12}^T & \mathcal{H}_{22} \end{pmatrix} < 0,$$ (17)

where $Y = \begin{pmatrix} Y_{11} & Y_{12} \\ Y_{12}^T & Y_{22} \end{pmatrix} > 0$.

Blocks $\mathcal{H}_{ij}$ are defined as follows:

$$\mathcal{H}_{11} = A_{11}Y_{11} + Y_{11}A_{11}^T + A_{12}Y_{12}^T + Y_{12}A_{12}^T + B_{11}KY_{11} + Y_{11}K^TB_{11}^T,$$ (18)

$$\mathcal{H}_{12} = A_{11}Y_{12} + A_{12}Y_{22} + Y_{11}A_{21}^T + Y_{12}A_{22}^T + B_{11}KY_{12},$$ (19)

$$\mathcal{H}_{22} = A_{21}Y_{12} + Y_{12}^TA_{21}^T + A_{22}Y_{22} + Y_{22}A_{22}^T.$$ (20)

Consider solution (17) in the class of block-diagonal matrices $Y = \begin{pmatrix} Y_{11} & 0 \\ 0 & Y_{22} \end{pmatrix}$. Then inequality (17) takes the form

$$\begin{pmatrix} (A_{11} + B_{11}K)Y_{11} + Y_{11}(A_{11} + B_{11}K)^T & A_{12}Y_{22} + Y_{11}A_{21}^T \\ (A_{12}Y_{22} + Y_{11}A_{21}^T)^T & A_{22}Y_{22} + Y_{22}A_{22}^T \end{pmatrix} < 0, \qquad (21)$$

where $Y_{ii} > 0$.

By applying Schur lemma [8] to (21) we can conclude that the solvability of this inequality for some matrices $Y_{11} > 0$ and $Y_{22} > 0$ can be ensured by the controller matrix. This means that solution (17) in the class of block-diagonal matrices Y is justified and conditions (13) and (15) allow us to solve the problem. Let us prove that the synthesis problem is convex and reduces to solving (16). As is shown in the first section, a static output feedback exists if and only if system (5) is solvable. Let us denote $Y^{-1} = X = \begin{pmatrix} X_{11} & 0 \\ 0 & X_{22} \end{pmatrix}$, $X_{ii} = Y_{ii}^{-1}$, and rewrite the first inequality of (5) as

$$\mathcal{N}_C^T \begin{pmatrix} X_{11}A_{11} + A_{11}^T X_{11} & X_{11}A_{12} + A_{21}^T X_{22} \\ (*)^T & X_{22}A_{22} + A_{22}^T X_{22} \end{pmatrix} \mathcal{N}_C < 0. \qquad (22)$$

Since $C = (C_p 0)$, then we can take $\mathcal{N}_C = \begin{pmatrix} 0 \\ I \end{pmatrix}$. As a result, the system (5) takes the form:

$$X_{22}A_{22} + A_{22}^T X_{22} < 0, \qquad (23)$$

$$\mathcal{N}_{B^T}^T \left(AY + YA^T \right) \mathcal{N}_{B^T} < 0.$$

Let us multiply the first inequality (23) right and left by $X_{22}^{-1} = Y_{22}$ and write the result

$$A_{22}Y_{22} + Y_{22}A_{22}^T < 0, \qquad (24)$$

$$\mathcal{N}_{B^T}^T \left(AY + YA^T \right) \mathcal{N}_{B^T} < 0.$$

The matrix B^T kernel can be divided into blocks as follows:

$$\mathcal{N}_{B^T} = \begin{pmatrix} \mathcal{N}_1 & 0 \\ 0 & I \end{pmatrix}, \qquad (25)$$

where $\mathcal{N}_1 = \begin{pmatrix} 0 \\ I \end{pmatrix}$.

After substituting (25) into the second inequality (24), we obtain

$$A_{22}Y_{22} + Y_{22}A_{22}^T < 0, \qquad (26)$$

$$\begin{pmatrix} \mathcal{N}_1^T(A_{11}Y_{11} + Y_{11}A_{11}^T)\mathcal{N}_1 & \mathcal{N}_1^T(A_{12}Y_{22} + Y_{11}A_{21}^T) \\ (*)^T & A_{22}Y_{22} + Y_{22}A_{22}^T \end{pmatrix} < 0.$$

Due to these computations, instead of the original nonlinear system (5), we obtain one linear matrix inequality, which is equivalent to (16). Therefore, if this inequality is unsolvable, then the problem is unsolvable and the static output controller does not exist.

Sufficiency. The solvability of (16) is equivalent to the existence of the static output feedback. The controller matrix K can be found from inequality (17) which for the given matrix $Y = Y^T > 0$ is linear matrix inequality. The theorem is proved.

The considered case allows us to transform the static output feedback synthesis problem to the problem of convex optimization. To do this, it is necessary to organize control and measurement and set the corresponding matrices in such a way that the controller matrix allows us to change the unstable diagonal block of the system matrix. If the input and output matrices are not block-homogeneous, but the controlled and measured variables partially or completely coincide, then using a linear transformation it is possible to choose a basis in which the input and output matrices are block-homogeneous. The matrix of the system in the new basis is similar to the matrix of the same name in the original basis [6].

Consider now the case when the input and output matrices are block-homogeneous, but the above conditions are not satisfied. Suppose that the input matrix has the form (6), and the output matrix has the form (9). We define the dimensions m and p as follows:

$$m = k, \tag{27}$$

$$p = n - k. \tag{28}$$

The corresponding matrix of the closed loop system takes the form

$$A_c = \begin{pmatrix} A_{11} & A_{12} + K \\ A_{21} & A_{22} \end{pmatrix}. \tag{29}$$

Consider the next auxiliary linear matrix inequality:

$$AY + YA^T + BKC + C^T K^T B^T < 0, \tag{30}$$

where $Y = Y^T > 0$.

Prove the theorem:

Theorem 2. *The problem of the matrix (29) stabilization is solvable in the terms of linear matrix inequalities.*

Proof. The inequality (30) is solvable if and only if the system

$$\mathcal{N}_C^T(AY + YA^T)\mathcal{N}_C < 0, \tag{31}$$

$$\mathcal{N}_{B^T}^T(AY + YA^T)\mathcal{N}_{B^T} < 0$$

is solvable.

Show that (31) is solvable. The kernels of B^T and C_0 can be the following matrices:

$$\mathcal{N}_{B^T} = \begin{pmatrix} 0 \\ I \end{pmatrix}, \tag{32}$$

$$\mathcal{N}_C = \begin{pmatrix} I \\ 0 \end{pmatrix}. \tag{33}$$

In accordance with dimensions of blocks of (29) divide the matrix $Y = Y^T > 0$ into the blocks of appropriate dimensions. Then, on the strength of (32) and (33) the system (31) takes the form

$$\begin{pmatrix} \mathcal{D}_{11} & 0 \\ 0 & \mathcal{D}_{22} \end{pmatrix} < 0, \tag{34}$$

where

$$\mathcal{D}_{11} = A_{11}Y_{11} + Y_{11}A_{11}^T + A_{12}Y_{12}^T + Y_{12}A_{12}^T,$$

$$\mathcal{D}_{22} = A_{21}Y_{12} + Y_{12}^T A_{21}^T + A_{22}Y_{22} + Y_{22}A_{22}^T.$$

We can see that under the conditions (27) and (28) the matrix space $\mathcal{D}$ is decomposed into two non-intersecting subspaces. Let us show that inequality (34) is solvable. Assume that A_{22} is Hurwits matrix. For $A_{12} \neq 0$ the inequality $\mathcal{D}_{11} < 0$ can be solved by the matrices $Y_{11} > 0$ and Y_{12}. Then the second inequality $\mathcal{D}_{22} < 0$ with respect to unknown matrix $Y_{22} > Y_{12}^T Y_{11}^{-1} Y_{12}$ and a scalar parameter $\gamma^2 \neq 0$ can be written as

$$\gamma^2 \mathcal{L}(Y_{22}) < \mathcal{M}, \tag{35}$$

where $\mathcal{L}(Y_{22}) = A_{22}Y_{22} + Y_{22}A_{22}^T$.

The matrix $\mathcal{L}(Y_{22})$ is a variable negative definite matrix, and $\mathcal{M}$ is a given symmetric matrix without sign definition. The parameter γ^2 allows increase the left-hand side of (35) and hence we can achieve the negative definiteness of (34). In order to find the controller matrix it is necessary to solve inequality (30). Find it from the next system

$$AY + YA^T + BK_1C + C^T K_1^T B^T < 0, \tag{36}$$

$$\mathcal{D}_{11} + K_1 Y_{22}^{-1} Y_{12}^T + Y_{12}Y_{22}^{-1}K_1^T < 0.$$

The system (36) is the linear matrix inequalities system with respect to K_1. Let us prove that the system is solvable. Assume that $K_1 = \mathcal{A}Y_{12}$, where $\mathcal{A}$ is the some square matrix, defined from the solvable inequality $\mathcal{A}\left(Y_{12}Y_{22}^{-1}Y_{12}^T\right) + \left(Y_{12}Y_{22}^{-1}Y_{12}^T\right)\mathcal{A}^T < 0$. Then the system (36) can be rewrite as follows

$$\begin{pmatrix} \mathcal{D}_{11} & \mathcal{D}_{12} + \mathcal{A}Y_{12} \\ \mathcal{D}_{12}^T + Y_{12}^T \mathcal{A}^T & \mathcal{D}_{22} \end{pmatrix} < 0, \tag{37}$$

where $D_{12} = A_{11}Y_{12} + A_{12}Y_{22} + Y_{11}A_{21}^T + Y_{12}A_{22}^T$.

The solvability of this inequality for some matrix $Y > 0$ can be ensured by the matrix $\mathcal{A}$. Since (37) is solvable, then the system (36) is also solvable. As result we get the negative defined matrix (37). Similar, the main inequality (4) with the same matrix $Y = Y^T > 0$ and an unknown matrix K_2 takes the form

$$\begin{pmatrix} D_{11} + K_2Y_{12}^T + Y_{12}K_2^T & D_{12} + K_2Y_{22} \\ D_{12}^T + Y_{22}K_2^T & D_{22} \end{pmatrix} < 0. \qquad (38)$$

If $K_2 = K_1 Y_{22}^{-1}$, then (38) is also the negative defined matrix. The theorem is proved.

It is necessary to note that the Theorem 2 is also valid for the case when the input matrix has the form (7), the output matrix has the form (8) and the corresponding dimensions equal

$$m = n - k,$$

$$p = k.$$

The closed loop system matrix in such case takes the form

$$A_c = \begin{pmatrix} A_{11} & A_{12} \\ A_{21} + K & A_{22} \end{pmatrix}.$$

The all results are guarantee valid if the matrices B and C are block-homogeneous. The some possible solutions of the synthesis problem when B and C are an arbitrary matrices are discussed in the next section.

4 Synthesis of Controllers with Arbitrary Input and Output Matrices

Let B and C are an arbitrary matrices. In is known [17] that the general formulation of the static output feedback synthesis problem is as follows: for given matrices A, B and C find a matrix K that ensures the stability of (3). In scope of this section consider two other problems. Let us at first consider the solution of the static output feedback synthesis problem supposing that the matrix C elements can be changed. We assume that the couple (A, B) is controllable. Without loss of generality, we will consider that the condition $rank(C) = p$ is satisfied. If not, then it can be corrected by a linear transformation. As before, we assume that one diagonal block of the system matrix is Hurwitz. For solving such problem let us prove the next lemma:

Lemma 1. *For the existence of a static output feedback ensuring the closed loop system matrix (3) stability, it is sufficient that linear matrix inequality (30) be solvable.*

Proof. Assume that the auxiliary inequality (30) is solvable. Denote $X = Y^{-1}$ and rewrite (30) in the form

$$AY + YA^T + BKC_XY + YC_X^TK^TB^T < 0, \qquad (39)$$

where $C_X = CX$.

The solvability of (39) implies that the closed loop system matrix

$$A_c = A + BKC_X$$

is stable. Consider the main inequality (4). Comparing it and (39), we come to the conclusion that if we substitute in the first one $C = C_X$, then it is equivalent to the second one. The lemma is proven.

Lemma 1 is only sufficient condition, i.e. if inequality (30) is unsolvable, then the question of the existence of a static output feedback is open. Next consider a somewhat different problem. If in the main inequality (4) to set of matrix K, then we obtain a inequality with respect to Y and C. Because the dimension of the C more than the dimension of the K, the problem of C calculating is simpler then the problem of K calculating. The mathematical formulation of the such problem is as follows: for given matrices A, B and dimension $m \leq p < n$ find a matrix C of minimal dimension that ensures the solvability of (4). It can be show that the problem is convex. In order to solve it to prove the theorem:

Theorem 3. *For the existence of a minimal dimension matrix ensuring the closed loop system matrix (3) stability, it is necessary and sufficient the controllability of the $C(A, B)$.*

Proof. Need. Rewrite (4) as the linear matrix inequality:

$$AY + YA^T + BZ + Z^T B^T < 0, \tag{40}$$

where $Y = Y^T > 0$.

If the controllability of the (A, B) is not fulfilled, then (40) as well as (4) are unsolvable.

Sufficiency. Let the controllability of the (A, B) is fulfilled. Then (40) is solvable. Consider the next equation

$$KCY = Z. \tag{41}$$

Rewrite it in the form

$$KC = ZY^{-1}. \tag{42}$$

Dimensions of right side of (41) and (42) equal $(m \times n)$. If $p \geq m$ this equation is solvable with respect to the unknown matrix C. Let $p = m$. We obtain the square matrix K of the minimal dimension. Then the unknown matrix C of the minimal dimension can be found as follows:

$$C = K^{-1}ZY^{-1}. \tag{43}$$

So if we get the output matrix in according with (43), then the inequality (4) is solvable. Note that the Eq. (42) is also solvable with respect to C for any another $p > m$. The theorem is proved.

Let's illustrate the Theorem 3 application. Consider the stabilization of a rigid rotor, rotating in electromagnets bearings [18, 19]. Electromagnetic bearings are electrome-chanical systems that use magnetic forces to levitate a rotor without physical contact. Magnetic forces are created by four pairs of electromagnets. The systems are of great interest for a number of industrial applications [20, 21]. One of the main advantages of such systems is frictionless operation. As a result, it allows significantly increase the service life and efficiency compared to the traditional mechanical counterparts. The most actual problem in electromagnetic bearings is the rotor control. The system matrix equals [18].

$$
A = \begin{pmatrix}
0 & 0 & 0 & 0 & 1 & 0 & 0 & 0 & 0 & 0 & 0 & 0 \\
0 & 0 & 0 & 0 & 0 & 1 & 0 & 0 & 0 & 0 & 0 & 0 \\
0 & 0 & 0 & 0 & 0 & 0 & 1 & 0 & 0 & 0 & 0 & 0 \\
0 & 0 & 0 & 0 & 0 & 0 & 0 & 1 & 0 & 0 & 0 & 0 \\
10.2 & 0 & 0 & 0 & -2 & 0 & 0 & -5.1 & -5.1 & 0 & 0 & 0 \\
0 & 10.2 & 0 & 0 & 2 & 0 & 0 & 0 & 0 & -5.1 & 5.1 \\
0 & 0 & 2 & 0 & 0 & 0 & 0 & 0 & 0 & -1 & -1 \\
0 & 0 & 0 & 2 & 0 & 0 & 0 & -1 & -1 & 0 & 0 \\
0 & 0 & 0 & -1 & 0 & 0 & -1 & -1 & 0 & 0 & 0 \\
0 & 0 & 0 & -1 & 0 & 0 & -1 & 0 & -1 & 0 & 0 \\
0 & 0 & 0 & 0 & 1 & 1 & 0 & 0 & 0 & -1 & 0 \\
0 & 0 & 0 & 0 & 1 & -1 & 0 & 0 & 0 & 0 & -1
\end{pmatrix}.
$$

The corresponding state space dimension equals $n = 12$. It is necessary to determine a matrix C of minimal possible dimension ensuring the solvability of (4). Let's the matrix B satisfy to the next view

$$
B = \begin{pmatrix} I_1 \\ 0_{11\times1} \end{pmatrix}.
$$

The matrix form means that the first state variable is measured, i.e. the angle of rotation of the rotor relative to the axis of the rotor mass center. It is easy to verify that the rank condition for the controllability of the (A, B) is satisfied. If $p = m$, then it follows from Theorem 1 that the synthesis problem of the static output feedback of minimum output dimension is solvable. Note that in this case we obtain a scalar controller, which, in general, is no obviously. Let's $K = 1$. Next, we solve (40) and then calculate the matrix according with (43). As a result we obtain the next matrix

$$C^T = 10^3 \times \begin{pmatrix} -0.0219 \\ -7.8549 \\ -9.6834 \\ 3.4886 \\ -0.0621 \\ -1.2457 \\ -5.2934 \\ 2.4632 \\ -1.0339 \\ -1.0119 \\ 4.8731 \\ 0.2526 \end{pmatrix}.$$

The spectral abscissa of the closed loop matrix equals is negative quantity:

$$\max_{1 \leq i \leq 12} \{Re\lambda_i(A + BC)\} = -0.4907.$$

So, the closed loop matrix is stable. Consider another approach to control. Let's the matrix B satisfies to the next form:

$$B = \begin{pmatrix} 0_{11 \times 1} \\ I_1 \end{pmatrix}.$$

In this case the current in the electromagnet circuit is the controllable variable. The system is also controllable. If $K = 1$, then we obtain the next matrix

$$C^T = 10^4 \times \begin{pmatrix} -2.3091 \\ -1.2361 \\ -1.2341 \\ 4.7760 \\ -0.1912 \\ -0.1351 \\ -0.6167 \\ 2.8764 \\ -1.8851 \\ -0.7126 \\ 0.6211 \\ -0.0038 \end{pmatrix}.$$

The spectral abscissa of the closed loop matrix is also negative quantity:

$$\max_{1 \leq i \leq 12} \{Re\lambda_i(A + BC)\} = -0.4774.$$

Thus we can make the conclusion that it is reasonably to set C as result of solving (40). Due to this the synthesis problem is solvable by the static output feedback of the minimal dimension. If C is some arbitrary matrix of full rank, then the synthesis problem will most probably be unsolvable, because, as rule, the dimension of the K less than the dimension of the C.

5 Conclusion

Based on the obtained results it can be concluded that the introduction of block-homogeneous input and output matrices simplifies the synthesis problem. In order to stabilize unstable matrices using the static output feedback in the class of block-homogeneous matrices, the dimensions of the input and output should be matched with the dimensions of the stable and unstable blocks in the system matrix. In all such cases it can be guaranteed that the static output feedback synthesis problem will be solvable and the sum of m and p will be no more than n. Also it is showed that the problem of C calculating is significantly simpler then the problem of K calculating and it can be reduces to solving a linear matrix inequality.

References

1. Syrmos, V.L., Abdallah, C.T., Dorato, P., Grigoriadis, K.: Static output feedback. a survey. Automatica **33**(2), 125–137 (1997)
2. Astolfi, A., Colaneri, P.: Static output feedback stabilization of linear and nonlinear systems. In: Proceeding 39th Conference on Decision and Control, Sydney, Australia, pp. 2920–2925 (2000)
3. Astolfi, A., Colaneri, P.: An algebraic characterization for the static output feedback stabilization problem. In: Proceedings American Control Conference, Arlington, VA, pp. 1408–1413 (2001)
4. Kimura, H.: Pole assignment by gain output feedback. IEEE Trans. Autom. Control **20**, 509–516 (1975)
5. Röbenack, K., Voswinkel, R., Franke, M., Franke, M.: Stabilization by static output feedback: a quantifier elimination approach. In: International Conference on System Theory, Control, Computing (ICSTCC), Sinaia, Romania (2018)
6. Mukhin, A.V.: About static output controller existing. Large-Scale Syst. Control **96**, 16–30 (2022)
7. Nemirovskii, A.A.: Several NP-hard problems arising in robust stability analysis. Math. Control Signals Syst. **6**, 99–105 (1994)
8. Malkin, I.G.: Theory of stability of motion. Moscow-Leningrad (1952)
9. Balandin, D.V., Kogan, M.M.: Sintez zakonov upravleniya na osnove lineinykh matrichnykh neravenstv (Design of Control Laws on the Basis of Matrix Inequalities). Fizmatlit, Moscow (2007)
10. Polyak, B.T., Khlebnikov, M.V., Rapoport, L.B.: Matematicheskaya teoriya avtomaticheskogo upravleniya (Mathematical Theory of Automatic Control). LENAND, Moscow (2019)
11. Balandin, D.V., Kogan, M.M.: Synthesis of controllers on the basis of a solution of linear matrix inequalities and a search algorithm for reciprocal matrices. Autom. Remote. Control. **66**, 74–91 (2005)
12. Hassibi, A., How, J., Boyd, S.: A path following method for solving BMI problems in control. In: Proceedings of American Control Conference, pp. 1385–1389 (199)

13. Henrion, D., Loefberg, J., Kocvara, M., Stingl, M.: Solving polynomial static output feedback problems with PENBMI. In: Proceedings IEEE Conference on Decision Control and European Control Conference, Sevilla, Spain (2005)

14. Cao, Y.-Y., Lam, J., Sun, Y.-X.: Static output feedback stabilization: an ILMI approach. Automatica **34**, 1641–1645 (1998)

15. El Ghaoui, L., Oustry, F., Aitrami, M.: A cone complementarity linearization algorithm for static output-feedback and related problems. IEEE Trans. Autom. Control **42**, 1171–1176 (1997)

16. Sadabadi, M.S., Peaucelle, D.: From static output feedback to structured robust static output feedback: a survey. Annu. Rev. Control **42**, 11–26 (2016)

17. Polyak, B.T., Scherbakov, P.S.: Hard problems in linear control theory: possible approach to solution. Autom. Remote Control **66**(5), 681–718 (2005)

18. Balandin, D.V., Kogan, M.M.: Motion control for a vertical rigid rotor rotating in electromagnetic bearings. J. Comput. Syst. Sci. Int. **50**, 683–697 (2011)

19. Mukhin, A.V.: Mathematical modeling of a rigid rotor stabilization process, rotating in electromagnetic bearings. Trans. NNSTU n.a. R.E. Alekseev **2**, 36–48 (2021)

20. Zhuravlev, Yu.N.: Active magnetic bearings. Theory, calculation, application. Politechnica, St. Petersburg (2003)

21. Schweitzer, G., Maslen, E.: Magnetic Bearings. Theory, Design, and Application to Rotating Machinery. Springer, Berlin (2009). https://doi.org/10.1007/978-3-642-00497-1

Weak Penalty Decomposition Algorithm for Sparse Optimization in High Dimensional Space

Kirill Spiridonov, Sergei Sidorov$^{(\boxtimes)}$ ⓘ, and Michael Pleshakov

Saratov State University, Saratov, Russian Federation
sidorovsp@sgu.ru
http://www.sgu.ru

Abstract. Many machine learning problems may be reduced to finding a sparse approximation to a loss function minimum, i.e. to finding the infimum of a given convex function among all elements of the search space that satisfy a cardinality constraint. One of the well-known iterative methods for solving such problems is the penalty decomposition algorithm (PDA), which uses two auxiliary optimization problems at each iteration. However, when applying PDA in practice, there is a difficulty associated with the computational complexity of the sub-problems being solved. To reduce it, we propose to add some weakening conditions to the algorithm. We study the properties of the proposed algorithm, which we called the Weak Penalty Decomposition Algorithm (WPDA), and we show that the sequence of points of the search space generated by WPDA converges to a stationary point that satisfies the first-order necessary optimality conditions. In real applications, the dimension of the search space can be extremely large, therefore, to study the independence of the properties of the applied algorithms from dimension, in this paper we assume that the search space is infinite-dimensional.

Keywords: Machine learning · Convex optimization · Penalty decomposition algorithm · Cardinality constraint · Sparse solution

1 Introduction

Finding sparse solutions to optimization problems is of interest in many applied areas. A sparse solution is a point in the search space that can be represented as a linear combination of a small number of basis vectors (or elements of some dictionary defined in the search space). For example, machine learning methods are often reduced to finding the minimum of a parametric loss function on a set of training examples, or to maximizing the expectation for the maximum

This work was supported by the Ministry of science and education of the Russian Federation in the framework of the basic part of the scientific research state task, project FSRR-2020-0006.

likelihood function [3,22,31]. At the same time, deep learning problems have a huge number of parameters, and for stable operation of deep neural networks on out-of-sample data, one should look for sparse solutions during the learning process. Sparse solutions are especially important in feature selection problems, and is also often used in the fight against overfitting.

Further, to select features in classification problems, a logistic regression model is used, the parameters of which are found as a sparse solution of the problem of minimizing the mathematical expectation of losses [14,17]. Similar problems are arisen in sparse inverse covariance selection [2,5,27,32] and in multivariate linear regression with a small number of training points. Another applications arise in portfolio investment, where it is necessary to find portfolio weights (shares of investments in assets), for which some risk evaluation function takes the smallest values, under certain restrictions on the return and on the number of assets included in the portfolio [12,26,28].

The most obvious way to obtain a sparse solution to an optimization problem is to use an additional constraint on its cardinality (which is the constraint on the number of basis vectors used in linear combinations representing approximate solutions to the problem).

At the same time, optimization problems with the cardinality constraint have non-polynomial complexity in many cases. In this regard, in real applications, data analysts prefer to use alternative methods. Perhaps one of the most common approaches to solving sparse optimization problems is l_1-regularization [8,13,20], i.e. the addition of the l_1-norm constraint on solutions. It turned out that l_1-norm relaxation methods can effectively find sparse solutions under certain assumptions for a fairly wide class of problems, including the compressed sensing. Recently, a different approach has been attracted interest, in which a constraint on the l_p-norm for some $p \in (0,1)$ is added. However, it seems that it is too early to judge its effectiveness. Another favorite of practitioners for finding approximate sparse solutions to convex conditional optimization problems in Euclidean spaces are greedy algorithms, one of which is the conditional gradient method. For convex optimization problems in Banach spaces, its analogue, for example, is the class of weak biorthogonal greedy algorithms. Convergence results for this class of algorithms were obtained in [7,15]. By design, greedy algorithms are capable of producing sparse solutions because exactly one basis vector (or dictionary element) is added at each iteration to the linear combination representing the solution.

In this paper, we will use the ideas of the paper [21], which proposed another approach for solving the problem with the cardinality constraint that uses the penalty decomposition. This technique of finding sparse solutions to optimization problems has been turned out to be very effective, which has been confirmed by the ever-growing interest in these methods in recent years [9,11,23,29,33], [16,20,24,25], [9,18,19,30]. The penalty decomposition algorithm (PDA) solves two auxiliary optimization problems at each of sub-iterations. However, when applying PDA in practice, there is a difficulty associated with the computational complexity of the sub-problems being solved. To reduce it, we propose to add some weakening conditions to the algorithm. We study the properties of the proposed method, which we called the Weak Penalty Decomposition Algorithm (WPDA), and we show that the sequence of points of the search space generated by the WPDA converges to a stationary point that satisfies the first-order necessary optimality conditions.

2 Weak Penalty Decomposition Method in Hilbert Spaces

2.1 The Cardinality Constrained Optimization Problem

In real applications, the dimension of the search space can be extremely large, therefore, to study the independence of the properties of the applied algorithms from dimension, in this paper we assume that the search space is an infinite-dimensional (Hilbert) space H endowed with norm $\|\cdot\|_H$. Denote $\mathcal{B} = \{e_1, e_2, \ldots\}$ the orthonormal basis in H.

Let us consider a convex differentiable function E defined on H. We are interested in solving the following problem:

$$E(x) \to \inf_{x \in \Sigma_m}, \tag{1}$$

where Σ_m is the set of all m-term polynomials with respect to $\mathcal{B}$:

$$\Sigma_m = \Sigma_m(\mathcal{B}) = \Big\{ x \in H \ : \ x = \sum_{i \in I} x_i e_i, \ \mathrm{card}(I) = m \Big\}, \tag{2}$$

where the index set $I = I(x)$ is such that $x_i \neq 0$ for all $i \in I$ and $x_i = 0$ if $i \notin I$.

The quantity $\|x\|_0 := \mathrm{card}(I)$ represents the number of non-zero coefficients in the expansion of x with respect to the basis $\mathcal{B}$ (note that l_0-norm $\|x\|_0$ does not satisfy norm properties). Thus, an equivalent for problem (1)–(2) can be presented as $E(x) \to \inf_{\|x\|_0 \leq m}$.

Necessary optimality conditions for cardinality constrained problems in Euclidean spaces have been proposed in paper [21]. The same conditions may be applied to analyze the convergence of the PD method in Hilbert spaces.

An element $x^* = \sum_{i \in I^*} x_i^* e_i$ of H is said to satisfy *the first order optimality conditions of Lu–Zhang type* [21] for problem (1)–(2) if $\|x^*\|_0 = m$, and $\nabla E(x^*)$ is orthogonal to the linear subspace $\mathrm{span}\{e_i : \ i \in I^*\}$.

It can be shown in paper [25] that if an element x^* of H is the solution to the problem (1)–(2), then x^* satisfies first order optimality conditions of Lu–Zhang type.

2.2 The Weak Penalty Decomposition Algorithm

Let us define the auxiliary function F_δ as follows

$$F_\delta(x, y) = E(x) + \frac{\delta}{2}\|x - y\|_H^2, \tag{3}$$

where the parameter $\delta > 0$ calibrates the importance of two penalty components.

An PD algorithm for solving problem (1)–(2) was considered in [25]. This algorithm is based on the method proposed in paper [21], but is designed to solve the problem in Hilbert space. At each main iteration k of the algorithm, a while–loop is started (which is executed until the gradient of the auxiliary function is less than ϵ_k), within which two auxiliary optimization problems are solved. The first of them finds the unconditional minimum of the function F_δ with respect to the first variable (while the second one is fixed). Then the minimum of the function F_δ is found with respect to the second variable (with the fixed first variable) over all elements that satisfy the cardinality constraint. Despite the fact that the first sub problem is a convex unconstrained optimization problem, for which there are a wide variety of sufficiently efficient methods, its exact solution in high-dimensional spaces is cost ineffective. For the second problem, there is a closed-form solution in Euclidean space, but in high-dimensional spaces, its exact finding can be difficult.

In this paper, we propose a development of this algorithm (Algorithm 1) that allows finding approximate solutions to these sub problems at each iteration. We use two weakening sequences $\{\xi_l\}$ and $\{\tau_l\}$ that describe the allowable error in solving these sub-problems at each iteration. The algorithm is called the *weak penalty decomposition algorithm* (WPDA), since the weakened conditions are used.

Algorithm 1: WPDA IN HILBERT SPACE

---

begin

· Input $x^0 = y^0 \in \Omega(F)$, s.t. $\|x^0\|_0 \le m$, $\{\delta_k\}$ s.t. $\delta_k \to \infty$ and $\delta_k > 0$, $\{\epsilon_k\}$ s.t. $\epsilon_k \to 0$ and $\epsilon_k > 0$, $\{\xi_l\}$ s.t. $\xi_l \to 0$, $\{\tau_l\}$ s.t. $\tau_l \to 0$, $\sum_l \xi_l + \sum_l \tau_l < B$, $A \ge \max\{E(x^0), \inf_x F_{\delta_0}(x, y^0)\} + B$;

for each $k \ge 0$ **do**

　· $l = 0$;

　· $u^0 = x^k$;

　· **if** $\inf_x F_{\delta_k}(x, y^k) \le A$ **then**

　　$v^0 = y^k$

　$v^0 = y^{k-1}$

　· **while** $\|\nabla_u F_{\delta_k}(u^l, v^l)\|_H > \epsilon_k$ **do**

　　· Find u^{l+1} s.t. $F_{\delta_k}(u^{l+1}, v^l) \le \min_u F_{\delta_k}(u, v^l) + \xi_l$;

　　· Find $v^{l+1} \in \Sigma_m$ s.t. $F_{\delta_k}(u^{l+1}, v^{l+1}) \le \min_{v \in \Sigma_m} F_{\delta_k}(u^{l+1}, v) + \tau_l$;

　　· $l := l + 1$;

　· $x^{k+1} = u^l$, $y^{k+1} = v^l$;

· Output $\{x^k\}$, $\{y^k\}$;

end

The sequence $\{x^k\}$ is generated as the output of the WPDA in the following way. The algorithm starts with any feasible element x^0 of H, i.e. such that $\|x^0\|_0 \le m$. At each subsequent iteration ($k = 1, 2, \ldots$), the algorithm finds x^k based on x^{k-1} by using the while loop of the block coordinate descent until the norm of the gradient for $F_{\delta_{k-1}}$ at a current element would not be bounded by ϵ_{k-1}. The while loop consists of two main steps, the first of which minimizes (with an error depending on ξ_l) the auxiliary penalty function $F_{\delta_{k-1}}(u^l, v^l)$ over u^l without any restriction on cardinality. The second finds a point u^l as the "nearest" element to v^l (with an error depending on τ_l) satisfying the cardinality constraint $\|v^l\|_0 \leq m$.

The notation $\nabla_u F(u, v)$ denotes the gradient of F over u with fixed v.

As the number of iteration increases, the penalty parameter δ_k tends to infinity to ensure that the generated elements $\{x_k\}$ are getting closer to a point satisfying the cardinality constraint. The sequence $\{\epsilon_k\}$ decreases to zero to obtain the acceptable level of the approximation to the minimum of the auxiliary function $F_{\delta_{k-1}}$.

Before the k-th iteration starts, we verify that the elements x^{k-1} and y^{k-1} generated on the previous iteration belong to a compact set. If it is true, the k-th iteration starts with x^{k-1} and y^{k-1}; otherwise, the iteration should use the pair x^{k-1} and y^{k-2} as its starting points.

2.3 Convergence Analysis

Theorem 1. *Let the function $F_\delta(x, y)$ be defined in (3) and suppose that E is differentiable and convex, and for any $x \in H$, $x \neq 0$, we have $E(\alpha x) \to \infty$ if $\alpha \to \infty$. Let $\epsilon_k \to 0$ and $\delta_k \to \infty$ as $k \to \infty$. Let $\{\xi_l\}_{l \geq 1}$ and $\{\tau_l\}_{l \geq 1}$ be two positive and decreasing to zero sequences such that $\sum_l \xi_l + \sum_l \tau_l < B$ for some positive B.*

Suppose that x^* is the limit points of sequence $\{x^k\}_{k \geq 0}$ generated by Algorithm 1, i.e. there is $K \subset \{1, 2, \ldots\}$, such that $x^k \to_K x^*$.
Then

1. *x^* is a feasible point for problem (1)–(2) .*
2. *x^* satisfies the first order optimality conditions of Lu–Zhang type for problem (1)–(2).*

Proof. First we will show that $\Omega(F_\delta, x^0, y^0, B) := \{x, y \in H : F_\delta(x, y) \leq F_\delta(x^0, y^0) + B\}$ is a compact set in H for any δ, x^0, y^0, $B > 0$. It follows from the definition of $F_\delta(x, y)$ that F_δ is differentiable and convex on $H \times H$, and for any $x \in H$, $x \neq 0$, we have both $F_\delta(\alpha x, y) \to \infty$ and $F_\delta(x, \alpha y) \to \infty$ as $\alpha \to \infty$. It follows from the coercivity and convexity of $F_\delta(x, y)$ on $H \times H$ that the set $\Omega(F_\delta, x^0, y^0, B)$ is compact.

Next we will show that the while loop of the algorithm can be terminated in a finite number of iterations for any k. Let us suppose the opposite, i.e. for some k, the k-th iteration generates an infinite sequence $\{u^l, v^l\}$. Then $F_{\delta_k}(u^l, v^l) \leq F_{\delta_k}(u^0, v^0) + B$, and therefore all $\{u^l, v^l\}$ belong to the compact $\Omega(F_{\delta_k}, u^0, v^0, B)$. Then there exists $L \subset \{0, 1, \ldots\}$ such that $u^l \to_L u^*$, $v^l \to_L v^*$. The while loop of the algorithm and the convergence of ξ_l and τ_l to zero implies that there is a subsequence $L_1 \subset L$ such that

$$F_{\delta_k}(u^{l+1}, v^l) \leq \frac{1}{2}\left(F_{\delta_k}(u^l, v^l) + \min_u F_{\delta_k}(u, v^l)\right),$$

$$F_{\delta_k}(u^{l+1}, v^{l+1}) \leq \frac{1}{2}\left(F_{\delta_k}(u^{l+1}, v^l) + \min_{v \in \Sigma_m} F_{\delta_k}(u^{l+1}, v)\right),$$

for all $l \in L_1$. In another words,

$$F_{\delta_k}(u^{l+1}, v^l) + F_{\delta_k}(u^{l+1}, v^l) \leq F_{\delta_k}(u^l, v^l) + \min_u F_{\delta_k}(u, v^l),$$

and

$$F_{\delta_k}(u^{l+1}, v^{l+1}) + F_{\delta_k}(u^{l+1}, v^{l+1}) \leq F_{\delta_k}(u^{l+1}, v^l) + \min_{v \in \Sigma_m} F_{\delta_k}(u^{l+1}, v),$$

and therefore

$$F_{\delta_k}(u^{l+1}, v^l) + \epsilon_k \leq F_{\delta_k}(u^l, v^l),$$

and

$$F_{\delta_k}(u^{l+1}, v^{l+1}) + \tau_k \leq F_{\delta_k}(u^{l+1}, v^l).$$

We get

$$F_{\delta_k}(u^{l+1}, v^{l+1}) + \gamma_k \leq F_{\delta_k}(u^l, v^l),$$

where $\gamma_k = \epsilon_k + \tau_k$ tends to zero as $k \to \infty$.

Then there exists $l_0 \in L$ such that for all $l \geq l_0$, $l \in L_1$ we get the contradiction to the terminal condition of the `while` loop $\|\nabla_u F_{\delta_k}(u^l, v^l)\|_H > \epsilon_k$.

Suppose that sequences $\{x^k\}$, $\{y^k\}$ are generated by the algorithm. It can be verified that points $\{x^k\}$ belong to the compact set $\Omega(F_{\delta_k}, x^0, y^0, B)$. Then there exists a subsequence of $\{x^k\}$ that has at least one limit (accumulation) point. It should be noted that this accumulation point may not be unique, in general.

From the definition of the algorithm we get the inequality $\|x^{k+1} - y^{k+1}\|_H^2 \leq 2(A - E(x^{k+1}))/\delta_k$, from which follows that points $\{y^k\}$ belong to a compact set as well. Thus, the sequence $\{y^k\}$ is bounded and has at least one accumulation point.

Now we should show that the limit point x^* satisfies the cardinality constraint. On k-th iteration the inequality $\|\nabla_x F_{\delta_k}(x^k, y^k)\|_H \leq \epsilon_k$ holds, and therefore,

$$\left\|\frac{\nabla_x E(x^k)}{\delta_k} + (x^k - y^k)\right\|_H \leq \frac{\epsilon_k}{\delta_k}.$$

Since $\delta_k \to \infty$, $\epsilon_k \to 0$ as $k \to \infty$ and $\{\nabla_x E(x^k)\}$ are bounded, we get $\|x^* - y^*\|_H = \lim_{k \in K, k \to \infty} \|x^k - y^k\|_H = 0$. We have $\|y^*\|_0 \leq m$ by its definition in the algorithm, and consequently, $\|x^*\|_0 \leq m$.

For a fixed $x \in H$ let us denote $J(x) \subset \{1, 2, \ldots\}$ the index set satisfying the following properties:

- card$(J(x)) \leq m$;
- if $i \in J(x)$ then $x_i \neq 0$;
- if $i \in J(x)$ then $|x_i| > |x_j|$ for all $j \notin J(x)$.

Let K_1 be any infinite subset of K such that $J(x^k) = J(x^*)$ for all $k \in K_1$.
It follows from $\|\nabla_x F_{\delta_k}(x^k, y^k)\|_H \leq \epsilon_k$ that for any $i \in J(x^*)$

$$\frac{\partial E(x^k)}{\partial x_i} + \delta_{k-1}(x_i^{(k)} - y_i^{(k)}) \to_{K_1} 0.$$

Since $x_i^{(k)} = y_i^{(k)}$ for every $i \in J(x^*)$, we get $\frac{\partial E(x^k)}{\partial x_i} \to_{K_1} 0$, for all $i \in J^*$, i.e.
the first order optimality condition of Lu–Zhang type holds. $\square$

Let X be a closed convex subset of Hilbert space H. Let us consider the
constrained optimization problem

$$E(x) \to \min,$$

under constraints

$$x \in X, \ g(x) \leq 0, \ h(x) = 0, \ \|x\|_0 \leq r,$$

for some fixed integer $r \geq 0$.
    Starting from this problem, let us denote

$$G_\delta(x, y) = E(x) + \frac{\delta}{2}\left(\||g(x)|^+\|_H^2 + \|h(x)\|_H^2 + \|x - y\|_H^2\right) \tag{4}$$

where $\delta > 0$ is a penalty parameter. Then the next proposition is an analogue
of Theorem 1.

Theorem 2. *Let the function $G_\delta(x, y)$ be defined in (4) and suppose that E is
differentiable and convex, and for any $x \in H$, $x \neq 0$, we have $E(\alpha x) \to \infty$ if
$\alpha \to \infty$. Let $\epsilon_k \to 0$ and $\delta_k \to \infty$ as $k \to \infty$. Let $\{\xi_l\}_{l \geq 1}$ and $\{\tau_l\}_{l \geq 1}$ be two
positive and decreasing to zero sequences such that $\sum_l \xi_l + \sum_l \tau_l < B$ for some
positive B.*

*Suppose that x^* is the limit points of sequence $\{x^k\}_{k \geq 0}$ generated by Algo-
rithm 1 with respect to the penalty function $G_\delta(x, y)$, i.e. there is $K \subset \{1, 2, \ldots\}$,
such that $x^k \to_K x^*$.*

Then

1. *x^* is a feasible point for problem (1)–(2).*
2. *x^* satisfies the first order optimality conditions of Lu–Zhang type for problem
(1)–(2).*

2.4   Empirical Results

2.4.1   The Compressed Sensing Problem To evaluate applicability of the
WPDA in practice, in this section we compare its performance with the basic
PD algorithm [21] on the compressed sensing problem. A sparse signal of a large
dimension is decoded by a signal of much smaller one, and in order to recover
it, one should find a sparse solution to the system of linear equalities [4, 6, 10].

Let A be a $p \times n$ matrix with $p < n$, $\xi \in \mathbb{R}^n$ be such that $\|\xi\|_0 \leq m$, $b := A\xi \in \mathbb{R}^p$. We obtain elements of matrix A using the Gaussian random generator. We recover ξ from b by solving the optimization problem $E(x) := \|Ax - b\|_2^2 \to \min$ with the cardinality constraint $\|x\|_0 \leq m$.

All experiments were performed on a personal computer with an Apple M1 CPU (3.2 GHz) and 8 GB memory, using a Python package.

We compare the WPD algorithm with the original PD algorithm for different p. The dimension $n = 5000$ was fixed and the cardinality of signals was fixed at level $m = 100$. We let $\epsilon_k = \frac{1}{k}$, $\delta_k = \log k$, $\tau_l = \xi_l = \frac{1}{l^2}$. For each $p = 500, \ldots, 1000$, the algorithms generated 100 solutions, and then they values were averaged. The averaged performance for the two algorithms are shown in Fig. 1. The results show that the WDP algorithm has a significant increase in solution generation speed, while the accuracy of solutions in terms of $MSE = \frac{1}{n}\|x^* - x\|_H$ remains at about the same level as for the original PD method.

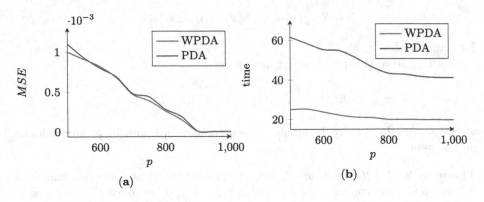

Fig. 1. Results of performance for the WPDA and the PDA for the compressed sensing problem with different p averaged over 100 independent experiments. (a) MSE over p; (b) CPU time over p.

2.4.2   The Index Tracking Problem Let n be the number of assets and let $R = (r_{ti})$ be the matrix of asset returns, where r_{ti} is the i-th asset return at time t, where $1 \leq i \leq n$ and $1 \leq t \leq m$. A portfolio is a vector composed of asset weights, i.e. $x = (x_1, \ldots, x_n)^T \in \mathbb{R}^n$. Denote I_t the index return at time t, $1 \leq t \leq m$, $I = (I_1, \ldots, I_t)^T \in \mathbb{R}^m$. We consider the simple version of index tracking problem in which the portfolio does not change over time. Moreover, we will assume that transaction costs are equal to 0, all weights x_i can be as positive and negative, and the constraint $x^T 1_n = 1$ holds, where 1_n denotes the vector from $\mathbb{R}^n$ with all its component equal to 1.

We will compare two algorithms for solving the index tracking problem with the cardinality constraint:

$$x^* = \arg\min \frac{1}{m}\|I - Rx\|_2 \quad s.t. \quad x^T 1_n = 1, \ \|x\|_0 \leq K, \tag{5}$$

where K is the maximum number of assets in the portfolio with non-zero weights.

We use a personal computer with an Apple M1 CPU (3.2 GHz) and 8 GB memory, using a Python package. In our empirical analysis we use publicly available data taken from the OR-Library [1]. In our experiments we use S&P 100 dataset (USA, $n = 98$), for which the number of time periods each is $m = 290$ (weekly data). We transformed the original data in form of absolute asset price into asset return matrix.

The comparison of the WPD algorithm with the original PD algorithm for different p is shown in Fig. 2. The dimension of 290 was fixed and the cardinality took its level from $6 \leq K \leq 25$. Again, we take $\epsilon_k = \frac{1}{k}$, $\delta_k = \log k$, $\tau_l = \xi_l = \frac{1}{l^2}$. For each $K = 6, \ldots, 25$, the algorithms generated 20 solutions, and then they values of error $E(x) = \frac{1}{m}\|I - Rx\|_2$ as well as their run times were averaged. The averaged performance for the two algorithms are visualized in Fig. 1. Figure 2 shows that the WDP algorithm has an increase in solution rate, while the accuracy of solutions in terms of $MSE = \frac{1}{n}\|x^* - x\|_H$ remains at about the same level as for the original PD method.

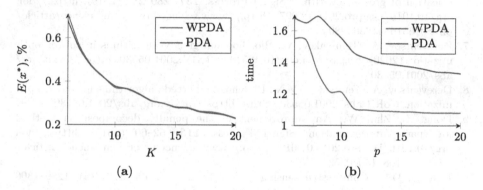

Fig. 2. Results of performance for the WPDA and the PDA for the index tracking problem with different p averaged over 100 independent experiments. (a) MSE over p; (b) CPU time over p.

3   Conclusion

This paper proposed a new penalty decomposition algorithm with relaxation constraints. These weakening conditions allow one to solve two sub-problems on while-loop inexactly. Despite this relaxation, Theorem 1 shows that the sequences generated by the WPDA converge to a stationary point satisfying the Lu–Zhang type necessary optimality conditions (that is, this important property, which is valid for the original algorithm, holds as well). Empirical results show that the WPD algorithm has a faster execution speed without losing the accuracy of the solutions obtained compared to the original PD algorithm.

References

1. Beasley, J.E., Meade, N., Chang, T.J.: An evolutionary heuristic for the index tracking problem. Eur. J. Oper. Res. **148**(3), 621–643 (2003). https://doi.org/10.1016/s0377-2217(02)00425-3, https://www.sciencedirect.com/science/article/pii/S0377221702004253

2. Bollhofer, M., Eftekhari, A., Scheidegger, S., Schenk, O.: Large-scale sparse inverse covariance matrix estimation. SIAM J. Sci. Comput. **41**(1), A380–A401 (2019). https://doi.org/10.1137/17M1147615

3. Bubeck, S.: Convex optimization: algorithms and complexity. Found. Trends Mach. Learn. **8**(3–4), 231–358 (2015). https://doi.org/10.1561/9781601988614

4. Chen, Z., Huang, C., Lin, S.: A new sparse representation framework for compressed sensing MRI. Knowl.-Based Syst. **188**, 104969 (2020). https://doi.org/10.1016/j.knosys.2019.104969, http://www.sciencedirect.com/science/article/pii/S0950705119303983

5. Dempster, A.P.: Covariance selection. Biometrics **28**(1), 157–175 (1972). https://doi.org/10.2307/2528966

6. Deng, Q., et al.: Compressed sensing for image reconstruction via back-off and rectification of greedy algorithm. Signal Process. **157**, 280–287 (2019). https://doi.org/10.1016/j.sigpro.2018.12.007, http://www.sciencedirect.com/science/article/pii/S0165168418303980

7. Dereventsov, A., Temlyakov, V.: Biorthogonal greedy algorithms in convex optimization (2020). https://doi.org/10.48550/ARXIV.2001.05530, https://arxiv.org/abs/2001.05530

8. Dereventsov, A., Temlyakov, V.N.: Biorthogonal greedy algorithms in convex optimization. CoRR abs/2001.05530 (2020). https://arxiv.org/abs/2001.05530

9. Dong, Z., Zhu, W.: An improvement of the penalty decomposition method for sparse approximation. Signal Process. **113**, 52–60 (2015). https://doi.org/10.1016/j.sigpro.2015.01.012, http://www.sciencedirect.com/science/article/pii/S0165168415000353

10. Donoho, D.L.: Compressed sensing. IEEE Trans. Inf. Theor. **52**(4), 1289–1306 (2006). https://doi.org/10.1109/TIT.2006.871582

11. Dou, H.X., Huang, T.Z., Deng, L.J., Zhao, X.L., Huang, J.: Directional l_0 sparse modeling for image stripe noise removal. Remote Sens. **10**(3) (2018). https://doi.org/10.3390/rs10030361, https://www.mdpi.com/2072-4292/10/3/361

12. Fan, J., Lv, J., Qi, L.: Sparse high-dimensional models in economics. Ann. Rev. Econ. **3**(1), 291–317 (2011). https://doi.org/10.1146/annurev-economics-061109-080451

13. Figueiredo, M.A.T., Nowak, R.D., Wright, S.J.: Gradient projection for sparse reconstruction: application to compressed sensing and other inverse problems. IEEE J. Sel. Top. Signal Process. **1**(4), 586–597 (2007). https://doi.org/10.1109/JSTSP.2007.910281

14. Gajare, S., Sonawani, S.: Improved logistic regression approach in feature selection for EHR. In: Abraham, A., Cherukuri, A.K., Melin, P., Gandhi, N. (eds.) ISDA 2018 2018. AISC, vol. 940, pp. 325–334. Springer, Cham (2020). https://doi.org/10.1007/978-3-030-16657-1_30

15. Jiang, B., Ye, P., Zhang, W.: Unified error estimate for weak biorthogonal greedy algorithms. Int. J. Wavelets Multiresolution Inf. Process. 2250010 (2022). https://doi.org/10.1142/S0219691322500102

16. Jin, Z.F., Wan, Z., Zhao, X., Xiao, Y.: A penalty decomposition method for rank minimization problem with affine constraints. Appl. Math. Model. **39**(16), 4859–4870 (2015). https://doi.org/10.1016/j.apm.2015.03.054

17. Kampa, K., Mehta, S., Chou, C.A., Chaovalitwongse, W.A., Grabowski, T.J.: Sparse optimization in feature selection: application in neuroimaging. J. Glob. Optim. **59**, 439–457 (2014). https://doi.org/10.1007/s10898-013-0134-2

18. Lapucci, M., Levato, T., Sciandrone, M.: Convergent inexact penalty decomposition methods for cardinality-constrained problems. J. Optim. Theory Appl. **188**, 473–496 (2020). https://doi.org/10.1007/s10957-020-01793-9

19. Leung, H.M.F., Dong, Z., Lin, G., Chen, N.: An inexact penalty decomposition method for sparse optimization. Comput. Intell. Neurosci. **2021**(9943519), 1–8 (2021). https://doi.org/10.1155/2021/9943519

20. Lu, Z., Li, X.: Sparse recovery via partial regularization: Models, theory, and algorithms. Math. Oper. Res. **43**(4), 1290–1316 (2018). https://doi.org/10.1287/moor.2017.0905

21. Lu, Z., Zhang, Y.: Sparse approximation via penalty decomposition methods. SIAM J. Optim. **23**(4), 2448–2478 (2013). https://doi.org/10.1137/100808071

22. Luo, X., Chang, X., Ban, X.: Regression and classification using extreme learning machine based on L1-norm and L2-norm. Neurocomputing **174**, 179–186 (2016). https://doi.org/10.1016/j.neucom.2015.03.112, http://www.sciencedirect.com/science/article/pii/S092523121501139X

23. Pan, L.L., Xiu, N.H., Fan, J.: Optimality conditions for sparse nonlinear programming. Sci. China Math. **60**(5), 759–776 (2017). https://doi.org/10.1007/s11425-016-9010-x

24. Patrascu, A., Necoara, I.: Penalty decomposition method for solving ? 0 regularized problems: application to trend filtering. In: 2014 18th International Conference on System Theory, Control and Computing (ICSTCC), pp. 737–742 (2014). https://doi.org/10.1109/ICSTCC.2014.6982506

25. Pleshakov, M., Sidorov, S., Spiridonov, K.: Convergence analysis of penalty decomposition algorithm for cardinality constrained convex optimization in hilbert spaces. In: Kononov, A., Khachay, M., Kalyagin, V.A., Pardalos, P. (eds.) MOTOR 2020. LNCS, vol. 12095, pp. 141–153. Springer, Cham (2020). https://doi.org/10.1007/978-3-030-49988-4_10

26. Pun, C.S., Wong, H.Y.: A linear programming model for selection of sparse high-dimensional multiperiod portfolios. Eur. J. Oper. Res. **273**(2), 754–771 (2019). https://doi.org/10.1016/j.ejor.2018.08.025, http://www.sciencedirect.com/science/article/pii/S0377221718307203

27. Scheinberg, K., Ma, S., Goldfarb, D.: Sparse inverse covariance selection via alternating linearization methods. In: Lafferty, J.D., Williams, C.K.I., Shawe-Taylor, J., Zemel, R.S., Culotta, A. (eds.) Advances in Neural Information Processing Systems, vol. 23, pp. 2101–2109. Curran Associates, Inc. (2010). https://ui.adsabs.harvard.edu/abs/2010arXiv1011.0097S

28. Sidorov, S.P., Faizliev, A.R., Khomchenko, A.A.: Algorithms for L1-norm minimisation of index tracking error and their performance. Int. J. Math. Oper. Res. **11**(4), 497–519 (2017). https://doi.org/10.1504/ijmor.2017.087743, https://ideas.repec.org/a/ids/ijmore/v11y2017i4p497-519.html

29. Teng, Y., Yang, L., Yu, B., Song, X.: A penalty palm method for sparse portfolio selection problems. Optim. Methods Softw. **32**(1), 126–147 (2017). https://doi.org/10.1080/10556788.2016.1204299

30. Wang, D., Jin, Z.F., Shang, Y.: A penalty decomposition method for nuclear norm minimization with L1-norm fidelity term. Evol. Eqn. Control Theory **8**(4), 695–708 (2019). https://doi.org/10.3934/eect.2019034
31. Wipf, D.P., Rao, B.D.: Sparse Bayesian learning for basis selection. IEEE Trans. Signal Process. **52**(8), 2153–2164 (2004). https://doi.org/10.1109/TSP.2004.831016
32. Xu, F., Deng, R.: Fast algorithms for sparse inverse covariance estimation. Int. J. Comput. Math. **96**(8), 1668–1686 (2019). https://doi.org/10.1080/00207160.2018.1506108
33. Zhu, W., Dong, Z., Yu, Y., Chen, J.: Lagrange dual method for sparsity constrained optimization. IEEE Access **6**, 28404–28416 (2018). https://doi.org/10.1109/ACCESS.2018.2836925

Mathematical Modelling and Optimization of Scheduling for Processing Beet in Sugar Production

Dmitry Balandin[ID], Albert Egamov[(✉)][ID], Oleg Kuzenkov[ID],
Oksana Pristavchenko[ID], and Vadim Vildanov[ID]

Lobachevsky University of Nizhny Novgorod, Gagarin Avenue,
23, Nizhny Novgorod 603950, Russia
dmitriy.balandin@itmm.unn.ru, albert810@yandex.ru, pristavchenko@unn.ru

Abstract. The article is devoted to the problems of mathematical modeling and optimization of the sugar beet processing schedule, taking into account the level of its sugar content which varies over time. The problem of finding the optimal processing schedule that ensures the maximum yield of the final product (sugar) is set. It is shown that the optimization problem reduces to a well-known assignment problem. Algorithms for solving the problem are discussed. Some special cases allowing to obtain the solution of the optimization problem in an analytical form are considered. Numerical experiments for the real parameters of sugar beet are carried out. Recommendations on the practical implementation of the obtained optimal solutions are proposed. These results are useful in the practical activities of sugar production enterprises. In addition, the results obtained can be used as a basis for the development of more general mathematical models describing the process of sugar beet processing, taking into account a number of other factors.

Keywords: Mathematical modelling · Sugar beet processing · Optimal schedule · Assignment problem

1 Introduction

The food supply of the population is among the most important factors determining the sovereignty of the country. The problem of the availability, sufficiency and accessibility of food for the population is an integral part of ensuring national

The article was carried out under the contract No SSZ-1771 dated 22.04.2021 on the implementation of R&D on the topic: "Creation of high-tech sugar production on the basis of JSC "Sergach Sugar Plant", within the framework of the Agreement on the provision of subsidies from the federal budget for the development of cooperation between the Russian educational organization of higher education and the organization of the real sector of the economy in order to implement a comprehensive project to create high-tech production No. 075-11-2021-038 of 24.06.2021 (IGC 000000S407521QLA0002).

security. Sustainable provision of the country's population with economically and physically accessible high-quality food products in volumes corresponding to scientifically based consumption standards is the most important task of the agricultural and industrial complex. An essential factor inherent in the production of agricultural products is the seasonality of this production, that is, the unevenness of production during the year associated with the season (time of year). This feature finds its expression in the rise, reduction or even complete cessation of production in certain periods of the year. As is well known, the seasonality of production is characteristic of many sectors of the economy, but the most distinct properties of the seasonality of production are manifested in agriculture. An important aspect of agricultural production, along with the problem of preservation in proper condition, is the efficient processing of agricultural products [1, 2].

Sugar production is an important industry in many countries of the world [3, 4], because sugar is of great importance not only as one of the most important food products, but also as a raw material for other industries. Sugar production is one of the most complex and energy-consuming, and therefore, at present, its indicators such as quality, energy consumption, and cost are coming to the fore. It is believed that one of the main ways to reduce the cost of production is to increase the duration of processing plants due to proper storage of agricultural products, more rational use and processing. Currently, the processing period of sugar beet is about 100 days after the start of harvesting. The introduction of modern resource-saving technologies for storage and processing allows both to increase the duration of the processing season and to improve its quality, preventing excessive loss of sucrose in root crops [5, 6]. Storage of root crops is often accompanied by excessive losses of beet pulp and sucrose. The reason for such losses is both the wrong choice of the storage mode of raw materials and an irrational schedule for processing raw materials. Usually, the greatest attention during processing is paid to ensuring that beet processing is completed as soon as possible immediately after harvesting. In principle, this is correct, since reducing the time of "laying" beets in the beet pile fields reduces the loss of sucrose during its storage. In addition, part of the beet remaining in the fields has not yet been harvested, manages to "grow up" somewhat and increase the sucrose content. However, an important role can also be played by the order of processing of certain batches of beets coming for processing from various beet producers.

This article describes a mathematical model of optimal processing of sugar beet, which takes into account the minimum number of factors affecting the result of processing [7]. This "minimality" of the model allows us to obtain, using mathematical methods, very important conclusions in practical terms about the nature and features of the optimal schedule for processing raw materials. The problem of optimal schedule for processing beet in sugar production was considered in papers [8–12]. It turns out that the stated problem is a special case of the classical linear programming problem, which can be solved by using the simplex algorithm. Moreover, this problem is one of the variants of the well-known assignment problem [13, 14].

The rest of this paper is organized as follows. In Sect. 2 the mathematical model of sugar beet processing is described and optimization problem is stated. In Sect. 3 the rigorous mathematical results concerning optimal scheduling for sugar beet processing age given. Section 4 is devoted to numerical experiments by using real data. The last Sect. 5 summarizes the contribution and results obtained.

2   Mathematical Model and Statement of the Problem

Let sugar beet of n varieties be harvested for further processing. The quantity (mass) M of beets of each variety is the same and is processed during one production cycle during a fixed period of time (one day). Accordingly, n processing periods are necessary to process the entire beet, individual batches of raw materials must be stored for a certain number of periods before being processed. We introduce the following notation: a_i is the sugar content (percentage of sugar content) of the i-th beet variety, $i = \overline{1, n}$, b_{ij} is the reduction coefficient of sugar content of the i-th beet variety as a result of storage for the j-th period of time, $0 < b_{ij} < 1$. The sugar content of the i-th beet variety changes as follows: $a_i b_{i1}$ is after the first period, $a_i b_{i1} b_{i2}...b_{ik-1}$ is by the beginning of the k-th processing period (unless, of course, it is processed before this moment). Denote p_{ij} as sugar content of the i-th beet variety by the beginning of the j-th processing period, then $p_{i1} = a_i$, $p_{i2} = a_i b_{i1}$, ..., $p_{in} = a_i b_{i1} b_{i2}...b_{in-1}$, $i = \overline{1, n}$. From the elements p_{ij} we will form a square matrix P so that

$$(P)_{ij} = p_{ij}. \tag{1}$$

In this $n \times n$-matrix, the column number determines the number of the processing stage, and the row number corresponds to the batch number of beets. The output of the finished product (sugar) at each processing period, other things being equal, is the greater, the greater the sugar content of the substance processed at this stage. Let a batch of raw materials for the j-th processing period be prepared as follows: all beet varieties are mixed in unequal proportions so that the total mass is M, the proportion of first grade beets is x_{1j}, the share of second grade beets is x_{2j}, share of i-th grade is x_{ij}, share of the n-th grade is x_{nj}. Obviously, these shares must satisfy the following conditions

$$\sum_{j=1}^{n} x_{ij} = 1 \text{ for } i = \overline{1, n}, \quad \sum_{i=1}^{n} x_{ij} = 1 \text{ for } j = \overline{1, n}. \tag{2}$$

The product yield for the entire processing time is proportional to

$$S = \sum_{i=1}^{n} \sum_{j=1}^{n} p_{ij} x_{ij}. \tag{3}$$

The optimization problem consists in choosing $x_{ij} \geq 0$, satisfying the conditions (2), under which the objective function (3) takes the maximum value S.

3   Optimal Scheduling

3.1   Does Mixing Beet Varieties Lead to Maximizing the Objective Function?

The answer to the question put in the heading of this subsection is given in the following theorem.

Theorem 1. *The largest value of the objective function (3) can always be achieved with arguments x_{ij}, taking the values 0 or 1 only.*

Proof. From the equalities (2) we can express

$$x_{in} = 1 - \sum_{j=1}^{n-1} x_{ij}, \ i = \overline{1, n-1}; \qquad x_{nj} = 1 - \sum_{i=1}^{n-1} x_{ij}, \ j = \overline{1, n-1}; \quad (4)$$

$$x_{nn} = 1 - \sum_{i=1}^{n-1} x_{in} = 1 - \sum_{i=1}^{n-1}\left(1 - \sum_{j=1}^{n-1} x_{ij}\right) = 2 - n + \sum_{i=1}^{n-1}\sum_{j=1}^{n-1} x_{ij}. \quad (5)$$

Therefore, it is possible to evaluate the presented expressions as follows

$$0 \le x_{ij} \le 1, \quad \sum_{j=1}^{n-1} x_{ij} \le 1, \ i = \overline{1, n-1}; \quad \sum_{i=1}^{n-1} x_{ij} \le 1, \ j = \overline{1, n-1}; \quad (6)$$

$$n - 2 \le \sum_{i=1}^{n-1}\sum_{j=1}^{n-1} x_{ij} \le n - 1. \quad (7)$$

Substituting expressions (4) and (5) into the objective function (3), we arrive at the *linear programming problem* of maximizing a linear function on the set of independent variables x_{ij}, $i = \overline{1, n-1}$, $j = \overline{1, n-1}$, in the $(n-1)^2$-dimensional domain given by inequalities (6) and (7). This domain is a polyhedron in $(n-1)^2$-dimensional space. Its faces satisfy one of the following equations:

$$x_{ij} = 0, \quad i = \overline{1, n-1}, \quad j = \overline{1, n-1}; \quad (8)$$

$$\sum_{j=1}^{n-1} x_{ij} = 1, \ i = \overline{1, n-1}; \quad \sum_{i=1}^{n-1} x_{ij} = 1, \ j = \overline{1, n-1}; \quad (9)$$

$$\sum_{i=1}^{n-1}\sum_{j=1}^{n-1} x_{ij} = n - 2. \quad (10)$$

Equations (9) correspond to cases where one of the variables x_{in} or x_{nj}, $i = \overline{1, n-1}$, $j = \overline{1, n-1}$, goes to zero, the Eq. (10) refers to the case when $x_{nn} = 0$. Thus, any face of this polygon corresponds to the set of variables x_{ij}, $i = \overline{1, n}$, $j =$

$\overline{1, n}$, that satisfy conditions (2), in which one variable vanishes. It is known that a linear function does not occur (because of the rare trivial case of degeneration into a constant) of the largest value at points inside the domain. This follows from the fact that the partial derivatives of a nontrivial linear function are constants not equal to zero together. Thus, the largest value of the objective function must be sought on one of the boundaries of the domain, which corresponds to zero at least one of the arguments. Then the original problem is reduced to the problem of maximizing the same linear function in a bounded set of lower dimension. For the reduced problem, all the properties of the original problem noted above are valid. Consequently, there are two possibilities – the objective function degenerates into a constant on this set (face), or its maximum is reached on one of the boundaries of this set. In the first case, any point of the set is optimal, while there is always a point whose components are equal only to zeros and units. The second case again corresponds to zero or unit of new arguments. Repeating these arguments a finite number of times, we come to narrowing the possible location domain of the objective function largest value to sets of arguments that are only zero (or unit). This completes the proof.

Thus, mixing beet varieties never leads to an increase in yield, the highest possible yield can always be achieved without mixing varieties in the process of sequential processing of each variety in one production period.

3.2 "Hungarian Algorithm" for Solving the Optimization Problem

Now, to solve the optimization problem, it is necessary to find the optimal sequence for processing different batches of beets without mixing them. We number the batches of raw materials in the order of their processing. Then the sugar yield after completion of all stages will be proportional to the value of the objective function

$$S^* = a_1 + a_2 b_{21} + a_3 b_{31} b_{32} + \ldots + a_n b_{n1} b_{n2} b_{nn-1} \tag{11}$$

The problem of searching for an optimal processing schedule is reduced to finding such a sequence of processing of raw materials for which the value of S will be maximal. Taking into account the matrix P that is defined in (1), the problem of choosing the optimal processing schedule can be reformulated as follows: from each row of the matrix P, select exactly one element so that each column contains only one of the selected elements, and the sum of the selected elements is maximum. This task can be considered as a special case of the well-known assignment problem [13–16]. The assignment problem is a fundamental problem of combinatorial optimization. To solve it, in 1955, Harold Kuhn developed an algorithm called the "Hungarian algorithm" [17], later it was proved that it has polynomial complexity $O(n^4)$. Sometime later, the Hungarian algorithm was modified to polynomial complexity $O(n^3)$. The Hungarian algorithm can find both the maximum and minimum value of the objective function, as well as the corresponding choice of matrix rows, that is, the extreme processing schedule.

3.3   Analytical Solutions

Under some assumptions regarding the parameters a_i and b_{ij}, it is possible to obtain accurate analytical solutions for optimal raw material processing schedules.

Optimal Plan A. Let the degradation coefficient not depend on the batch of raw materials, but depend only on the processing period, that is

$$b_{ij} = \bar{b}_j, \quad i = \overline{1,n}, \quad j = \overline{1,n}. \tag{12}$$

In this case, the following theorem takes place.

Theorem 2. *Let the conditions (12) be fulfilled, then the highest yield of the finished product (sugar) can be obtained if batches of raw materials are processed in descending order of the coefficients* a_i, $i = \overline{1,n}$.

Proof. Let the batches of raw materials be numbered in the optimal processing order that provides the maximum yield of the product. This yield must be greater than the corresponding yield for any other batch processing sequence. Then the objective function (11) with the optimal processing order has the following form

$$S^* = a_1 + a_2\bar{b}_1 + a_3\bar{b}_1\bar{b}_2 + \ldots + a_n\bar{b}_1\bar{b}_2\ldots\bar{b}_{n-1}, \tag{13}$$

where S^* is the maximal value of the function S. Consider the processing order, which differs from the optimal one by interchanging the k-th and $k+1$-th batches of raw materials, where $k = \overline{1, n-1}$. In this case, we have the inequality

$$S^* = a_1 + a_2\bar{b}_1 + \ldots + a_k\bar{b}_1\bar{b}_2\ldots\bar{b}_{k-1} + a_{k+1}\bar{b}_1\bar{b}_2\ldots\bar{b}_k + \ldots + a_n\bar{b}_1\bar{b}_2\ldots\bar{b}_{n-1}$$
$$\geq a_1 + a_2b_1 + \ldots + a_{k+1}\bar{b}_1\bar{b}_2\ldots\bar{b}_{k-1} + a_k\bar{b}_1\bar{b}_2\ldots\bar{b}_k + \ldots + a_n\bar{b}_1\bar{b}_2\ldots\bar{b}_{n-1},$$

from which the equivalent inequalities follow

$$a_k\bar{b}_1\bar{b}_2\ldots\bar{b}_{k-1} + a_{k+1}\bar{b}_1\bar{b}_2\ldots\bar{b}_k \geq a_{k+1}\bar{b}_1\bar{b}_2\ldots\bar{b}_{k-1} + a_k\bar{b}_1\bar{b}_2\ldots\bar{b}_k$$

$$\iff a_k\bar{b}_1\bar{b}_2\ldots\bar{b}_{k-1}(1 - \bar{b}_k) \geq a_{k+1}\bar{b}_1\bar{b}_2\ldots\bar{b}_{k-1}(1 - \bar{b}_k) \iff a_k \geq a_{k+1}.$$

The proof is completed.

Thus, the *optimal plan A* corresponds to beet processing in the order of descending the parameter a_i characterizing the content of sucrose in the various batches of raw material, i.e. $a_1 \geq a_2 \geq \ldots \geq a_n$.

Optimal Plan B. Let us consider another particular case. It is assumed that

$$b_{ij} = \tilde{b}_i, \quad i = \overline{1,n}, \quad j = \overline{1, n-1}. \tag{14}$$

In other words, the parameters b_i depend only on the batch number of the raw material and do not depend on the stage of processing. In addition, it is assumed that all varieties of beets have the same initial sugar content, that is

$$a_i = a, \quad i = \overline{1,n}. \tag{15}$$

Theorem 3. *Let the conditions (14), (15) and the inequality*

$$\beta = \min_i \tilde{b}_i \geq \frac{n-2}{n-1} \tag{16}$$

be fulfilled. Then the highest yield of the finished product (sugar) can be obtained if batches of raw materials are processed in ascending order of coefficients b_i.

Proof. Let the batches of raw materials be numbered in the optimal processing order that provides the maximum yield of the product. This yield must be greater than the corresponding yield for any other batch processing sequence. Then the objective function (11) presumably with the optimal processing order has the following form

$$S^* = a\tilde{b}_1^0 + a\tilde{b}_2 + a\tilde{b}_3^2 + \ldots + a\tilde{b}_n^{n-1}, \tag{17}$$

where S^* is the maximal value of the function S. Consider the processing order, which differs from the optimal one by interchanging the k-th and $(k+1)$-th batches of raw materials, where $k = \overline{1, n-1}$. In this case, we have the inequality

$$S^* = a + \ldots + a\tilde{b}_k^{k-1} + a\tilde{b}_{k+1}^k + \ldots + a\tilde{b}_n^{n-1} \geq a + \ldots + a\tilde{b}_{k+1}^{k-1} + a\tilde{b}_k^k + \ldots + a\tilde{b}_n^{n-1},$$

from which the equivalent inequalities follow

$$\tilde{b}_k^{k-1} + \tilde{b}_{k+1}^k \geq \tilde{b}_{k+1}^{k-1}\tilde{b}_k^k \iff \tilde{b}_k^{k-1}(1-\tilde{b}_k) \geq \tilde{b}_{k+1}^{k-1}(1-\tilde{b}_{k+1}) \iff f(\tilde{b}_k) \geq f(\tilde{b}_{k+1}).$$

Here, the function $f(x)$ looks like $f(x) = x^{k-1} - x^k$. Its first derivative has the form $f'(x) = (k-1)x^{k-2} - kx^{k-1}$. It vanishes at the point $x^* = (k-1)/k$. If $x \geq x^*$, then this derivative is negative, and the function decreases monotonically; a larger function value corresponds to a smaller argument value. Since

$$\tilde{b}_k \geq \beta = \frac{n-2}{n-1} \geq x^* \quad \text{and} \quad \tilde{b}_{k+1} \geq \beta = \frac{n-2}{n-1} \geq x^*,$$

it what follows that $\tilde{b}_k \leq \tilde{b}_{k+1}$. Therefore, it is necessary that the inequalities $\tilde{b}_k \leq \tilde{b}_{k+1}$ must be satisfied for an optimal processing sequence which was required to be proved.

Thus, the *optimal plan B* corresponds to beet processing in the order of increasing the degradation parameter $\tilde{b}_i$, i.e., in the following order of the degradation parameters $\tilde{b}_1 \leq \tilde{b}_2 \leq \ldots \leq \tilde{b}_n$.

4 Numerical Results

In practice, when processing sugar beets not all the numerical parameters of the optimization problem are known and have exact numerical values. If the parameters corresponding to the values of sugar content for different varieties of beets can be measured relatively accurately, then the parameters characterizing the degree of beet wilting and loss of sugar content and depending on poorly predicted weather conditions cannot be specified in advance before processing the entire harvested beet crop. In addition, in order to apply the Hungarian algorithm, it is necessary to know what the degradation factors of the batches would be before they have been processed. This can only be predicted by some empirical means, therefore, the optimal plan, in practice, generally speaking, is not achievable. The question arises, how, in this case, to correctly organize the process of beet processing. Further, it is proposed to discuss and evaluate some processing strategies.

Despite the simplicity of the above optimal plans A and B, their exact implementation in practice is rather problematic, due to the lack of reliable information on the degradation parameters b_{ij} at the beginning of the raw material processing season. In this regard, approximate solutions can be proposed, which will take into account the current information on the residual sugar content of beet varieties that have not yet been processed to this stage. Based on the optimal plan A the following processing strategy can be proposed: at the next stage, raw materials with the highest residual sugar content are supplied for processing. This strategy will be called the *greedy algorithm* [18,19], meaning that the best of the remaining varieties of raw materials is sent to the next stage of processing. Another processing strategy can be associated with the optimal plan B: at the next stage, raw materials with the lowest residual sugar content enter the processing. Such a strategy can be called a *thrifty algorithm*, emphasizing by this name that the processing of the best varieties should be carried out to the last stages, and the processing of less valuable beet varieties (before they have completely lost their production value) should be carried out at the beginning of the season.

The initial data of sugar content and the rate of its loss by different batches of beets were recorded during many years of measurements carried out at the Sergachsky sugar plant. In the computational experiments below, it is assumed that there are a total of 100 batches and sugar beet processing takes place at a hundred stages ($n = 100$). The sugar content parameters are set randomly on some interval (empirical observations). The coefficients b_{ij} are set as random variables, from interval $(\beta, 1)$, in other words, the distribution of parameters a_i and b_{ij} are obtained in accordance with the law on the uniform distribution of a random variable on the corresponding segments.

Figures 1 and 2 show the resulting graphs of the dependence of the objective function S on the number of processed batches when implementing the optimal, *greedy* and *thrifty* algorithms for different input data. The graphs in Fig. 1 correspond to the situation when the coefficients b_{ij} are uniformly distributed in the interval $(0.97, 1)$, the coefficients a_i are uniformly distributed in the interval $(0.16, 0.20)$. The graphs in Fig. 2 correspond to the situation when the coefficients a_i are uniformly distributed in the interval $(0.16, 0.2)$, the coefficients b_{ij} are chosen from the interval $(0.95, 1)$, with restriction $|b_{ij} - b_{ik}| \leq 0.01$ and the distribution for each fixed i is uniform.

The blue line marks the change in the objective function when implementing the optimal processing algorithm, the green line shows the change in the

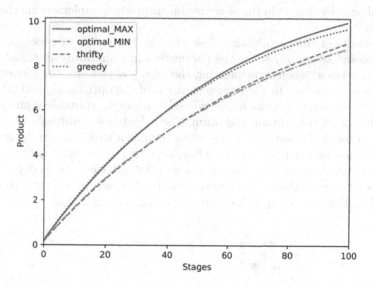

Fig. 1. Product dynamics versus the number of stages for three algorithms in the case when *greedy* algorithm is better than *thrifty* one (Color figure online)

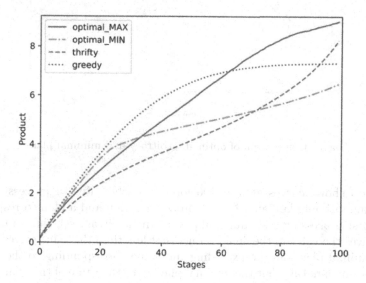

Fig. 2. Product dynamics versus the number of stages for three algorithms in the case when *thrifty* algorithm is better than *greedy* one (Color figure online)

objective function when implementing the *thrifty* algorithm, and the orange dotted line shows the change in the objective function when implementing the *greedy* algorithm.

The results of the calculations shown in Fig. 1 allow us to conclude that in the considered situation, when the parameters are uniformly distributed over the given sections, when implementing the *greedy* algorithm, the losses will be insignificant in relation to the implementation of the optimal algorithm. In this case, the *greedy* algorithm can be considered as a simple, convenient and efficient approximation of the optimal algorithm. The calculation results shown in Fig. 2 allow us to conclude that in the case when there is a small scatter in the values of the degradation coefficients in each batch relative to the periods of processing and the scatter of the initial sugar content is also small, then the *thrifty* algorithm gives a better result than the it greedy one. In this case, it is better to use the *thrifty* algorithm as an approximation of the optimal algorithm.

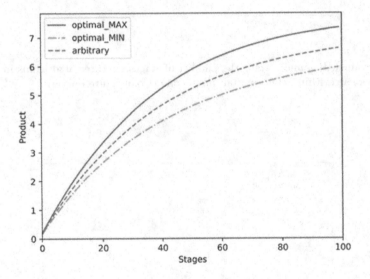

Fig. 3. Comparison of optimal, arbitrary and minimal plans

Figure 3 shows curves at $n = 100$ for three scheduling of processing beet, namely: an optimal plan, an arbitrary processing plan and a plan corresponding to the worst processing case (minimal plan). Any arbitrary schedule of beet processing gives the value of the final product, not less than the value corresponding to the minimal plan and not exceeding the value corresponding to the optimal plan. It is worth noting that the minimal plan gives the value of the final product (sugar) at the end of the processing season by 20% less than the optimal plan allows.

5   Conclusion

This article deals with the problem of the optimal processing schedule for sugar beet. An upper bound for the objective function is given. In the case when all the coefficients of the stage-by-stage degradation of beets are known, the task is reduced to the known assignment problem, therefore, the exact solution of the problem posed can theoretically be obtained by using the Hungarian algorithm or its varieties, but is practically not feasible in real life. In practice, all the exact coefficients of the stage-by-stage degradation of beets during the season are not known in advance. Therefore, it is more justified to use simpler solutions that are quasi-optimal and based on some estimates of the degradation rates of raw materials.

Two strategies for beet processing under uncertainty are proposed. The first strategy based on the use of a greedy algorithm, relies on measurements of the sugar content of each variety at the beginning of each stage. According to this strategy at the next stage one must choose raw materials with the highest residual sugar content. The second strategy is so called thrifty algorithm. According to this strategy at the next stage one must choose raw materials with the lowest residual sugar content. The numerical results obtained show that by the end of the sugar beet processing season, the relative losses of the final product for these two quasi-optimal strategies in comparison with the optimal plan do not exceed 8.5%.

The conducted studies allow us to make the following recommendations to sugar producers. If different beet varieties that differ significantly in sugar content are stored under approximately the same conditions, i.e. the degradation parameter depends mainly on the processing stage and weakly depends on the beet variety, then a processing strategy similar greedy algorithm may be recommended. On the contrary, if different beet varieties slightly differing in sugar content are stored in significantly different conditions, then a processing strategy similar thrifty algorithm may be recommended.

References

1. Sapronov, A.R.: Technology of sugar production. Kolos, Moscow (1999). [in Russian]
2. Anichin, V.L.: Theory and practice of production resources management in the beet sugar subcomplex of the agro-industrial complex. Publ. House of the BelGSHA, Belgorod (2005). [in Russian]
3. Jiao, Z., Higgins, A.J., Prestwidge, D.B.: An integrated statistical and optimisation approach to increasing sugar production within a mill region. Comput. Electron. Agric. **48**, 170–181 (2005)
4. Junqueira, R., Morabito, R.: Modeling and solving a sugarcane harvest front scheduling problem. Int. J. Prod. Econ. **231**(1), 150–160 (2019)
5. Shirokov, E.P.: Technology of storage and processing of fruits and vegetables with the basics of standardization. Agropromizdat, Moscow (1988). [in Russian]
6. Manzhesov, V.I.: Technology of storage, processing and standardization of crop production. Troitsky Bridge, St. Petersburg (2010). [in Russian]

7. Kukhar, V.N., Chernyavsky, A.P., Chernyavskaya, L.I., Mokanyuk, Y.A.: Methods for assessing the technological properties of sugar beet using indicators of the content of potassium, sodium and α-amine nitrogen determined in beetroot and its processing products. Sugar **1**, 18–36 (2019). [in Russian]
8. Balandin, D.V., Vildanov, V.K., Kuzenkov, O.A., Egamov, A.I.: Optimal schedule of sugar beet processing in conditions of uncertainty. In: Actual Problems of Applied Mathematics, Computer Science and Mechanics: Proceedings of the International Scientific Conference, pp. 328–334. Voronezh (2022). [in Russian]
9. Balandin, D.V., Vildanov, V.K., Kuzenkov, O.A., Zakharova, I.V., Egamov, A.I.: Strategy of processing sugar beet batches with close parameters of its withering. In: Proceedings of the Second All-Russian scientific and Practical Seminar "Mathematical and Computer Modeling and Business Analysis in the Conditions of Digitalization of the Economy". Lobachevsky University of Nizhny Novgorod, pp. 10–18. N.Novgorod (2022). [in Russian]
10. Balandin, D.V., Kuzenkov, O.A., Vildanov, V.K.: A software module for constructing an optimal schedule for processing raw materials. Mod. Inf. Technol. IT Educ. **17**(2), 442–452 (2021). [in Russian]
11. Balandin, D.V., Kuzenkov, O.A.: Optimization of the schedule of processing of raw materials in the food industry. Mod. Eng. Innov. Technol. **17**(1), 59–66 (2021). [in Russian]
12. Balandin, D.V., Kuznetsov, Y.A.: The problem of optimizing the schedule of processing perishable agricultural products. Econ. Anal. Theory Pract. **20**(11), 2134–2150 (2021). [in Russian]
13. Bunday, B.: Basic linear programming. London (1984)
14. Sigal, I.H., Ivanova, A.P.: Introduction to Applied Discrete Programming: Models and Computational Algorithms. Textbook, Moscow (2002). [in Russian]
15. Lelyakova, L.V., Kharitonova, A.G., Chernyshova, G.D.: Applied assignment problems (models, solution algorithms). Bull. Voronezh State Univ. **2**, 22–27 (2017). [In Russian]
16. Rainer, B., Mauro, D., Silvano, M.: Assignment problems. Society for Industrial and Applied Mathematics, Philadelphia, USA (2009)
17. Kuhn, H.: The Hungarian Method for the assignment problem. Naval Res. Logist. Q. **2**, 83–97 (1955)
18. Asanov, M.O., Baransky, V.A., Racin, V.V.: Discrete Mathematics: Graphs, Matroids, Algorithms, 2nd edn. Textbook. St. Petersburg (2010). [in Russian]
19. Roughgarden, T., Algorithms Illuminated. Part 3: Greedy Algorithms and Dynamic. Soundlikeyourself Publishing, LLC, New York (2019)

Integration of Information Systems Data to Improve the Petroleum Product Blends Quality

Viacheslav Kuvykin[1], Artem Kolpakov[2], and Mikhail Meleshkevich[3](✉)

[1] National Research Lobachevsky State University of Nizhny Novgorod,
Nizhny Novgorod, Russia
vkuvykin@yandex.ru
[2] PJSC LUKOIL, Moscow, Russia
kolpakovav@list.ru
[3] LLC LUKOIL-Nizhegorodnefteorgsintez, Kstovo, Russia
meleshkevichma@mail.ru

Abstract. The article deals with innovative solutions to improve the preparation of mixtures with the data integration of advanced planning, material balance and laboratory information management systems. The problem of developing models for calculating the kinematic viscosity of blending petroleum products was considered. To assess the discrepancy between the optimal planned recipes and real recipes when blending oil products, the cosine similarity was used. If the discrepancy was large, the optimal planning model should be updated. To calculate the viscosity of blends, a number of formulas was considered that are most widely used in modern NLP-models. The results were compared with the experimental data. To optimize blending, it was necessary to adjust the parameters of the calculation formulas depending on the current state of production. The parametric analysis of the NLP-model of a real refinery was carried out based on numerical calculations of the optimal plan. It is shown that the change in the parameter significantly influences the optimal utilization of process units and the magnitude of marginal profit. The economic evaluation of the proposed solutions during the operation of the refinery was performed.

Keywords: Cyber-physical systems · Mathematical modeling · Optimization · Viscosity · Blending · Nonlinear programming · Automation in industry · Systems integration · Data reconciliation · Material balance · Refinery

1 Introduction

The relationship between refinery optimal production planning and the cyber-physical production environment is a new question posed in the study of process systems engineering [1]. To solve the optimization problems of refineries have been successfully used methods of mathematical programming with the application of advanced planning system (APS), such as RPMS (Honeywell) [2], PIMS

D. Balandin et al. (Eds.): MMST 2022, CCIS 1750, pp. 239–250, 2022.
https://doi.org/10.1007/978-3-031-24145-1_20

(Aspen Technology) [3], GRTMPS (Haverly Systems) [4], Plan and Schedule (Aveva) [5] and others. One of the features of modern NLP-models of refineries is their high dimensionality. At the same time, the mathematical model of an enterprise should be as simple as possible, not clogged with a mass of secondary details, since their consideration complicates the economic analysis and further increases the dimensionality. Although modern planning systems allow modeling non-linear dependencies, in practical applications for refineries are widely used various indices that allow to linearize the problem. However, the use of blending indices with some pre-set coefficients often leads to unacceptable errors [6–8]. The parameters of the model calculation formulas should not be taken as the prescriptive ones, but should be adjusted in accordance with the existing state of production [9–11].

According to dynamical systems theory, for practical applications it is important that the mathematical model is structurally stable, namely the conclusions do not vary significantly with a small change in the parameters and functions describing the model. Such systems are called crude [12]. It is known that optimization of plan parameters can result in incorrect solutions because of the instability arising due to optimization [13]. The concept of crude models can be extended to the problems of mathematical programming. Although in NLP-models the solutions vary step-wise, but there are parameter values at which the variations in solutions is so great that it is physically unrealizable and the model at these points does not reflect actual production and requires detailed study and adjustment [14].

Production planning process is included in a system with supply chain planning, inventory management, scheduling, compounding and dynamic process control systems [15–17]. Improper modeling can lead to incorrect management decisions while developing production and product shipment schedules, setting up management systems. For example, one of the main technical and economic indicators is the volume of hydrocarbon feed conversion, which determines the plan for the feed supply, volume of the shipped products, utilization of process units, consumption of energy resources, and auxiliary materials, etc. At the same time, the practice shows that the calculating methods of a blend viscosity have a significant impact on the economic results of the refinery [6].

Data reconciliation is widely used in the petrochemical industry as a method of error reduction measuring side with instrument redundancy [18,19]. Large dimension of modern production models production inevitably requires appropriate information systems. Enterprise digitalization contributed to the active use Manufacturing execution systems (MES) such as Production Balance (Honeywell) [20], ROMeo Material Balance (AVEVA) [21], I-DRMS (IndaSoft) [22], Sigmafine (Pimsoft) [23], Aspen Operations Reconciliation and Accounting (Aspen-Tech) [24], etc.

The paper [25] considers the ways to improve the quality of oil refinery process models in information systems at various management levels. The integration of such systems today is an important problem [26,27].The authors demonstrate the necessity to introduce a universal (basic) model for all management systems in order to ensure their effective interaction and integration [25]. The methodology for the automated refinery material balance reconciliation, based on the technology of reverse balancing and integration of data reconciliation and optimal planning systems has been developed [28].

In order to set up planning models correctly, it is important to use reliable data that can be obtained by balancing the material balance of an enterprise. Due to the constant increase in the complexity of modern technological processes, production chains are changing and becoming more complex, which ultimately leads to the need to use special information systems to correctly solve the problem of data reconciliation. Planning models and material balance models should be integrated into a single information space for comparative analysis of recipes, "plan-fact" analysis of deviations and statistical analysis [28]. At the same time, the integration of planning systems and material balance systems brings production automation to a qualitatively new level: instead of using a set of separate unrelated systems, a new cyber-physical system appears with constant feedback between all subsystems [1]. This approach makes it possible to use intelligent planning, namely, the timely updating of constraint sets for the NLP model, the operational adjustment of "on-line" mixture control systems, as well as a reliable forecast for making effective management decisions.

Viscosity is one of the most important physical and chemical indicators of oil and petroleum products. Rheological properties play an important role in pumping through pipelines, drain-fill operations and fuel system operations. When blending hydrocarbons, the actual task is to predict the viscosity of the blend, which is a non-linear function of the components. The analysis of various formulas for calculating the viscosity of binary blends was carried out in a review papers [29, 30], and it was noted that there are no universal calculation rules suitable for all types of oil and petroleum products. Formulas with blending indices are widely used, they are additive in terms of the volume or weight basis at a constant temperature [31, 32].

The study primarily focuses on the data integration of production planning and material balance reconciliation systems to improve the petroleum product blends quality.

2 Experiment

The rheological characteristics of an oil products blend were studied with different percentages of the components in the laboratory information management system (LIMS) of LLC LUKOIL-Nizhegorodnefteorgsintez. LUKOIL has been operating in Nizhny Novgorod Region since 2002. Today the Nizhny Novgorod region is one of the key areas for the Company. The region is home to LLC LUKOIL-Nizhegorodnefteorgsintez, one of the largest petroleum refineries in Russia [33].

The kinematic viscosity of liquid petroleum products was measured in accordance with GOST 33-2000 [34]. The density of liquid petroleum products was measured in accordance with ASTM D 4052 [35].

The blend compositions, which where the most similar to the actual blending of heating oil, the content of the components varied. In addition to the viscosity, we consider the density of the blend, since it is often necessary to meet the requirements for the density of fuel oil, for example, according to the standard ISO 8217 [36].

The viscosity was measured at the temperature of 100°C, the density was adjusted to values at 15°C in accordance with standards for the production of heating oil GOST 10585-2013 [37]. Seven blends of six components ($B1$, $B2$, $B3$, $B4$, $B5$, $B6$) were used.

The following notations have been introduced: component $B1$ is combined product of vacuum residue visbreaking unit, $B2$ is vacuum residue of crude distillation unit, $B3$ is light gasoil of FCC unit, $B4$ is vacuum fraction of crude distillation unit, $B5$ is heavy residue of FCC unit, $B6$ is asphalt of vacuum residue deasphalting unit. The main part of the blend (from 50% to 75%) is the combined product of the tar visbreaking unit.

The kinematic viscosity measurement results v^e of the blend at 100°C are shown in Table 1.

Table 1. Results of kinematic viscosity measurement v^e at 100°C.

Blend No	$B1$,%	$B2$,%	$B3$,%	$B4$,%	$B5$,%	$B6$,%	v^e, cSt
$B1$	100						118.4
$B2$		100					909.1
$B3$			100				1.0
$B4$				100			3.7
$B5$					100		10.4
$B6$						100	3379.0
1	55	25	20	0	0	0	35.0
2	65	15	10	10	0	0	37.9
3	60	20	15	0	2	3	46.8
4	50	25	10	10	2	3	46.4
5	65	20	10	5	0	0	51.7
6	55	22	7	10	3	3	55.0
7	75	15	10	0	0	0	63.1

The composition of the blends was varied in such a way that the viscosity of the blended product was around 50 cSt at 100°C. The kinematic viscosity of the blend varied from 34 cSt to 63 cSt with the median of 47 cSt and the boundaries of the first quartile of 38 cSt and the third of 55 cSt, the density was in the range from 998 kg/m^3 to 1011 kg/m^3, median was of 1003 kg/m^3, the first and third quartiles were 1001 kg/m^3 and 1008 kg/m^3.

3   Results and Discussions

Now we will turn to the analysis of optimal planning issues related to the constraints in the model and nonlinear dependencies. We will consider the mathematical programming problem for variables x_1, x_2, ..., x_n, that would ensure the maximum of the objective function

$$L(x_1, x_2, \ldots, x_n) \tag{1}$$

and satisfy the set of constraints

$$\begin{cases} g_i(x_1, x_2, \ldots, x_n, C_l) = b_i, & i = 1, 2, \ldots, m_1 \\ g_j(x_1, x_2, \ldots, x_n, C_l) \geq b_j, & j = m_1 + 1, m_1 + 2, \ldots, m_2, \\ g_k(x_1, x_2, \ldots, x_n, C_l) \leq b_k, & k = m_2 + 1, m_2 + 2, \ldots, m \end{cases} \tag{2}$$

where x_1, x_2, ..., $x_n \geq 0$, b_i, b_j, b_k are constants (describe the limitations of plant utilization, the quality of blends, flows, marketable products, etc.), $g(x_1, x_2, \ldots, x_n, C_l)$ are linear and nonlinear functions containing l parameters C_l [6, 25].

With respect to an industrial enterprise, the $L(x_1, x_2, \ldots, x_n)$ is to be maximized up to L_{max}. The refinery model is described by a system of Eqs. (1) and constraints (2), which number m is several thousand. It should be noted that the dependence of the marginal profit L_{max} on the amount of feed processing in the general case is a non-monotonic function, and can have several local maxima. At the same time, the parameters of the problem, including the parameters of the model for calculating the rheological characteristics, significantly influence the magnitude of the marginal profit and the position of the extremum.

In problems of mathematical programming, when calculating the viscosity of a blend, blending indices of the following type are usually used

$$I_i = \log \log(v_i + C), \tag{3}$$

$$I_b = \sum_i w_i I_{il}, \tag{4}$$

where v_i is the measured kinematic viscosity of i-component, w_i is the volume (or weight) fraction of i-component in the blend, C is the parameter of the calculation formula.

The inverse transformation allows calculating the viscosity of the blend v_i^{blend}

$$v^{blend} = 10^{10^{I_b}} - C.$$

Let us consider some formulas that are widely used in practical calculations.

Walther equation [38] for calculating the viscosity of the blend v^{blend} based on the known values of the kinematic viscosity of i-component can be written as follows

$$\log \log(v^{blend} + C) = \sum_i V_i \log \log(v_i + C), \tag{5}$$

where C is empirical coefficient, V_i is volume fraction of i-componentin the blend.

The Eq. (5) can be easily represented in the form of (3)–(4).

In Refutas equation the blending indices

$$I_i = 23.097 + 33.469 \log \log(v_i + 0.8), \tag{6}$$

are calculated for the weight fractions of the blend w_i, $C = 0.8$ [32].

The RPMS software calculates the kinematic viscosity (which is recommended at 50°C) as per the formula [31]:

$$I_i = 41.10743 - 49.08258 \log \log(v_i + 0.8), \tag{7}$$

for volume fractions of the blend. Other formulas are also available in RPMS in particular such as (3–4).

In should be noted that formulas (3) with parameter $C = 0.7$ are used for the viscosity calculation in accordance with ASTM D7152 [39]

$$\log \log(v^{blend} + 0.7) = \sum_i V_i \log \log(v_i + 0.7), \tag{8}$$

both for the weight and volumetric methods. The summands and multipliers in formulas (6–7), as it is shown in the work [6], do not influence the solution of the problem and serve for the convenience of numerical calculations.

Let us compare the calculating results for the blend viscosity v_j^{blend} according to various formulas (5)–(8) with experimental viscosity v_j^e (j is the experiment number) and choose the optimal parameter C for calculating the viscosity of blends considered in the article. For this purpose, we estimate the absolute relative error RE_j expressed as percentage

$$RE_j = \left| \frac{v_j^e - v_j^{blend}}{v_j^e} \right| \times 100. \tag{9}$$

The Mean absolute percentage error ($MAPE$) for N blends is used as a criterion, when comparing different algorithms for calculating the viscosity with experimental data

$$MAPE = \frac{1}{N} \sum_{j=1}^{N} RE_j. \tag{10}$$

The optimal parameter C in Walther equation, which ensures the minimum $MAPE$ (9), (10) for blends, corresponds to the value $C = 1.06$.

The results of calculations using Eqs. (9) and (10) are given in Table 2. As it follows from Table 2, the best convergence of the calculated and experimental data is achieved using Walther equation with bulk blending. Standard methods with parameters of 0.7 and 0.8 show the worst results.

Formulas (3) and (4) with different parameter values in mathematical programming problems lead to a solution with different optimal blending formulas. If you use incorrect formulas in the planning of blending, then as a result of compounding, either substandard products are obtained, or there is a quality giveaway, which leads to the lost profits.

Table 2. Comparison of $MAPE$ (Mean absolute percentage error) calculating results with the use of various methods for a blend of petroleum products.

Parameter	$MAPE$	Max RE_i
0.70	12.9	17.9
0.80	8.9	12.5
0.97	3.2	6.6
1.06	1.8	4.6

We estimate the effect of the parameter C of the calculation formula (4) on the enterprise's economy. For clarity, we consider the model of a refinery, where the functional dependence on the plant utilization has one extremum. We introduce a dimensionless function, where $L_* = L/L_0$, where L_0 is characteristic value of the marginal profit, rated value of the feed $M_* = M/M_0$, M_0 is the characteristic value of the throughput. The results of the calculation using RPMS system are shown in Fig. 1. The mathematical programming problem (1), (2) was solved in the optimal planning system RPMS (Honeywell) using the XPRESS software package. In the XPRESS, the high-dimensional problem is solved by the Primal-Dual Barrier method. If this method fails to obtain a solution with the required accuracy, then the found solution can be considered as a starting point for solving by the simplex method [40].

Variation of the parameter values C leads to a change in the marginal profit L and the optimal refinery utilization.

The model for viscosity calculating has an effect on the optimal feed utilization of the refinery. As the parameter value increases, the maximum marginal profit $L_{*max} = L_{max}/L_*$ decreases (Fig. 2), upon that the optimal utilization of the refinery increases. The function $M(C)$, describing the dependence of the optimal throughput volume on the parameter C near $C = 1.3$, has discontinuity (in the dimensional variables of the annual plan the discontinuity value is around 100 tons). When using such parameter values, the model ceases to describe the behavior of the real production and requires a detailed analysis of the scenario conditions and constraints in the model (2).

To assess the discrepancy between the optimal planned recipes and real recipes when blending oil products, the angular distance was used. The component composition of the blending recipe from optimal APS can be represented as a vector $\mathbf{B}_{pl}$, and the real recipe from Process Data Reconciliation (PDR) [19] system by the vector $\mathbf{B}_f$. One of the methods for estimating the deviation of the planned recipe from the reconciliation recipe is to compare the angle α between the vectors $\mathbf{B}_{pl}$ and $\mathbf{B}_f$.

The Angular distance α is determined from the ratio

$$\cos \alpha = \frac{(\mathbf{B}_{pl}, \mathbf{B}_f)}{|\mathbf{B}_{pl}||\mathbf{B}_f|}.$$

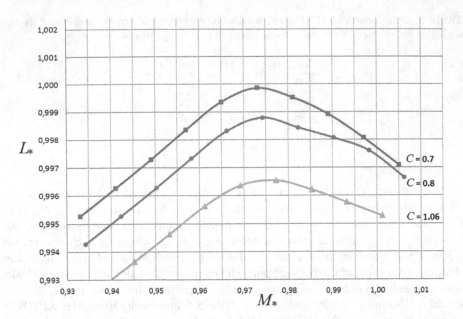

Fig. 1. Dependence of the objective function L_* on the refinery utilization M_*.

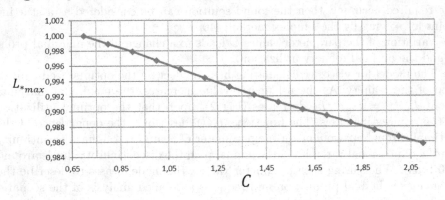

Fig. 2. Dependence of the maximum margin profit L_{*max} on parameter C in the formula for viscosity calculating.

When the value of the angle $\alpha = 0°$, the recipes completely coincide, an increase in the angle α corresponds to an increase in the deviation of the recipes.

If $\alpha > \alpha_*$, where α_* is the critical value, then the deviation from the planned v_i^{pl} and actual values v_i^f of the kinematic viscosity was compared (Fig. 3).

If the deviations are insignificant $\beta_i \leq \beta_*$, where

$$\beta_i = \left| \frac{v_i^{pl} - v_i^f}{v_i^{pl}} \right| \times 100,$$

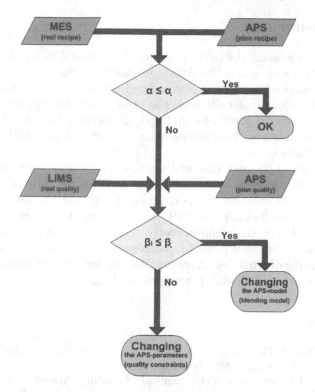

Fig. 3. Blending model improvement algorithm.

then the empirical coefficient C in formula (5) was changed (Fig. 3) in accordance with the above algorithm. To improve the model, it is necessary to use laboratory information management system data or special experimental data. It was necessary to adjust the parameters of the calculation formulas depending on the current state of production. It was found that in a certain variation range of the viscosity index, the value of the refinery optimal utilization varies step-wise, the model was structurally unstable.

The application of the methodology described in the article made it possible to reduce the angular distance when mixing fuel oil by more than two times.

The selection of the relevant parameters improved the accuracy of the production and economic planning of the refinery, made it possible to optimize the costs while compounding petroleum products. For a refinery with the throughput of 16 million tons of hydrocarbon feed per year, the change in the coefficient from 0.8 to 1.1 will result in the increase of the optimal utilization by 100 tons per year. At the same time, the economic results will differ by 1\$ per ton of petroleum fuel.

4   Conclusion

It is proposed to monitor the deviation of data integrated into a single information system of the system of optimal planning and material balance. To assess the discrepancy between the optimal planned recipes and real recipes when blending oil products, the angular distance was used. If the discrepancy is large, the optimal planning model should be updated. To improve the model, it is necessary to use as a laboratory information management system (LIMS) and special experimental studies.

The adequacy of the mathematical models of a refinery with various formulas for calculating the blend viscosity on the basis of the full-scale experiment was verified. The calculation of the marginal profit and utilization of refinery process units showed its high sensitivity to the change in the parameters in the formulas for predicting the blend viscosity. The selection of appropriate formulas and parameters makes it possible to increase the accuracy of production and economic planning of refineries, make calculations for capacity management and optimize costs for compounding petroleum products.

References

1. Joly, M., Odloak, D., Miyake, M.Y., et al.: Refinery production scheduling toward Industry 4.0. Front. Eng. **5**(2), 202–213 (2018)
2. Honeywell Refinery and Petrochemical Modeling System (RPMS). www.processonline.com.au/content/software-it/product/honeywell-refinery-and-petrochemical-modeling-system-rpms-1335653382. Accessed 11 Oct 2021
3. Aspen PIMS. www.aspentech.com/en/products/pages/aspen-pims. Accessed 11 Oct 2021
4. Haverly GRTMPS. www.haverly.com/planning. Accessed 11 Oct 2021
5. Aveva. Plan and Schedule. www.aveva.com/en/ solutions/operations/refinery-planning-scheduling. Accessed 11 Oct 2021
6. Kuvykin, V.I., Bryukhanov, M.V., Kuvykina, E.V., Piskunov, I.V., Sychev, A.G.: Updating the viscosity calculations for blends of heavy petroleum products in the production planning system of the refinery. World Pet. Prod. Bull. Oil Co. **9**, 25–31 (2017)
7. Kuzora, I.E., Dubrovskiy, D.A., Semenov, I.A., et al.: Reduction of the quality giveaway of petroleum fuels based on the results of the precise assessment of viscosity parameters of heavy components. World Pet. Prod. Bull. Oil Co. **9**, 25–31 (2017)
8. Centeno, G., Sánchez-Reyna, G., Ancheyta, J., et al.: Testing various mixing rules for calculation of viscosity of petroleum blends. Fuel **90**, 3561–3570 (2011)
9. Essien, G., Kuye, A.: Development and validation of linear programming models for gasoline and fuel oil blending. Int. J. Sci. Eng. Res. **7**(7), 1204–1209 (2016)
10. Stratiev, D., et al.: Dependence of visbroken residue viscosity and vacuum residue conversion in a commercial visbreaker unit on feedstock quality. Fuel Proces. Technol. **138**, 595–604 (2015)
11. Piskunov, I.V., Bashkirceva, N.Y., Emelyanycheva, E.A.: The mathematical modeling of bitumen properties interrelations (Review). J. Chem. Technol. Metall. **57**(3), 464–479 (2022)

12. Andronov, A.A., Pontryagin, L.S.: Crude systems. Rep. Sci. Acad. USSR **14**(5), 247–250 (1937)
13. Arnold, V.I.: Hard and Soft Mathematical Models. Moscow Center for Continuous Mathematical Education, Moscow (2004)
14. Kuvykin, V.I., Kuvykina, E.V., Petukhov, M.Y.: Analysis of optimal solutions in problems of nonlinear programming. Bull. Lobachevsky State Res. Univ. Nizhny Novgorod **4–5**, 2285–2286 (2011)
15. Joly, M.: Refinery production planning and scheduling: the refining core business. Braz. J. Chem. Eng. **29**(02), 371–384 (2012)
16. Kuvykin, V.I.: Optimal planning and analysis of continuous production models. Autom. Remote. Control. **79**(2), 384–390 (2018). https://doi.org/10.1134/S0005117918020170
17. Piskunov, I.V., Shamanin, M., Bashkirtseva, N.Y.: Development of It-systems for planning and control of technological processes in oil refining. Bull. Technol. Univ. **24**(10), 62–71 (2021)
18. Câmara, M.M., et al.: Numerical aspects of data reconciliation in industrial applications. Processes **5**(4), 56 (2017)
19. Narasimhan, S., Jordache, C.: Data Reconciliation and Gross Error Detection: An Intelligent Use of Process Data. Golf Publishing Company, Houston (2000)
20. Production Balance PIN - Honeywell Process Solutions. http://paperzz.com/doc/6820600/production-balance-pin-honeywell-process-solutions. Accessed 11 Oct 2021
21. ROMeo Material Balance. www.utitech.com.tw/download/DataSheet/AVEVA/O&O/AVEVA%20Proces%20Optimization/%E7%89%A9%E6%96%99%E5%B9%B3%E8%A1%A1.pdf. Accessed 11 Oct 2021
22. Data Reconciliation Management System - I-DRMS. http://indusoft.ru/en/products/indusoft/ Accessed 11 Oct 2021
23. Sigmafine. https://sigmafine.pimsoftinc.com/ Accessed 11 Oct 2021
24. Aspen Operations Reconciliation and Accounting, www.aspentech.com/en/products/msc/aspen-operations-reconciliation-and-accounting Accessed 11 Oct 2021
25. Kuvykin, V.I., Petukhov, M.Y.: Improving the quality of process models in oil refinery information systems. Int. J. Qual. Res. **13**(3), 539–552 (2019)
26. Chu, H., You, F.: Model-based integration of control and operations: overview, challenges, advances, and opportunities. Comput. Chem. Eng. **83**(1), 2–20 (2015). https://doi.org/10.1016/j.compchemeng.2015.04.011
27. Martinez, M.E.C., Aranda, D.A., Gutierrez, L.G.: IT integration, operations flexibility and performance: an empirical study. J. Ind. Eng. Manage. **9**(3), 684–707 (2016). https://doi.org/10.3926/jiem.1869
28. Kuvykin, V.I., Logunov, P.L.: Data reconciliation for refinery material balance. Autom. Inf. Fuel Energy Complex **2**(583), 41–48 (2022)
29. Centeno, G.: Testing various mixing rules for calculation of viscosity of petroleum blends. Fuel **90**(12), 3561–3570 (2011)
30. Piskunov, I.V., Kuvykin, V.I., Tankov, D.Y., Tychkin, A.A., et al.: Methods for calculation of the viscosity of oil and residualpetroleum products. Oil Refining Petrochemistry **1**, 25–38 (2022)
31. Antonchenkov, V.P.: Calculation of fuel viscosity in the linear programming model. Chem. Technol. Fuels Oils **41**(4), 323–324 (2005)
32. Maples, R.E.: Petroleum refinery process economics. PennWell Corp (2000)
33. LLC LUKOIL-Nizhegorodnefteorgsintez. https://nnos.lukoil.ru Accessed 11 Oct 2021

34. GOST 33–2000 (ISO 3104–94): Petroleum products. Transparent and opaque liquids. Determination of the kinematic viscosity and calculation of dynamic viscosity
35. ASTM D4052: standard test method for density and relative density of liquids by digital density meter
36. ISO 8217:2010: petroleum products – Fuels (class F) – Specifications of marine fuels (MOD)
37. GOST 10585–2013: petroleum fuel. Mazut
38. Walther, C.: The evaluation of viscosity data. Erdol und Teer **7**, 382–384 (1931)
39. ASTM D7152: standard practice for calculating viscosity of a blend of petroleum products
40. Tsodikov, Y.M.: Successive linear programming method efficiency in solving the problems of production planning at oil refinery. Control Sci. **6**, 55–61 (2018)

Supercomputer Simulation

High-Performance Graph Coloring on Intel CPUs and GPUs Using SYCL and KOKKOS

Anastasia Kurnikova[✉], Anna Pirova, Valentin Volokitin, and Iosif Meyerov

OneAPI Center of Excellence, Lobachevsky State University of Nizhny Novgorod, Nizhny Novgorod, Russia
alxndrvna@icloud.com

Abstract. A variety of high-performance computing devices opens up many opportunities for numerical simulation in science and industry. Therefore, there is a growing need to investigate approaches for heterogeneous programming that allow developing and using a single code for different devices, sometimes with minor modifications to improve performance. In this paper, we explore the efficiency of using SYCL (Data Parallel C++) and KOKKOS to develop portable codes solving the graph coloring problem. We employ the commonly used Catalyurek algorithm, implement it on C++/OpenMP, SYCL (DPC++) and KOKKOS, and evaluate performance on Intel CPUs and GPUs. The paper discusses approaches of optimization performance that allow getting portable codes that run efficiently on both CPUs and GPUs. It is shown that performance portability takes place for this algorithm. We hope that our results can be used by researchers who implement graph algorithms for high-performance computing devices of various architectures.

Keywords: Graph coloring · SYCL · KOKKOS · OpenMP · CPU · GPU · Performance analysis and optimization

1 Introduction

Great progress has been made in the area of heterogeneous computing in the last decade. So, nowadays various languages, libraries and frameworks for heterogeneous programming are developed and actively used. They make it possible to use processors and co-processors of various architectures for computations. At the same time, the problem of code portability between different hardware platforms is still not totally resolved. On the one hand, we have powerful frameworks and toolkits (OpenCL [1], OpenACC [2], OneAPI [3], KOKKOS [4], Alpaka [5], HIP [6], HPX [7] and others) that allow us to write a single code for different devices. On the other hand, performance portability is an open question. So, we cannot be sure that code originally optimized for x86 CPUs run with at least acceptable performance on GPUs without several rounds of fine-tuning. It is also interesting how difficult are the improvements that we need to make for each specific device.

In this paper we discuss these questions for one of the key graph theory problems. The area of graph theory applications is rapidly expanding, the number of problems

D. Balandin et al. (Eds.): MMST 2022, CCIS 1750, pp. 253–265, 2022.
https://doi.org/10.1007/978-3-031-24145-1_21

that can be solved by means of graph theory methods is increasing. At the same time, many algorithms in this area are very difficult to optimize for modern high-performance systems. So, graph algorithms are characterized by low arithmetic intensity, irregular memory access, imbalance of the computational load in decomposing into subtasks and distributing them among threads. These and other problems lead to insufficient use of the modern HPC systems potential [8]. This paper considers the problem of finding the chromatic number of a graph that is the minimal number of colors needed to mark all the vertices.

The graph coloring problem is NP-hard [9], and in practice greedy algorithms [10], contraction algorithms [11], polynomial algorithms [12] and others are used to solve it. High-performance applications use parallel graph coloring algorithms for computing systems with shared and distributed memory. In this paper we consider the Catalyurek algorithm that is one of the best algorithms for systems with shared memory [13]. At first, we implement this algorithm using C++ with OpenMP considering this implementation as the baseline. Next, we develop portable codes using the KOKKOS library and the SYCL language, compare their performance on Intel CPUs with the baseline implementation and move on to experiments on Intel GPUs. We pose the following key questions:

1. Is there any overhead when switching to KOKKOS and SYCL?
2. What code optimization techniques on KOKKOS and SYCL can speed up calculations on Intel CPUs?
3. What performance can be achieved when switching to Intel GPUs? Is fine-tuning required?
4. Do we end up with an implementation that works well on both CPUs and GPUs?

The paper is organized as follows. In Sect. 2 we review related work and developments in the area of solving the graph vertex coloring problem. In Sect. 3 we present implementations of the Catalyurek parallel graph coloring algorithm using OpenMP, SYCL and KOKKOS, and discuss performance optimization techniques. Section 4 presents the numerical results. Section 5 concludes the paper.

2   Background and Related Work

At first, we consider the graph vertex coloring problem. Assume that a graph $G = (V, E)$ is given with a set of vertices V and a set of edges E. It is required to find such partition of the vertex set V into L non-intersecting subsets when each subset does not contain adjacent vertices. If now each of these subsets is marked with a certain color (an integer number), then all vertices inside one subset can be marked with only this color. Now we get the *Distance* $- 1$ vertex coloring of the graph (the type of the vertex coloring when the correctness of the resulting coloring is checked by colors of vertices that are adjacent through one edge to this vertex) [14]. The minimal possible number of colors L is called the chromatic number of the graph. Finding a correct optimal coloring is an NP-hard problem, so in practice heuristic algorithms are applied for large graphs.

The most preferable sequential graph coloring algorithm is the greedy algorithm. It makes locally optimal decisions at each step assuming that the final solution is also optimal. Parallelization of this algorithm raises lots of problems. For example, the simplest version of the parallel greedy algorithm requires $O(n)$ synchronization iterations in the worst case [10].

The pseudocode of the sequential greedy algorithm is as follows:

Algorithm 1 Sequential greedy coloring algorithm

---

1. Input: $G(V, E)$

2. $U = V$ **>** uncolored set

3. $for\ \forall v \in U$:

4. $v_i \to w(v_i)$ **>** give a priority

5. $\forall u \in adj(v_i): w(u) > w(v_i)\ C = C \backslash c(u)$ **>** available colors

6. $c(v_i) = \min C$ **>** minimal available color

7. $U = U \backslash v_i$ **>** remove from uncolored

At first, all vertices are marked as uncolored (Algorithm 1, line 2). Next, they are reordered according to the priority function (Algorithm 1, line 4), and each vertex in order of its priority gets a minimal available color (Algorithm 1, lines 5–6). After that, the colored vertex is removed from the uncolored set (Algorithm 1, line 7).

Jones and Plassmann suggested a parallel vertex coloring algorithm based on the choice of independent sets [15]. The vertices in each set are colored together. The Jones-Plassmann algorithm requires a large number of synchronization iterations, but allows the use of priority functions to improve the quality of the coloring (to reduce the number of colors).

The pseudocode of the parallel Jones-Plassmann algorithm is as follows:

Algorithm 2 The parallel graph coloring Jones-Plassmann algorithm

---

1. Input: $G(V, E)$

2. $\forall v \in V \to w(v)$ **>** give a priority

3. $U = V$ **>** uncolored set

4. while $U \neq \emptyset$:

5. $I = \{z: w(z) > w(u)\ for\ \forall u \in adj(z) \cap U\}$ **>** independent set

6.   parallel for $z \in I$: $c(z)$ **>** minimal available color

7. $U = U \backslash I$ **>** remove from uncolored

Here, in the first step we assign to each vertex a priority according to the priority function (Algorithm 2, line 2) and mark all vertices as uncolored (Algorithm 2, line

3). Next, we choose an independent set of vertices which are not connected to each other by any edge (Algorithm 2, line 5), and mark them in parallel with the minimal available color (Algorithm 2, line 6). After that the colored vertices are removed from the uncolored set (Algorithm 2, line 7), and we repeat the iteration, so the neighbors of already colored vertices become colored in order of their priority. The actions are repeated until the uncolored set becomes empty (Algorithm 2, line 4).

Catalyurek suggested a parallel algorithm based on conflict solving. At first, all vertices are divided into equal blocks and colored in parallel by threads. Next, threads are synchronized and we detect the conflicts of coloring. On the next iteration the previously identified conflict vertices are recolored by threads in parallel. The Catalyurek algorithm is suitable for any shared-memory system including multicore platforms and allows to achieve good performance [13].

The pseudocode of the parallel Catalyurek algorithm is as follows:

Algorithm 3 The parallel graph coloring Catalyurek algorithm

---

1. Input: $G(V, E)$

2. $N / p: V_i \in thread_i$ > divide into threads

3. $U = V$ > uncolored set

4. while $U \neq \emptyset$:

5.   parallel for $\forall v \in V_i \cap U : c(v)$ > minimal available color

6.   barrier   > synchronize

7. $R = \emptyset$ > conflicts set

8.   parallel for $\forall v \in V_i \cap U$:   > detect conflicts

9.   if $\{(w, v) \in E : c(v) == c(w), v > w\} : R = R \cup \{v\}$

10.   barrier   > synchronize

11. $U = R$ > uncolored set = conflicts

According to it, we pre-divide all vertices into equal blocks (Algorithm 3, line 2) and mark them by threads in parallel with the minimal available colors (Algorithm 3, line 5) taking into account vertex connection to each other (we perform the sequential greedy coloring on each thread). After that, the threads are synchronized (Algorithm 3, line 6), and we detect vertices that are incorrectly colored at the blocks junctions (Algorithm 3, lines 8–9). These vertices are marked as a conflicts set and we recolor them on the next iteration (Algorithm 3, line 11). The actions are repeated until the conflicts set after synchronization becomes empty (Algorithm 3, line 4).

Nowadays, some modifications of this algorithm are known. For example, Rokos, Gorman and Kelly presented an algorithm [16] that reduces the number of synchronizations in the basic Catalyurek algorithm. The modified algorithm does improve performance on sparse graphs, but in general case the basic Catalyurek algorithm shows quite balanced results, so we consider it below.

3 Implementation of Coloring Algorithms on OpenMP, SYCL and KOKKOS

3.1 Overview

Parallel algorithms can have two strategies of coloring vertices depending on the requirements for the resulting coloring. If it is required to mark the graph with the minimal possible number of colors, then the algorithm is usually based on a choice of the minimal color that is not yet used in coloring neighbor vertices. If the goal of the coloring is a balanced vertex coloring, then the algorithm chooses the minimally used available color to balance a number of each color vertices. We consider the Catalyurek algorithm focused on the first strategy.

The Catalyurek algorithm is already parallel and can be easily implemented for systems with shared memory. However, we still should care about how to implement the code in the most effective way and think out what tools and frameworks we need to provide performance and portability between different hardware architectures. Nowadays, there are several languages and open-source software libraries for parallel programming, whose developers claim the possibility of writing a single code for various computing devices including CPUs and GPUs. That is an indisputable advantage of such parallelization tools.

This paper explores the developing a portable code that runs with good performance on Intel CPUs and GPUs. To do this, we use the SYCL programming language [17] (we consider the Intel OneAPI implementation which is Data Parallel C++, hereinafter referred as DPC++) and the KOKKOS library [18]. The main interest is whether the code that is originally optimized for CPUs runs efficiently on GPUs, and whether it requires any GPU-specific optimizations.

3.2 Implementation on C++

At first, we implement the basic Catalyurek algorithm (Algorithm 3) using C++ and OpenMP. For storing graphs, we use an adjacency list structure. Note that we implement the code using KOKKOS and SYCL with the same data structure. The parallel algorithm for multicore CPUs is based on OpenMP. We implement it using the "parallel for" construct (Algorithm 3, lines 5–6, 8–10). Such mean of parallelization is straightforward, but, as it is shown below, quite effective.

3.3 Implementation on SYCL

SYCL [3] is a high-level programming model for improving performance on different devices. It is based on the idea of programming a single code for all types of devices. Data Parallel C++ is the oneAPI Implementation of SYCL. SYCL borrowed the concepts of portability and efficiency from OpenCL [1]. It allows the CPU to rely on the runtime without any specific compiler. SYCL extends the power of C++ by freeing the programmer from explicitly passing data between the host and devices. SYCL provides single-source programming where C++ template functions can contain both host and device code, and allows them to be reused in source code for different data types.

The implementations using KOKKOS and SYCL are organized in other way, in the style of GPU programs. To port the code to KOKKOS and SYCL, we need to answer two main questions: how to work with memory and how to implement kernel functions responsible for performing one portion of work. Now we show how these problems are resolved.

To work with memory in SYCL (DPC++), we use the Unified Shared Memory (USM) extension. It supports memory management through pointers. USM has two models: automatic data movement between host and device in shared memory and manual data movement in and out of separate device memory. Here we use them both.

To modify the Catalyurek algorithm for the SYCL model we employ the NDRange loop model (many loops become one) and the single-loops parallelization. SYCL, as well as the KOKKOS model, independently allocates the amount of work to the available physical resources. The transfer of computations is implemented through C++ functors. The template takes the amount of work and distributes it among the functors. Additionally, we refine mechanisms for allocating and freeing memory to obtain better datary use and improve performance (Listing 4). We also develop two special kernel functions for resolving conflicts and coloring the sub-graphs, respectively.

3.4 Implementation on KOKKOS

KOKKOS [4] is a model for parallel programming on C++. As well as SYCL, it is based on the idea of programming a single code for all types of devices. It is based on the C++ language library with added parallelism. As well as SYCL, KOKKOS extends the power of C++ by freeing the programmers from passing data between the host and devices and does it inside its structure.

To modify the Catalyurek algorithm for the KOKKOS model, we use the MDRange-Policy model (the KOKKOS ranking tool for parallelizing densely nested loops) and the single-loop parallelization (Listing 3, line 1). The main difference between the implementation on KOKKOS and the implementation on C++ with OpenMP is that KOKKOS independently distributes workload among execution resources. For a single loop, each iteration is taken as a unit of work. It is identified by the loop iteration number. The cycle range determines the total amount of work. We need to indicate the range of iterations and the body of calculations. Further, KOKKOS determines how the work is distributed according to the available resources. Computational bodies are given as C++ functors. The template takes the work items and distributes them over the functors one by one. For densely nested loops, we specify the dimension of the loops space, the lists of initializers for the start and the end of each loop, and the functors take the appropriate number of indexes.

3.5 Optimizations

We also implement technical modifications to improve the performance. First, for the conflict resolving step we change the array of indexes of vertices that have the same color as the neighbor vertex and the array of conflict detection flags to a bit shift by the conflict index number [19]. Listing 1 and Listing 2 show the structure of the conflict detection step before and after the modification.

Hereby, we manage to get by with one auxiliary element (Listing 2, line 8) instead of two as it was before (Listing 1, lines 8–9), and also change the bool array (Listing 1, line 8) to the bit shift (Listing 2, line 8).

```
1.for (int i = 0; i < nv; i++) {
2.     c = colors[i];
3.     for (int j = gr.Xadj[i]; j < gr.Xadj[i + 1]; j++)
4.         if (colors[col[j]] == c) {
5.             if (i < col[j]) conflictIndex = i;
6.             else conflictIndex = col[j];
7.             ... // check if index is already handled
8.             isDetected[conflictIndex] = true;
9.             out[index++] = conflictIndex;
10.        }
11. }
```

Listing 1. The standard implementation of the conflict detection step

```
1.for (int i = 0; i < nv; i++) {
2.     c = colors[i];
3.     for (int j = gr.Xadj[i]; j < gr.Xadj[i + 1]; j++)
4.         if (colors[col[j]] == c) {
5.             if (i < col[j]) conflictIndex = i;
6.             else conflictIndex = col[j];
7.             ... // check if index is already shifted
8.             out >> conflictIndex;
9.        }
10. }
```

Listing 2. The modified implementation of the conflict detection step

```
1.h.parallel_for<...>(range<1>(N), [=](id<1> i) {
2.    c = setMinAvailableColor(...);
3.    colors[i] = c;
4.});
```

Listing 3. The standard implementation of the part of coloring step

```
1.range<1> glo(N);
2.range<1> loc(B);
3.auto g = local_accessor<int, 1>(range<1>(B), h);
4.h.parallel_for<...>(nd_range<2>(glo, loc), [=](nd_item<2> item) {
5.    i = item.get_global_id();
6.    j = item.get_local_id();
7.    for (int k = 0; k < N/B; k++) {
8.        ... // load graph parts into local memory
9.        item.barrier(access::fence_space::local_space);
10.        for (int r = 0; r < B; r++) {
11.            c = setMinAvailableColor(...);
12.            colors[i] = c;
13.        }
14.        item.barrier(access::fence_space::local_space);
15.    }
16. });
```

Listing 4. The modified implementation of the part of coloring step

Secondly, we should think about smart memory management and implement caching for the primary coloring and recoloring steps to improve efficiency of working with memory. In this case, it is necessary to take into account the difference in memory architecture in CPUs and GPUs. On CPUs, we must ensure that the data loaded into the cache is reused as much as possible. On GPUs, the local memory needs to be used efficiently. At first glance, it may seem that these differences require the development and subsequent support of two implementations. However, it turns out that this can be avoided. To do this on KOKKOS and SYCL in this particular case, we can focus on the architecture of GPUs. For this we need to make an auxiliary local block, due to that the data is prepared for loading into local memory of GPU in advance. As it is shown below, such an implementation works efficiently on CPUs as well because the data loaded into the cache is reused in further calculations.

Listing 3 and Listing 4 show the structure of the step where we create a local buffer for data caching. Hereby, the graph is divided into parts (Listing 4, line 2) that we gradually load into local memory inside the loop (Listing 4, lines 7–9) and they place there entirely. We achieve better performance than in previous case by experimental graph parts size selection.

4 Numerical Results

4.1 Preliminaries

The experiments are performed on three graphs with the number of vertices from 8.22 million to 14.76 million from the Suite Sparse collection (Table 1) on the nodes of the Intel DevCloud cluster with Intel Xeon Gold 6338 CPU (2400 MHz, 192 GB, 2S, 32 cores) and Intel Iris XE Max GPU (1650 MHz, 4 GB, 68 GB/s, 96 EUs) [20].

Table 1. Test graphs characteristics

	Number of vertices	Non-sparseness percentage ($|E|/\left|V^2\right|$), %
dersame	8 222 012	50.15
mianse2	11 054 532	46.97
nu16ddk	14 758 344	45.14

4.2 Experiments on CPU

Figure 1 shows the running time of several implementations of the Catalyurek algorithm for three test graphs on Intel Xeon Gold CPU. We found that for all implementations on CPU and GPU the number of colors (the quality) of the coloring is exactly the same. Therefore, we need to compare only the computation time.

The results show that the implementation based on OpenMP scales well at least up to 32 CPU cores. We observe 87% of strong scaling efficiency on all three test graphs. We also found that in CPU runs the KOKKOS and SYCL implementations ("*default kokkos*" and "*default sycl*", respectively) are not worse in performance than the OpenMP implementation ("*default openmp*") with the same number of colors (Fig. 1). It means that in this case KOKKOS and SYCL add almost no overhead while allowing us to build and run the code on the GPU. Then we compare the performance with the implementation of the algorithm from the KOKKOS graph library ("*kokkos lib*") [19] and found that there is a room for optimizations, described above in Sect. 3. The experiments show that the bit shift optimizations ("*bit openmp*", "*bit kokkos*" and "*bit sycl*" for the OpenMP, KOKKOS and SYCL implementations, respectively) lead to much better performance. We also found that cache-friendly implementations on SYCL and KOKKOS ("*bitcache sycl*" and "*bitcache kokkos*") achieve almost the same performance as the implementation from the KOKKOS graph library.

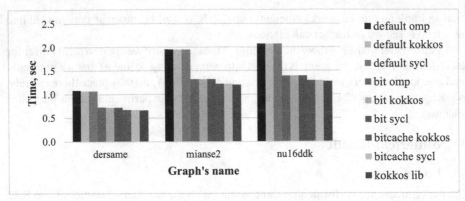

Fig. 1. Running time of several implementations of the Catalyurek algorithm on Intel Xeon Gold 6338 CPU. Codes are parallelized with OpenMP, KOKKOS, and SYCL. The *"default"* prefix in the name of the code corresponds to the reference implementations, the *"bit"* prefix corresponds to implementations with bit shifts optimizations, the *"bitcache"* prefix denotes final implementations with caching and bit shifts. The results are compared with coloring function from the KOKKOS graph library (*"kokkos lib"*).

4.3 Experiments on GPU

Then we run the experiments on Intel Iris XE Max GPU (Fig. 2). We start from the reference implementations of the Catalyurek algorithm using KOKKOS and SYCL (*"default kokkos"*, *"default sycl"*) and the coloring function from the KOKKOS library (*"kokkos lib"*) on the GPU to check how the reference codes perform on GPU. As expected, the *"kokkos lib"* implementation outperforms the others. After that we run the implementations optimized with bit shifts (*"bit kokkos"*, *"bit sycl"*) to compare the optimization results. Finally, we test cache-friendly implementations (*"bitcache kokkos"*, *"bitcache sycl"*).

We found that with the bit operations it becomes possible to get about 1.5x speedup compared to the reference implementations and to get closer to the performance of the coloring function from the KOKKOS library. After adding in loops data caching that reduces the number of expensive data copies (*"bitcache kokkos"*, *"bitcache sycl"*), we manage to get a speedup compared to the previous implementation of about 1.3x, compared to the coloring function from the KOKKOS library of about 1.2x. This speedup is likely due to more accurate fine-tuning of cache-blocking parameters.

Eventually, the final versions with SYCL and KOKKOS make it possible to achieve the same number of colors and to provide good performance on both CPU and GPU.

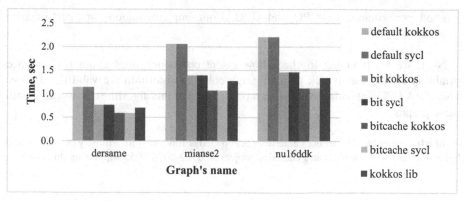

Fig. 2. Running time of several implementations of the Catalyurek algorithm on Intel Iris XE Max GPU. Codes are parallelized with KOKKOS and SYCL. The *"default"* prefix in the name of the code corresponds to the reference implementations, the *"bit"* prefix corresponds to implementations with bit shifts optimizations, the *"bitcache"* prefix denotes final implementations with caching and bit shifts. The results are compared with coloring function from the KOKKOS graph library (*"kokkos lib"*).

5 Conclusion

In this paper we considered the Catalyurek parallel algorithm for graph vertex coloring and implemented it on C++ with OpenMP, KOKKOS, and SYCL (DPC++) to compare the performance on Intel CPUs and GPUs and understand if this algorithm can be implemented as a performance portable code across these hardware architectures.

We developed several implementations on OpenMP, KOKKOS, and SYCL, starting from straightforward versions, and gradually improving them by employing bit shifts for the conflict resolving step and cache-friendly memory management. We tested the implementations on Intel Xeon Gold 6338 CPU and Intel Iris XE Max GPU and compared the performance of the developed codes with each other and also with the algorithm implementation from the KOKKOS graph library. Our main findings are the follows:

1. All the implementations achieve the same number of colors, therefore we can focus only on the performance comparison;
2. The KOKKOS and SYCL implementations of the Catalyurek algorithm do not add any overhead compared to our reference OpenMP implementation;
3. Bit shifts and cache-friendly memory management greatly improve the performance of KOKKOS and SYCL codes both on CPU and GPU. The final results are in good agreement with the implementation from the KOKKOS graph library, outperforming it by 20% on Intel Iris XE Max GPU;
4. A direct comparison of the CPU and GPU performance is difficult here because these devices have different architectures and computing capabilities. However, the GPU results meet the expectations that are based on the hardware specifications;
5. The memory management scheme was developed for GPU but worked also well on CPU. Finally, we found that both KOKKOS and SYCL implementations achieve

good performance on CPU and GPU. Both implementations are performance portable.

Next, we are planning to check how extent our conclusions about performance portability for the Catalyurek parallel graph coloring algorithm are valid if we move to NVIDIA GPUs, and also to consider modifications of this algorithm for working with sparse graphs.

Acknowledgements. The authors gratefully acknowledge funding from Ministry of Science and Higher Education of the Russian Federation project № 0729-2020-0055 supporting this work.

References

1. Stone, J.E., Gohara, D., Shi, G.: OpenCL: a parallel programming standard for heterogeneous computing systems. Comput. Sci. Eng. **12**(3), 66 (2010)
2. Farber, R.: Parallel programming with OpenACC. Newnes (2016)
3. Nozal, R., Bosque, J.L.: Exploiting co-execution with OneAPI: heterogeneity from a modern perspective. In: Sousa, L., Roma, N., Tomás, P. (eds.) Euro-Par 2021. LNCS, vol. 12820, pp. 501–516. Springer, Cham (2021). https://doi.org/10.1007/978-3-030-85665-6_31
4. Edwards, H.C., Trott, C.R.: Kokkos: enabling performance portability across manycore architectures. In: Extreme Scaling Workshop, pp. 18–24. IEEE (2013)
5. Zenker, E., et al.: Alpaka--an abstraction library for parallel kernel acceleration. In: IEEE International Parallel and Distributed Processing Symposium Workshops, pp. 631–640 (2016)
6. HIP C++ Runtime API and Kernel Language. https://github.com/ROCm-Developer-Tools/HIP. Accessed 18 Sep 2022
7. HPX C++ Standard Library for Concurrency and Parallelism. https://github.com/STEllAR-GROUP/hpx. Accessed 18 Sep 2022
8. TOP500 Modern HPC systems. https://www.top500.org/lists/top500/2022/06/. Accessed 15 July 2022
9. Dandashi, A., Al-Mouhamed, M.: Graph coloring for class scheduling. In: ACS/IEEE International Conference on Computer Systems and Applications, pp. 1–4 (2010)
10. Culberson, J.: Iterated greedy graph coloring and the difficulty landscape (1992)
11. Gabow, H.N., Galil, Z., Spencer, T.H.: Efficient implementation of graph algorithms using contraction. J. ACM **36**(3), 540 (1989)
12. Gutjahr, W., Welzl, E., Woeginger, G.: Polynomial graph-colorings. Discret. Appl. Math. **35**(1), 29 (1992)
13. Çatalyürek, Ü.V., et al.: Graph coloring algorithms for multi-core and massively multithreaded architectures. Parallel Comput. **38**(10–11), 576 (2012)
14. Gebremedhin, A.H., et al.: Colpack: software for graph coloring and related problems in scientific computing. ACM Trans. Math. Softw. **40**(1), 1 (2013)
15. Jones, M.T., Plassmann, P.E.: A parallel graph coloring heuristic. SIAM J. Sci. Comput. **14**(3), 654 (1993)
16. Rokos, G., Gorman, G., Kelly, P.H.J.: A fast and scalable graph coloring algorithm for multi-core and many-core architectures. In: Träff, J.L., Hunold, S., Versaci, F. (eds.) Euro-Par 2015. LNCS, vol. 9233, pp. 414–425. Springer, Heidelberg (2015). https://doi.org/10.1007/978-3-662-48096-0_32
17. SYCL Overview - The Khronos Group Inc. https://www.khronos.org/sycl/. Accessed 16 July 2022

18. Kokkos C++ Performance Portability Programming EcoSystem. https://github.com/kokkos. Accessed 16 July 2022
19. Kokkos Graph Library Coloring Function. https://github.com/kokkos/kokkos-kernels/blob/master/src/graph/KokkosGraph_Distance1Color.hpp. Accessed 30 July 2022
20. Specifications of Intel DevCloud. https://software.intel.com/content/www/us/en/develop/tools/devcloud.html. Accessed 17 Mar 2022

Automated Debugging of Fragmented Programs in LuNA System

Victor Malyshkin[1,2,3] , Andrey Vlasenko[1,2(✉)] , and Mihail Michurov[2]

[1] Institute of Computational Mathematics and Mathematical Geophysics SB RAS,
Novosibirsk, Russian Federation
[2] Novosibirsk State University, Novosibirsk, Russian Federation
`a.vlasenko@g.nsu.ru`
[3] Novosibirsk State Technical University, Novosibirsk, Russian Federation
`https://www.nsu.ru/n/information-technologies-department/`

Abstract. The LuNA system, which was created in ICMMG SB RAS, follows the approach of fragmented programming. The LuNA-program runs in parallel, but the programmer does not specify the behaviour of individual processes or threads when creating it. Instead, the user defines the content of computational fragments that may have dependencies on each other. Then, during the execution of the LuNA-program, the runtime system allocates independent computational fragments and distributes them to computing nodes and cores of the multicomputer.Some properties of the system play significant role, e.g. LuNA is the single assignment language and the execution order of operators in the subprogram body is undefined in general case. That is why LuNA-programs are characterized by specific errors. They are not peculiar neither to sequential programs, nor to parallel in classical technologies (MPI, OpenMP etc.) The paper contains classification of semantic errors that are specific for fragmented programs. The analysis of the various approaches applicability to automated debugging in the LuNA system is given. The paper also describes the operation principle of the tool created by the authors for detecting some popular fragmented program errors. The work of the tool is shown on the example of a test programs with different errors. Since the debugging tool is based on a "post-mortem" analysis, it is important to evaluate overhead. The evaluation results are also given in the paper. The directions of further work are described.

Keywords: Automated debugging · Post-mortem analysis · Fragmented programs · LuNA system

1 Fragmented Programs and the LuNA System

The problem considered in the article is the detection of semantic errors in fragmented programs on the example of the LuNA (Language for Numerical Algorithms) parallel programming system.

This work was carried out under state contract with ICMMG SB RAS 0251-2021-0005.

The system of fragmented programming LuNA [1, 2] was created in the Institute of Computational Mathematics and Mathematical Geophysics (Siberian Branch of the Russian Academy of Sciences). "System of fragmented programming" means that a program for the system is represented in the form of a fragmented algorithm, which consists of data fragments (DFs) and computational fragments (CFs). One can draw some analogy with conventional programming languages (C/C++, Fortran, etc.) by mapping DFs to variables, and CFs to calls of procedures. CFs accept as inputs some data fragments and produce others (called output DFs). In other words, a fragmented algorithm may be described by the Fig. 1.

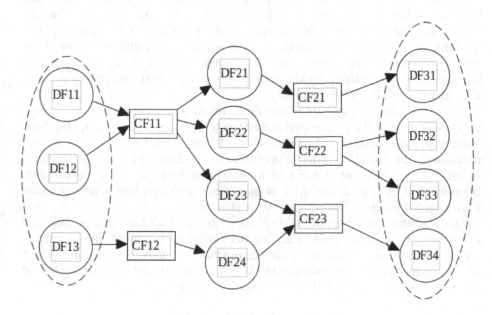

Fig. 1. Diagram of a fragmented algorithm.

The elements «DF xy»on the diagram are data fragments. Thus, the ovals on the left and on the right encircle the sets of input and output DFs for the program. Elements «CF xy»are computational fragments. The number of DFs that are input to each CF, as well as the number of output DFs, does not have a strict limit and depends on the program specificity.

CFs may be either subprograms written in the LuNA or sequential program modules, such as a C/C++-functions. In the first case CFs are called "structured computational fragments" because the LuNA system can distinguish individual structures of these fragments. In the second case CFs are "imported atomic computational fragments". The LuNA takes entire function "as is" and doesn't know anything about the content. DFs are variables of integer (int), floating (real), or string (string) type. DF may also be of "custom value" type.

DFs also may be indexed (e.g. *M[3]*, *M[i]*). These are something like arrays. But in conventional programming languages an array has a fixed length, which is defined statically or dynamically. The memory is allocated for an entire array. In the LuNA there is no explicit operation of memory allocation and the access to two indexed data fragments *x[2]* and *x[5]* is equal to the access to two differently named data fragments (e.g. *a* and *b*). Also there is no explicit boundary for indexed data fragments. It is not described anywhere and nothing stops the user from accessing data fragment with any integer index.

When the LuNA-program is compiled, it passes the number of intermediate phases and at finish results in a dynamic library, containing binary code of fragments. After that, the runtime system of the LuNA is started, it takes this library (*.so*-file) as an argument and executes CFs. During operation, the runtime system allocates independent and "ready-to-execute" CFs, running them on different computing nodes of the multi-computer or on different cores of the multi-processor. "Ready-to-execute" CFs are those for which all the input DFs have received their values. So the LuNA is suited for both computing systems with shared and distributed memory.

The user may also specify some directives in a program code to optimize implementation of the fragmented algorithm by reusing memory, mapping fragments to computational nodes, deleting no longer needed DFs etc. As already mentioned, the execution order of operators in the LuNA-program is undefined in general case. The only thing that limits this order is the data dependencies between the CFs. This is one of the reasons why there are many ways to execute a fragmented program. Directives reduce this set.

As for other programming technologies, for the LuNA-programs there are some specific semantic errors. Existing debugging tools cannot discover them. That's why the task of classifying such errors was set, analyzing the applicability of various approaches for debugging the LuNA-programs, and creating a tool that can help the user to detect the most common errors.

2   Errors in the LuNA-Programs

Syntactical errors discovered by the LuNA or C++-compilers are not considered in this paper as they are relatively easy for automated detecting and the user correction. The following categories of semantic errors may be distinguished (the meaning of these errors is given below):

1. Erroneous directives [3]:
   (a) One directive puts a DF on computational node №1 while later another directive requests it from node №2.
   (b) Incorrectly specified number of DF uses after which this fragment is deleted.
   (c) The DF is used after the directive that deletes it.
   (d) Too many deletions of DFs.
   (e) Some others.

2. Errors in the imported C/C++-code fragments
3. Errors in the fragmented algorithm:
   (a) DF is produced by several non-mutually exclusive CFs.
   (b) CF requires a DF that is not produced by any other CF.
   (c) DF is initialized with a value of one type, but after that is used as another type.
   (d) A cycle in the graph representing a fragmented algorithm (deadlock). Such a graph without cycles is shown on Fig. 1.

This classification may be not absolutely complete, but it contains most types of errors that LuNA users encountered during the testing and the operation of this system.

Directives are not obligatory elements of the LuNA-program. Correct directives do not affect the correctness of the program results. The purpose of using directives is to help a runtime system in optimizing the program execution. But an invalid directive may lead to a program malfunction.

However, one of the main directions of the LuNA system development is the reduction of the number of directives. The work that is currently performed by the user with the help of directives should be performed automatically by the system in the future. From version to version, the directives are modified and their level is raised. That's why semantic errors associated with incorrect directives are not discussed in detail in the paper.

Another type of errors comes from imported code fragments in C/C++. The compiler of the LuNA-programs does not analyze those functions and puts them in the final C++-file "as is". To analyze errors in the C/C++-code there are many high-quality, efficient, and widely-used tools that include static analyzers and dialog debuggers (Microsoft Visual Studio Debugger, gdb). For this reason, our paper will not consider such errors.

The last category of errors – "errors in the fragmented algorithm" – requires a detailed analysis and development of tools for their detection. There may be numerous situations in which the user may commit such an error.

The LuNA is a language of a single assignment. Every DF may have only one value throughout the entire program. This relates to the type of error *3a* in the classification above. The LuNA-program in *Listing 1* should initialize even elements of indexed DF *x* with zeros and odd elements – with ones. However, the user has made an error and written the subprogram *init_2* with *(ind - 1)* instead of *ind*.

Listing 1. Two CFs initialize one and the same DF

```
import c_init( name, int) as init; //code fragments init and
                                    //print are imported
import c_print( value) as print;    // C/C++-functions

sub init_1( name arr, int ind) { //LuNA-subprogram init_1 with
   init(arr[ind], 0);                //2 arguments imported code
```

```
                                   //fragment that performs
                                   //initialization
                                   //of first argument by second
}

sub init_2(name arr, int ind) {
  init(arr[ind-1], 1);             //an error was made here
}

sub main() {                       //LuNA-program starts here
  df x;                            //definition of Data Fragment

  for i = 0..10 {                  //the cycle from 0 to 10
    init_1(x, 2*i);                //call of subprogram init_1
    print(x[2*i])@{                //call of imported code fragment
      delete x[2*i];
    };

    init_2(x, 2*i+1);
    print(x[2*i+1])@{
      delete x[2*i+1];
    };
  }
}
```

The essence of the error is that a DF *x[2*i]* is computed twice for each *i*. The result of executing such a program is undefined. The reason is that the order of execution of the LuNA-program operators is generally undefined. After initializing some DF *x[2*i]*, it can be displayed with the CF *print*, and then deleted due to the delete directive. In this case, the *init_2* procedure will create a new DF with the same name and index. The program will work successfully and no conflicts will occur. But on the other hand, *init_2* can be run after the CF *init_1* immediately. Then the runtime system LuNA will detect an attempt to reinitialize DF and crashes all processes with an error.

The second reason for non-deterministic behavior of the program in *Listing 1* is the distribution of CFs across the nodes of multicomputer. If CFs *init_1* and *init_2* are executed on different nodes, then there will be no such report in any case.

A serious problem for the user may come from errors of the type *3b* "A CF requires some DF, which is not produced by any other CF". Such errors cause the program hang. In the example below (*Listing 2*) an error is evident - when the indexed DF *y* is initialized the second loop boundary is set as '*size-1*' while a screen output with the *print* CF requires every element up to and including *size*. Being unable to get the value *y[10]* CF print hangs up.

Listing 2. The lack of a required DF

```
import c_init( name, int) as init;
import c_print( value) as print;
#define SIZE 10        //''define'' like in C/C++

sub main() {
  df x, y, product;    //definition of 3 data fragments
  let size = \$SIZE {  //operator of assigning a value to variable
    for i = 1..size
      init( x[i], i);
    for i = 1..size-1
      init( y[i], i);
    for i = 1..size
      print( x[i] + y[i]);
  }
}
```

Errors of the following type (3c) are relatively easy to detect, and the LuNA runtime system throws an exception in case of an attempt to transmit an instrument of another type than what it should be.

The last type of errors that will be considered here is the occurrence of dependency cycle in algorithm graph (*3d*). This error also causes the program hang. The reason here is not that a required DF is not produced by any CF in its code but that CFs of the cycle cannot run in order to produce their DFs. An elementary example of program with a dependency cycle is given in *Listing 3*.

Listing 3. Fragment of the program with a dependency loop of CFs

```
sub main() {
  df x, y;
  init(x, y);   // x := y
  init(y, x);   // y := x
  print(x);
  print(y);
}
```

The two CFs *init* must initialize the DFs x and y respectively with a value of each other but, obviously, fail to do so.

3 Methods and Tools of Automated Debugging. The Analysis of Applicability to the LuNA

The LuNA is a system that 'accompanies' the user's program from the original code to the final result. After the user gives a command "*luna myprog.fa*" (myprog.fa - source code file in the LuNA programming language) his program runs through several stages of processing and a C++-code is produced. This code goes to a dynamic library that connects to the runtime system LuNA and works under its command. Thus, during development of the LuNA system we have various methods for automated debugging. It should be mentioned that the typical errors vary depending on the chosen programming technology, such as MPI, OpenMP etc. But there are not so many methods of automated debugging. And often for finding errors in programs of different languages and of different technologies the same methods are suitable. For example, [4] provides an overview of the applicability of the following approaches to the analysis of MPI-programs.

During the compilation of the LuNA program the **static analysis** is applied for relatively simple errors, which are not hard to find in the source code. Examples may be the following: invoking a subprogram with several arguments that do not correspond to the specification; invoking a subprogram that is not defined in the code; attempting to use an undefined DF; etc. For C/C++-programs there several debugging tools exist that use the method of the static analysis. The very first tool of this group was analyzer Lint [5]. At this moment there static analyzers PC-lint, Splint [6], PVS-Studio [7], Clang Static Analyzer [8] exist and are being developed. A significant problem for this category of tools is "false positives". This approach can be at least partially applied to analyze the LuNA programs (modules or generated C++ code). Many static analysis features are already included in C++ compiler, which is employed in the LuNA as a compilation step. Some static analysis is performed by the LuNA compiler.

Further on, during the program runs, the runtime system may **analyze the program online** (runtime analysis). The system processes the events of the creation and the completion of each CF with actual parameters as well as the creation and the deletion of each DF. A separate thread (or several threads) may gather this information in service structures and analyze it for the semantic errors from classification above. The problem of this approach in general case is high overheads. At each computation node, the runtime system of the LuNA tries employing all processor cores for the parallel execution of computational fragments. In the case of runtime-analysis, at least one core will be busy with these service operations. There may be a situation when a big program will be quickly generating and deleting many objects (CFs and DFs). In this case the service thread may become a performance bottleneck and fail to run all the assigned checks as soon as needed. The work on a program analysis may be parallelized but then the service function will involve several processor cores instead of one. Of course, the analysis work can be performed not in the service threads, but in the main threads that are busy by computations. In this case after each analyzed event the thread should stop and do analysis, but this will cause even greater overhead. The method of analysis for the occurrence of semantic errors during execution is used, in particular, in MUST [9–11], built on the basis of the experience in creating Marmot [12] and Umpire [13] to discover errors specific to MPI programs. Since an MPI program can run on different nodes of a computing system with distributed memory, the architecture of these systems has a separate service process or service thread within one of the worker processes, to which the processes transmit information about the MPI-calls, and it analyzes for errors from a certain list.

Another approach would be **«post-mortem analysis»** - looking for semantic errors over a collected trace. For this method, overheads are considerably lower than for the runtime-analysis. Roughly speaking, the information about events, which in the case of runtime-analysis goes to the service thread, here simply puts into a file. Of course frequently recording to the hard drive is a negative factor for performance. However, the situation is smoothed out, firstly, by the spooling mechanism of the operating system. And, secondly, by an opportunity to set up an intermediate buffer in the tracing tool. Event data can be dumped to the trace file from this intermediate buffer when a certain number of these events is reached, by buffer saturation, or by a timer. This approach is also used in some software tools for automated debugging of MPI programs, including, for example, Intel Trace Analyzer and Collector (ITAC). The error detection tool for the LuNA programs described below also uses this method.

Another approach for automated error detection in the LuNA-programs is **comparative debugging**. The main essence of the method is running two versions of the same program or to have the same program run under different conditions. One version of the program (one run) is called "original", another is "development". The original version, for example, may be running the program at one node (without MPI). A development version may be running several copies of the runtime system LuNA on different nodes communicating with each other. At certain "control points" of the program, data about the values of variables and other objects is collected during these two runs and compared with each other. A great difference is considered the error and the user gets informed about it. Comparative debugging has been implemented in systems such as the Guard [14] debugger and the DVM system [15].

One more method of automated debugging is **model checking**. The basic principle of the operation of systems using this method is as follows. The user specifies the properties that the program should satisfy using the formulas of temporal logic. An abstract formal model is formed in some pseudo-language, containing only those elements of the analysed program that are significant for checking the specified properties. During verification, a search is conducted along different traces for those states in the graph in which the specified properties are not fulfilled. This graph is built implicitly according to the model. If the test results are negative, then the user is provided with a trace containing an error (the so-called "counterexample").

4   The Tool for Analysis Hangups of the LuNA-Programs

The authors of the article have developed a module for backtracing the LuNA-programs (trace-module) as well as the *luna_ trace* utility for analysis of collected trace files. This tool provides the user with information that helps to detect the causes of semantic errors in the LuNA-programs.

First of all the *luna_trace* was developed for dealing with situations, when the LuNA-programs hang. Hangups may be caused by errors in a fragmented algorithm related to the categories 2) and 3) of the classification above. If one is to draw an analogy with programs in standard imperative languages (C, Fortran, Pascal, ...), the hangup effect may involve the following errors:

- Deadlock between threads executing different procedures.
- Violation of the array boundary.
- Trying to use noninitialized values.
- Some others.

With standard languages some of these errors are discovered by a compiler, some lead to a program crash during operation, and some to other consequences. Thus, the hangup-problem is very relevant for the LuNA-programs and the user needs the information, which can help him to explain the cause of the hangup.

Since the *luna_trace* uses the method of the post-mortem analysis, during the operation of the user program, a special module (trace-module) of the LuNA system only collects the trace from each process. If the program hangs, it must be completed in some way so that the *luna_trace* starts an analysis to identify semantic errors. The completion of a hung LuNA-program can be performed by two methods: automatically and by a SIGINT signal.

The automatic hang detection has a lot in common with solving the problem of stopping a distributed computing system. In the LuNA system, there is a module responsible for stopping the runtime system after performing all the CFs. It is based on the Dijkstra—Scholten algorithm [16].

The algorithm implies that each process can be in one of two states — "there is some work" and "there is no work". In short, the algorithm is as follows. The whole set of processes organized in a ring bypasses the integer counter. Initially, it is located in some process where there is work. When the work on the process ends, the counter is incremented and passed to the next process along the ring. If there is some work on this process, where the counter is transferred, then it is reset to 0. If there is no work, then it is incremented and transmitted further. When the value of the counter is equal to twice the number of processes (the counter made 2 rounds without finding work anywhere), then the module stops the LuNA-program.

To stop the hung program, the module was modified as follows. A process is considered to be in the "no work" state not only if all its CFs are completed, but also when the existing CFs have sent requests for the delivery of input DFs and cannot be executed until these DFs are delivered. And the state "there is work" means that at least one CF has received the necessary DFs or when it is already being executed.

If all the CFs in the system are waiting for input DFs, then the LuNA-program is hanging. In this case, according to the algorithm, the system will be stopped. After stopping, it is possible to determine whether the LuNA-program has hung up, or finished normally by the presence of unfinished CFs on the entire set of processes.

The feature of automatic hang detection can be disabled. The user may wish to disable this feature, because the Dijkstra-Scholten algorithm is not universal - it can give "false-positive" and the program will be terminated, despite the fact that the CFs could continue to be executed.

Thus, the decision to automatically force termination by the LuNA runtime system is better suited for the case when the user has no opportunities to interact with his program (e.g. during work on a supercomputer with the batch system).

In this regard, the second method of terminating a hung program is by a SIGINT signal. This signal is sent to the application by pressing the key combination "Ctrl+C". If the user can interact with the program during an execution, then forcibly interrupt of the program is best left to him.

When the LuNA-program stops by module of the automatic termination or intercepts SIGINT-signal, trace-module for each MPI-process finals writing 2 files. The first one - a json-file with information on uncompleted CFs with identifiers of needed DFs for them and a call stack. Another one is a file with matching identifiers of DFs, their names and relevant CF numbers.

It is worth noting that the trace-module writes information to these files throughout the entire operation of the program. As MPI-processes generate and perform CFs, the information about this gets into the intermediate buffer in memory. When the buffer size exceeds the value set by a certain parameter, its contents are written to a file and the buffer is cleared. Thus, by varying the buffer size using the parameter, the user has the opportunity to influence the performance of the program as a whole.

After stopping the LuNA-program the user starts up the analysis of the *luna_trace* utility. It parses these 2 files and for every hanging CF determines the element of call history that corresponds to the earliest subprogram, where the base name for an absent DF was declared. The *luna_trace* prints out the call history starting with this element and for every absent DF, it gives the local name and the expression corresponding to the local name in the code as well as the name of the subprogram where this DF was declared. The printout of the call history contains the lines of the original code that correspond to calls of subprograms, cycle notifications, etc. This is done with the help of service files generated by the compiler.

Let's analyze the output of the *luna_trace* on the example of the program from *Listing 2*. After starting, the program outputs the sums «$x[i] + y[i]$» for all i from 0 to 9. Without having a data fragment $y[10]$, the program hangs on the call «$print(x[10] + y[10])$». After interrupting the program using the combination "Ctrl+C" and running the *luna_trace* utility, the user will see the following output (Fig. 2).

The utility outputs a line of code of the hung CF, as well as the entire stack of procedure calls preceding the hung fragment (in this program, the hang occurred in the main procedure, which is called first, so the stack consists of only one element). As you can see, the utility shows a certain value of index for the missing piece of data. In addition, the *luna_trace* makes an assumption about which part of the code should initialize the missing indexed DF.

```
reworu@DESKTOP-CPJS18K:/mnt/c/Users/misha/Projects/luna/Programs/paper-listings/listing-2$ luna_trace
Following CFs appear to be the root cause of the system hang:

print( x[i] + y[i]) [./main.fa:13] never finished
main
    in for i = 1..size [./main.fa:12]
    in sub main() [./main.fa:5]

    Awaited DF y[10] (y[i])
        full name y[10], y declared in sub main
    Maybe it should have been initialized here:
        init( y[i], i) [./main.fa:11]
        in for i = 1..size-1 [./main.fa:10]
        in sub main() [./main.fa:5]
```

Fig. 2. Output of the *luna_ trace* utility on the program with lack of required DF.

The developed tool is able to detect cyclic dependencies between hung up CFs (the category *3c* from the classification above), giving a message about all fragments involved in the cycle.

Among all the hung CFs, there are often dependencies of one on the other. For example, it could be that fragments CF2, CF3, and CF4 hang and can not be executed because of hanging. Displaying all the hung CFs will make it more difficult for the user to find and analyze the cause of the error. That's why the created tool outputs only the "root cause" hung fragment of calculations (in the considered case - CF1) and the total number of hung fragments.

In addition, the *luna_ trace* is able to detect errors of multiple initialization of DF (category of errors *3a*). In the trace files for the CFs created during the operation of the program, there is in-formation about the identifiers of the input and output DFs. The *luna_ trace* iterates through pairs of CFs that have at least one output DF, and determines whether there are those among them where both elements of the pair initialize the same DF.

So, for the program from *Listing 1*, the output of the utility will be as follows (Fig. 3):

The user is provided with information about the DF initialized several times, as well as about the expressions performing initialization, with reference to the LuNA-program code.

5   Overhead Evaluation

An important characteristic of any automated debugging system is the overhead when running user programs under the control of the system. The most interesting is the evaluation of the overhead costs in the program's running time and the memory usage. For the tools of "post-mortem analysis", the working time of the trace analyzer is also important.

To evaluate overheads, a program in the LuNA language was used, implementing an algorithm for block multiplication of square matrices. Further, the size of the matrix and the number of blocks will be understood as these values along one of the axes.

```
reworu@DESKTOP-CPJS18K:/mnt/c/Users/misha/Projects/Luna/Programs/paper-listings/listing-1$ luna_trace
Following DFs are initialized multiple times:

x[0] (declared in sub main) in
    init(arr[ind-1], 1) [./main.fa:9]
        in sub init_2(name arr, int ind) [./main.fa:8]
        in init_2(x, 2*i+1) [./main.fa:19]
        in for i = 0..10 [./main.fa:14]
        in sub main() [./main.fa:12]

    init(arr[ind], 0) [./main.fa:5]
        in sub init_1(name arr, int ind) [./main.fa:4]
        in init_1(x, 2*i) [./main.fa:15]
        in for i = 0..10 [./main.fa:14]
        in sub main() [./main.fa:12]
```

Fig. 3. Part of output of the *luna_trace* utility on the program with multiple initialization of DF.

The tests were conducted on a computer with an Intel Core i7-3537U@ 2.00 GHz processor, 12 Gb DDR3, 2 MPI-processes, 4 threads per process, OS Ubuntu 20.04 LTS. Since an important parameter of the trace-module in the LuNA system which can affect overhead costs is the buffer size ("–log-buffer-size" key), a series of measurements was carried out depending on the value of this parameter. Table 1 shows the results for block size 50 and number of blocks 40.

Table 1. The operating time and the memory usage depending on the log-buffer-size value

The size of the trace file buffer, KB	The program operating time, sec.	The memory usage, MB
1	91, 48	2 679, 6
10	49, 54	2 678, 1
100	46, 23	2 680, 5
1000	35, 8	2 683, 1
10000	30, 01	2 688, 4

According to Table 1, it can be concluded that the choice of the buffer size can have a very significant effect on the total operating time while collecting the trace of a "sufficiently large program". However, the memory consumption increases slightly at the same time.

As for the operating time of the *luna_trace* utility itself, a significant part of it takes the search for re-initialization of the DFs. The *luna_trace* operating time for the same task of the block matrix multiplication of 2000 by 2000 with a variable block size (2004 by 2004 for block size 167) was estimated in the presence and absence of the "–no-double-init" key for a different number of CFs generated during LuNA-program run. Depending on this key, the data reinitialization check is performed or disabled.

The same computer was used for the tests. The buffer size was set to 1 MB.

Table 2. The *luna_ trace* operating time for different number of CFs

The block size	The number of blocks	The number of CFs	The program operating time, sec	The operating time of *luna_ trace* without the key "–no-double-init", sec	The operating time of *luna_ trace* with the key "–no-double-init", sec
500	4	247	47, 25	0, 49	0, 37
250	8	1483	17, 93	0, 68	0, 37
167	12	4479	15, 67	3, 4	0, 44
125	16	10003	14, 05	12, 24	0, 63
100	20	18823	12, 38	45, 7	0, 64

Thus, with a large number of CFs, the time to search for cases of repeated initialization, and, as a result, the running time of the *luna_ trace* can be quite long, and even exceed the running time of the program itself. In this regard, the use of the "–no-double-init" key may be justified (Table 2).

6   Conclusion and Future Work

Fragmented programming technology is characterized by its own errors that do not have a one-to-one mapping to errors in imperative languages, as well as in parallel programs that use MPI, OpenMP, and other technologies. That's why specific to fragmented programs debugging tools are needed. In addition, searching for errors in the C++-code that is generated from the source program by the LuNA-compiler does not make much sense. The reason is that the resulting C++-program is difficult for the user to match with the original LuNA-program. In this regard, the well-known dialog debugging tools (gdb, TotalView [17], Arm DDT [18]) are not suitable for the task of debugging fragmented programs. In addition, in the case of large computing programs, an automated tool is most convenient for the user, providing him with a list of errors found without his participation at all or with minimal participation.

At the moment, an automated debugging tool based on the "post-mortem" analysis approach has been developed. It is able to give the user information about those "primary" CF that caused hanging of all the rest. For these CFs, the names of the missing DFs are output according to the original LuNA-program. The tool is also able to detect dependency cycles between CFs. In addition, a module for automatic completion of the hung LuNA-program was added to the runtime system.

Work on the detection of semantic errors in LuNA-programs will be continued. In our plans, there are improving static analysis algorithms in order to detect errors that are specific to fragmented programs.

All errors listed in the classification above can also be found using the model checking method. As it known, the feature of this method is the ability to detect errors caused by the non-determinism of the behavior of a parallel program. In this regard, another area of our work is the development of an automatic model construction and a verifier that can detect errors by the model.

References

1. Malyshkin, V.: Technology of fragmented programming. Bulletin of YuUrGU, Series "computational Mathematics and Informatics", vol. 46(305), pp. 45–55 (2012)
2. Akhmed-Zaki, D., Lebedev, D., Malyshkin, V., Perepelkin, V.: Automated construction of high performance distributed programs in LuNA system. In: Malyshkin, V. (ed.) PaCT 2019. LNCS, vol. 11657, pp. 3–9. Springer, Cham (2019). https://doi.org/10.1007/978-3-030-25636-4_1
3. Malyshkin, V., Perepelkin, V.: Optimization methods of parallel execution of numerical pro-grams in the LuNA fragmented programming system. J. Supercomput. 61(1), 235–248 (2012). https://doi.org/10.1007/s11227-011-0649-6. Special issue on Enabling Technologies for Programming Extreme Scale Systems
4. Vlasenko, A.Y., Gudov, A.M.: The use of erratic behavior templates in debugging parallel programs by the automated validity verification method. J. Comput. Syst. Sci. Int. 56(4), 708–720 (2017). https://doi.org/10.1134/S1064230717040153
5. Kunst, F.: Lint, a C Program Checker. Vrije Universiteit, Amsterdam (1988)
6. Evans, D., Larochelle, D.: Improving security using extensible lightweight static analysis. IEEE Softw. 19(1), 42–51. IEEE Computer Society Press, Washington (2002). https://doi.org/10.1109/52.976940
7. Introduction to the PVS-Studio Static Code Analyzer on Windows (2020). https://pvs-studio.com/ru/m/0007/. Accessed 25 June 2022
8. Clang Static Analyzer. https://clang-analyzer.llvm.org/. Accessed 21 June 2022
9. Hilbrich, T., Schulz, M., de Supinski, B.R., Muller, M.S.: MUST: a scalable approach to runtime error detection in MPI programs. In: Müller, M., Resch, M., Schulz, A., Nagel, W. (eds.) Proceedings of the 3rd International Workshop on Parallel Tools for High Performance Computing, pp. 53–66. Springer, Berlin (2009). https://doi.org/10.1007/978-3-642-11261-4_5
10. Cramer, T., Münchhalfen, F., Terboven, C., Hilbrich, T., Müller, M.S.: Extending MUST to check hybrid-parallel programs for correctness using the OpenMP tools interface. In: Knüpfer, A., Hilbrich, T., Niethammer, C., Gracia, J., Nagel, W.E., Resch, M.M. (eds.) Tools for High Performance Computing 2015, pp. 85–101. Springer, Cham (2016). https://doi.org/10.1007/978-3-319-39589-0_7
11. Protze, J., Hilbrich, T., Schulz, M., de Supinski, B.R.: MPI runtime error detection with MUST: a scalable and crash-safe approach. In: 43rd International Conference on Parallel Processing Workshops, pp. 206–215. Minneapolis, MN, USA (2014)
12. Krammer, B., Müller, M.S., Resch, M.M.: MPI application development using the analysis tool MARMOT. In: Bubak, M., van Albada, G.D., Sloot, P.M.A., Dongarra, J. (eds.) ICCS 2004. LNCS, vol. 3038, pp. 464–471. Springer, Heidelberg (2004). https://doi.org/10.1007/978-3-540-24688-6_61

13. Vetter, J., de Supinski, B. R.: Dynamic software testing of MPI applications with umpire. In: Proceedings of the ACM/IEEE Conference on Supercomputing (SC 2000), pp. 70–79. Article No 51. Dallas, USA (2000)

14. Abramson, D., Watson, G., Dung, L.P.: Guard: a tool for migrating scientific applications to the .NET framework. In: Sloot, P.M.A., Hoekstra, A.G., Tan, C.J.K., Dongarra, J.J. (eds.) ICCS 2002. LNCS, vol. 2330, pp. 834–843. Springer, Heidelberg (2002). https://doi.org/10.1007/3-540-46080-2_88

15. Bakhtin, V., Zakharov, D., Ermichev, A., Krukov, V.: Comparative debugging of parallel DVMH programs. In: Scientific Service on the Internet: Proceedings of the XXI All-Russian Scientific Conference, 23–28 Sep 2019, Novorossiysk, pp. 91–104 (2019). https://doi.org/10.20948/abrau-2019-37

16. Fokkink, W.: Distributed Algorithms: An Intuitive Approach. MIT Press, Cambridge (2013)

17. TotalView   Tutorial.   https://hpc.llnl.gov/training/tutorials/totalview-tutorial. Accessed 29 May 2022

18. Get started with DDT. https://developer.arm.com/documentation/101136/2101/ DDT/Get-started-with-DDT. Accessed 29 May 2022

Multi-GPU GEMM Algorithm Performance Analysis for Nvidia and AMD GPUs Connected by NVLink and PCIe

Yea Rem Choi[1]($\boxtimes$) and Vladimir Stegailov[1,2,3]

[1] HSE University, Moscow, Russia
echoj@hse.ru
[2] Joint Institute for High Temperatures of RAS, Moscow, Russia
[3] Moscow Institute of Physics and Technology, Dolgoprudny, Russia

Abstract. Modern types of multi-GPU servers combine up to 8 A100 GPUs connected by NVLink 3.0 links through NVSwitch. This connectivity provides unprecedented capabilities for multi-GPU algorithms. In this work, we analyze the performance of matrix-matrix multiplication algorithm developed by us previously. Tuning principles and limits for maximum performance are discussed. Algorithm performance for much more affordable 4 AMD Radeon RX 6900 XT based server with PCI 4.0 working under ROCm HIP is described for comparison.

Keywords: Parallel computing · CUDA · HIP · GEMM · High-speed GPU interconnect · Multi-GPU programming

1 Introduction

Modern high performance computing shows an evident trend of the growing use of specialized computational elements. Among the top 10 systems in the current Top500 list (June 2022) there is only one CPU-only supercomputer Fugaku with the number 2. The number 6 is the Sunway TaihuLight supercomputer based on the special Sunway SW26010 processors (each with 260 computing cores) requiring special programming techniques. The number 9 is the Tianhe-2A supercomputer combining Intel Xeon CPUs with the Matrix-2000 accelerators. All other systems (the numbers 1, 3–5, 7, 8 and 10) provide the major share of computing power via GPU accelerators: the numbers 1 (Frontier), 3 (Lumi), 4 (Summit), 5 (Sierra), 7 (Perlmutter), 8 (Selene) and 10 (Adastra).

After the long period of Nvidia dominance in GPU computing technologies, now the real competition of vendors is developing. New GPUs made by AMD can be found in several largest supercomputers of the current top 10 systems (Frontier, Lumi and Adastra). The Aurora supercomputer that is to be commissioned soon in Argonne National Laboratory will be based on the new GPUs made by Intel. Each of these major GPU vendors proposes its own programming framework. In 2007 Nvidia pioneered the CUDA technology. In 2015 AMD

D. Balandin et al. (Eds.): MMST 2022, CCIS 1750, pp. 281–292, 2022.
https://doi.org/10.1007/978-3-031-24145-1_23

proposed the ROCm infrastructure and HIP for its GPUs as a nearly complete substitute for CUDA. In 2019 Intel announced the oneAPI infrastructure that uses the DPC++ cross-architecture language based on the SYCL standard for GPU programming. Interoperability of CUDA and the new technologies of AMD and Intel is crucial developing portable HPC software [1–3]. Many specific middle layers focused on performance portability are under active development (e.g. OpenMP, OpenACC, KOKKOS, Alpaka). Porting of linear algebra libraries is among the first priorities for the proper introduction of new technology. For example, the ROCm framework provides the hipBlas library that is a very close analogue of cuBlas and has similarly high efficiency [4].

Computing nodes of GPU-based supercomputers have multiple GPUs per node. The systems with AMD GPUs Frontier, Lumi and Adastra have very similar design with 4 GPUs/node. The systems with Nvidia GPUs differ: Summit has 6 GPU/node, Sierra and Perlmutter have 4 GPU/node and Selene has 8 GPU/node. In all these systems GPUs are interconnected by ultrafast communication links (Nvidia NVLink or AMD Infinity Fabric). For example, one A100 accelerator has 12 NVLink 3.0 links with 50 GB/s peak bandwidth each. One MI250X accelerator has 8 Infinity Fabric links with 100 GB/s peak bandwidth each. Such connectivity between GPUs within a supercomputer node opens unprecedented opportunities for parallel computations.

In this paper, we analyse the performance of the multi-GPU matrix-matrix multiplication algorithm developed and implemented by us previously [5] for two systems: a node with 8 A100 GPUs connected by NVLink 3.0 links through NVSwitch and a node with much more affordable AMD Radeon RX 6900 XT GPUs connected by PCIe 4.0. The SGEMM variant of the algorithm is considered. The accuracy of the previously proposed theoretical model [6] for performance tuning is validated. The performance influence of the tensor cores available in A100 [7,8] is described. The peculiarities of porting the algorithm from CUDA to HIP and running it on the AMD GPUs are described.

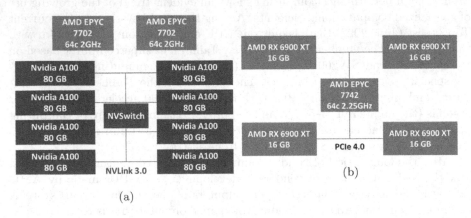

Fig. 1. The topology of the A100-equipped node of the cHARISMa supercomputer with two CPUs and eight Nvidia A100 GPUs by NVLink 3.0 (a) and the server with one CPU and four AMD RTX 6900 XT connected by PCIe 4.0 (b).

2   Related Work

Parallel algorithms for matrix multiplication evolve together with the development of the parallel computing technologies. The MPI algorithm for parallel matrix multiplication has been published soon after the MPI standard was introduced [9]. A runtime system called SuperMatrix that parallelize matrix operations for SMP and/or multi-core architectures was described in [10]. SuperMatrix introduced the concept of a set of tiles for work distribution among multiple threads. The PaRSEC framework [11] introduced the direct acyclic graph scheduling for dense algebraic operations. The subsequent PaRSEC implementation of the GEMM algorithm showed very high efficiency [12]. The near optimal parallel matrix-matrix multiplication algorithm COSMA was introduced based on the red-blue pebble game ideas [13]. Matrix multiplication is a generic test case for evaluation new programming models in HPC, e.g. the use for the Rust programming language [14].

Along with the academic projects SuperMatrix, PaRSEC and COSMA, there is a commercial multi-GPU Level-3 BLAS library cuBlas-XT developed by Nvidia. It was shown, however, that cuBlas-XT provides sub-optimal performance [15]. The multi-GPU level-3 BLAS library BLASX with improved scheduling was developed by Wang et al. [15]. The problem of communication optimal partitioning of a square computation domain over three heterogeneous processors has been considered recently [16].

PaRSEC, BLASX and COSMA are complex and multipurpose software projects. These projects (as well as cuBlas-XT) allow making calculations for matrices that are stored in the CPU memory (via special scheduling of CPU-GPU data transfers). The aim of this work is to analyze a much simpler matrix-matrix multiplication algorithm for matrices stored in GPU memory only [5,6]. Such an algorithm suits better for the purpose of benchmarking different types of GPUs and GPU-GPU interconnects.

3   Performance Model Overview

Here, we give a brief overview of the performance model developed in our previous work [5,6].

The GEMM algorithm is solving the equation

$$C = \alpha A * B + \beta C.$$

The uniqueness of the developed algorithm is that it uses only the resources of GPUs for its work, avoiding the necessity to wait for the data from the host CPU during the algorithm execution. This makes it possible to deploy such high bandwidth links as Nvlink between GPUs available in GPU servers nowadays (e.g., in DGX-like systems). Also, the asynchronous data transfers and computation overlap have been organized in the algorithm, providing the high performance rate.

For the case when we work with big matrix sizes the algorithm performance is limited by the computational abilities of GPUs. In the observed experiments the sizes of the tiles could be expected to be optimal for algorithm performance if they match the following conditions [6]

$$\begin{cases} N_i > 4k_{BW}N/(N - 2k_{BW}), & N > 2k_{BW} \\ N_i > 2(Num_{GPUs} - 1)BW_{math}/BW_{transfer}, \\ N_i > 2(Num_{GPUs} - 1)BW_{math}/BW_{transfer}, \end{cases}$$

where N and N_i are the sizes of original matrices and tile matrices, $k_{BW} = BW_{math}/BW_{mem}$ is the bandwidth coefficient, BW_{math} and BW_{mem} are the mathematical and memory bandwidths of a GPU device, and Num_{GPUs} is the number of implemented in computation GPUs.

4   Testing Platforms

The results reported in this study are obtained on the nodes of the cHARISMa supercomputer at HSE University [17,18]. The nodes are based on the 8x Nvidia A100 GPU "Delta" platform with NVSwitch (Fig. 1a). Each GPU has 80 Gb of HBM2 memory, and eight GPUs are connected by NVLINK 3.0 via NVSwitch.

The benchmarking studies on the A100 equipped node are carried out using the standard HPC software stack based on CentOS Linux release 7.9.2009, GNU compilers 8.3.0, and CUDA Version 11.7.64 with the driver ver. 515.43.04.

The second platform is the server with 4 AMD RX 6900 XT GPUs connected by PCIe 4.0 (Fig. 1b). Each GPU has 16 Gb of GDDR6 memory. The server is based on the ASRock ROMED8-2T single socket motherboard with one AMD EPYC 7742 CPU. The benchmarking studies on this server are carried out using Ubuntu 20.04 Linux with AMD ROCm 5.2.1. RX 6900 XT GPUs have RDNA2 architecture that is a close relative of CDNA2 architecture of MI250X GPUs used in Frontier, Lumi and Adastra supercomputers.

Table 1 summarizes the key features of two type of GPUs considered in terms of the parameters used in the performance model proposed in [6].

Table 1. Test platforms parameters

Hardware parameters	Nvidia A100	AMD RX 6900 XT
Peak FP32 performance (TFLOPS)	19.5	22.5
Real FP32 performance (TFLOPS)	18.4	21.4
Peak FP32 tensor core performance (TFLOPS)	156	–
Real FP32 tensor core performance (TFLOPS)	124	–
Peak GPU memory bandwidth (GB/s)	2039	512
Peak GPU-GPU bandwidth (GB/s)	300	32
Real GPU-GPU bandwidth (GB/s)	281	25.5

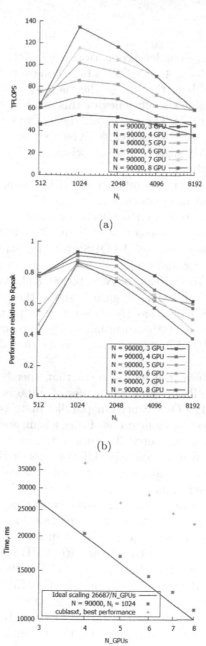

Fig. 2. Graph of the multi-GPU SGEMM operation on 3, 4, 5, 6, 7, and 8 A100 GPUs by tile size (N_i) for $(N = 90000)$ elements in a row (column) of matrices without tensor core in absolute TFLOPS (a) and in relative data (b). The graph (c) is the time dependency by number of GPUs for tile size $(N_i = 1024)$. The matrices A, B, and C are stored in devices 2, 1, 0 respectively.

5 Results

The possibility of reaching high levels of performance of the matrix multiplication algorithm considered ($C = \alpha A * B + \beta C$) is based on overlapping of computation and communication [5]. It is the size of the tiles N_i subdividing the matrices A, B and C that regulates the efficiency of this overlap, and hence the overall performance of the algorithm. Figure 2 shows the benchmark results for the A100 equipped node without using tensor cores. The results are presented for one of the largest possible matrix sizes $N = 90000$ where $N \times N$ is the size of the square matrices considered in this work.

We see that the higher is the number of GPUs N_{GPUs} the stronger is the dependence of the algorithm performance on the tile size N_i. This behavior is quite reasonable since the role of GPU-GPU communications increases for higher N_{GPUs}. For all $N_{GPUs} = 3$–8 we see the optimum $N_i = 1024$. At this tile size the efficiency of the algorithm (attained FLOPS over theoretical peak performance) reaches more than 80% for all N_{GPUs}.

The strong scaling for this case of $N = 90000$ is shown on Fig. 2 too and it is pretty close to the ideal scaling. Figure 2c shows the performance of cuBlas-XT for the exactly same problem ($N = 90000$, the points correspond to the minimum execution times at the variation of the tile size). One can see that the performance of cuBlas-XT is significantly worse that the performance of our algorithm.

A100 accelerators have the tensor cores that speed up multiplications of small matrices. While FP32 peak performance of A100 is 19.5 TFLOPs, tensor cores boosts it to 156 TFLOPs. Since our parallel matrix-matrix multiplication algorithm uses cuBlas for multiplications of tiles within each GPU, the algorithm can benefit from using tensor cores. The use of tensor cores in single precision can be switched on and switched off using CUDA calls (in double precision tensor cores can not be switched off and are deployed automatically whenever possible).

Figure 3 shows the benchmark results for our SGEMM algorithm with tensor cores switched on. One can see that the optimum values of N_i move to larger values. Despite significant acceleration in absolute values, the level of efficiency with tensor cores becomes lower (even lower than 40% $N = 8$).

Figure 4 shows the results for the server with AMD RX 6900 XT GPUs. The CUDA code of our algorithm has been ported to HIP using the Perl-based hipify tool available in the ROCm framework. hipBlas GEMM function calls are used instead of cuBlas. The results show surprisingly modest performance and low efficiency that is lower than 40% for 3 and for 4 GPUs. It is a strange fact since even for the quite old Nvidia GTX1070 GPUs the similar benchmark showed efficiency over 50% (see [6]).

6 Discussion

The algorithm had to be improved to manage with any size of the matrices. In the ending part of the algorithm the storing A and B devices exchange and

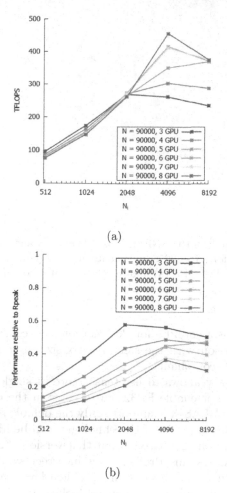

(a)

(b)

Fig. 3. Graph of the multi-GPU SGEMM operation on 3, 4, 5, 6, 7, and 8 A100 GPUs by tile size (N_i) for ($N = 90000$) elements in a row (column) of matrices with tensor core in absolute TFLOPS (a) and in relative data (b). The matrices A, B, and C are stored in devices 2, 1, 0 respectively.

compute the left matrices part after division them into bands. This way has been chosen firstly because we had an idea to improve the algorithm which requires it. Secondly, it is the one of less computationally expensive solution to manage with left parts at the same time. However, in some cases the effect of this step is too strong to be able to move the performance maxima. Moreover, for small block sizes the GPUs have to multiply tall-and-skinny matrices. Improvement of this issue has not been implemented in the algorithm yet.

Figure 5 shows the profiles of the algorithm execution with and without tensor cores. While we use tensor core, we achieve reasonably fast computational speed, but also we moderately, but sensibly lose the accuracy. The observed performance on each computation kernels are unduly low from peak for example in comparison

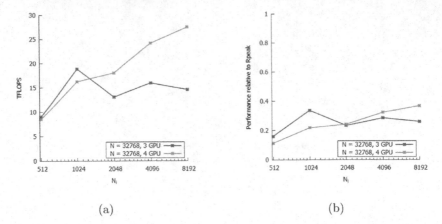

(a)   (b)

Fig. 4. Graph of the multi-GPU SGEMM operation on 3 and 4 Radeon RX 6900XT GPUs by tile size (N_i) for ($N = 32768$) elements in a row (column) of matrices without tensor core in absolute TFLOPS (a) and in relative data (b). The matrices A, B, and C are stored in devices 2, 1, 0 respectively.

with case without tensor core. In double precision the accuracy loss is much less detectable in comparison with the same matrix size in single precision (FP64 tensor cores are IEEE-compliant).

The cuBLAS library allows to use tensor core in single precision on A100 GPUs, but we could not enable FP32 tensor cores in the cuBLAS-XT library calls. Probably, the cuBLAS-XT supports only the default settings in this issue.

The difficulties in using new GPU technologies can be illustrated by the following fact. From experiments it comes out that version 2021.3.2.4-027534f and earlier Nvidia Nsight Systems Profiler gives incorrect synchronization profiles for multiple GPUs. In newer versions this problem has been corrected. Due to this problem a lot of efforts have been spend looking for the possible reasons of improper synchronization on A100 GPUs (while no problems were observed on V100 in our previous work).

We could not achieve the peak performance for RX 6900 XT GPUs during the algorithm. However, in a single launch of SGEMM in one GPU we do achieve the value very close to the peak. We marked (in Fig. 6) the time needed to compute a band multiplication with $2^{32} \approx 4.3$ GFLOPS, thus, the performance is 8 TFLOPS from the peak 22.5 TFLOPS. Figure 6 shows also considerable delays between data transfers. We suppose that these problems might be explained by the difficulty for the GPU (that is a consumer GPU, not a server grade GPU) to perform computations and data transfers simultaneously. This observation points to the fact that the multi-GPU performance of the algorithm considered can not be transferred easily to the consumer-grade GPU systems.

The results of the benchmarks of the A100-equipped node give us the possibility to test the performance model developed previously [6]. Table 2 summarizes the empirical values and the predictions. We show the threshold values "theoretical threshold" and the next larger $N_i = 2^n$ "theoretical N_i". The overall

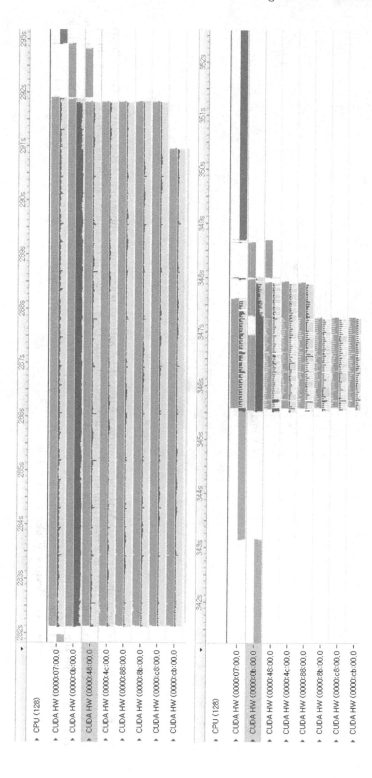

Fig. 5. The profiles of the multi-GPU SGEMM operation on 8 A100 GPUs without tensor core (upper) and with tensor core (lower) usage. Number of elements ($N = 90000$) in a row (column) of matrices and tile size ($N_i = 1024$) in case without tensor core and ($N_i = 4096$) in case with tensor core. The matrices A, B, and C are stored in devices 2, 1, 0 respectively. The green bars are the data supply from the host to devices for execution, the red bars are the resulting data supply to the host, the blue bars are the computations in GPUs, and the brown bars are the peer-to-peer data transfer operations between devices. (Color figure online)

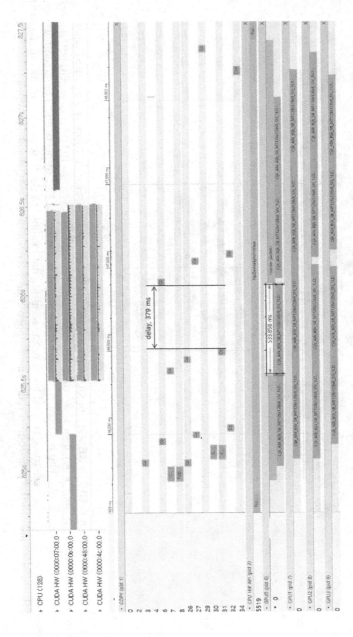

Fig. 6. The profiles of the multi-GPU SGEMM operation on 4 A100 GPUs without tensor core (upper) and 4 RX 6900 XT GPUs (lower). Number of elements ($N = 32768$) in a row (column) of matrices and tile size ($N_t = 1024$) in case on A100 GPUs and ($N_t = 8192$) in case on RX 6900 XT GPUs. The matrices A, B, and C are stored in devices 2, 1, 0 respectively. The green bars (upper, in lower omitted) are data supply from the host to devices for execution, the red bars (upper, in lower omitted) are result data supply to the host, the blue bars (upper) and purple bars (lower) are computations in GPUs, and the brown bars (upper) and pink bars (lower) are peer-to-peer data transfer operations between devices. (Color figure online)

accuracy of the predictions is higher than 50%. There are two factors that limit this accuracy: 1) the model does not take into account the possible last stage of the algorithm with poor load balancing, 2) the GPU performance for smaller tile sizes can be much lower than the maximum performance for larger tiles.

Table 2. The optimal tile sizes (N_i): the empirical values from the benchmarks of A100 system (see Fig. 2 and Fig. 3) and the predictions of the model [6].

A100 with NVLink		N_{GPUs}					
		3	4	5	6	7	8
Without tensor cores	Empirical	1024	1024	1024	1024	1024	1024
	Theoretical threshold	262	393	524	655	786	917
	Theoretical N_i	512	512	1024	1024	1024	1024
With tensor cores	Empirical	2048	4096	4096	4096	4096	4096
	Theoretical threshold	1766	2648	3531	4413	5296	6178
	Theoretical N_i	2048	4096	4096	8192	8192	8192

In the future, it would be interesting to see how the performance of the our algorithm compares with the performance of the SuperMatrix, ParSEC and COSMA implementations. Deployment of our algorithm in such a standard test as, for example, High Performance Linpack (HPL) would be another interesting problem for further study.

7   Conclusions

The empirical optimum parameters for our multi-GPU SGEMM algorithm obtained for 8 A100 based server with NVLink are compared with the theoretical model predictions. It is shown how using tensor cores changes the balance between communication and computation. The benchmark results obtained using the server with 4 AMD RDNA2-type GPUs connected by PCI 4.0 reveals certain peculiarities of porting our multi-GPU SGEMM algorithm from CUDA to HIP.

Acknowledgment. The article was prepared within the framework of the HSE University Basic Research Program. This research was supported in part through resources of supercomputer facilities provided by HSE University. This research was supported in part through computational resources of HPC facilities at HSE University [18].

References

1. Kondratyuk, N., Nikolskiy, V., Pavlov, D., Stegailov, V.: GPU-accelerated molecular dynamics: state-of-art software performance and porting from Nvidia CUDA to AMD HIP. Int. J. High Perform. Comput. Appl., p. 10943420211008288 (2021)
2. Williams-Young, D.B., et al.: Achieving performance portability in gaussian basis set density functional theory on accelerator based architectures in NWChemEx. Parallel Comput. **108**, 102829 (2021)

3. Cojean, T., Tsai, Y.H.M., Anzt, H.: Ginkgo-A math library designed for platform portability. Parallel Comput. **111**, 102902 (2022)
4. Brown, C., Abdelfattah, A., Tomov, S., Dongarra, J.: Design, optimization, and benchmarking of dense linear algebra algorithms on AMD GPUs. In: 2020 IEEE High Performance Extreme Computing Conference (HPEC), pp. 1–7. IEEE (2020)
5. Choi, Y.R., Nikolskiy, V., Stegailov, V.: Matrix-Matrix Multiplication using multiple GPUs connected by NVLink. In: 2020 Global Smart Industry Conference (GloSIC), pp. 354–361. IEEE (2020)
6. Choi, Y.R., Nikolskiy, V., Stegailov, V.: Tuning of a matrix-matrix multiplication algorithm for several GPUs connected by fast communication links. In: Sokolinsky, L., Zymbler, M. (eds.) PCT 2022. CCIS, vol. 1618, pp. 158–171. Springer, Cham (2022). https://doi.org/10.1007/978-3-031-11623-0_12
7. Markidis, S., Der Chien, S.W., Laure, E., Peng, I.B., Vetter, J.S.: Nvidia tensor core programmability, performance and precision. In: 2018 IEEE International Parallel and Distributed Processing Symposium Workshops (IPDPSW), pp. 522–531. IEEE (2018)
8. Dakkak, A., Li, C., Xiong, J., Gelado, I., Hwu, W.m.: Accelerating reduction and scan using tensor core units. In: Proceedings of the ACM International Conference on Supercomputing, pp. 46–57 (2019)
9. Van De Geijn, R.A., Watts, J.: SUMMA: scalable universal matrix multiplication algorithm. Concurrency Pract. Experience **9**(4), 255–274 (1997)
10. Chan, E., et al.: SuperMatrix: a multithreaded runtime scheduling system for algorithms-by-blocks. In: Proceedings of the 13th ACM SIGPLAN Symposium on Principles and Practice of Parallel Programming, pp. 123–132 (2008)
11. Wu, W., Bouteiller, A., Bosilca, G., Faverge, M., Dongarra, J.: Hierarchical DAG scheduling for hybrid distributed systems. In: 2015 IEEE International Parallel and Distributed Processing Symposium, pp. 156–165. IEEE (2015)
12. Herault, T., Robert, Y., Bosilca, G., Dongarra, J.: Generic matrix multiplication for multi-GPU accelerated distributed-memory platforms over PaRSEC. In: 2019 IEEE/ACM 10th Workshop on Latest Advances in Scalable Algorithms for Large-Scale Systems (ScalA), pp. 33–41. IEEE (2019)
13. Kwasniewski, G., Kabić, M., Besta, M., VandeVondele, J., Solcà, R., Hoefler, T.: Red-blue pebbling revisited: near optimal parallel matrix-matrix multiplication. In: Proceedings of the International Conference for High Performance Computing, Networking, Storage and Analysis, SC 2019, pp. 1–22. Association for Computing Machinery, New York, NY, USA (2019)
14. Bychkov, A., Nikolskiy, V.: Rust language for supercomputing applications. In: Voevodin, V., Sobolev, S. (eds.) RuSCDays 2021. CCIS, vol. 1510, pp. 391–403. Springer, Cham (2021). https://doi.org/10.1007/978-3-030-92864-3_30
15. Wang, L., Wu, W., Xu, Z., Xiao, J., Yang, Y.: BLASX: a high performance level-3 BLAS library for heterogeneous multi-GPU computing. In: Proceedings of the 2016 International Conference on Supercomputing, pp. 1–11 (2016)
16. Malik, T., Lastovetsky, A.: Towards optimal matrix partitioning for data parallel computing on a hybrid heterogeneous server. IEEE Access **9**, 17229–17244 (2021)
17. Kondratyuk, N., et al.: Performance and scalability of materials science and machine learning codes on the state-of-art hybrid supercomputer architecture. In: Voevodin, V., Sobolev, S. (eds.) RuSCDays 2019. CCIS, vol. 1129, pp. 597–609. Springer, Cham (2019). https://doi.org/10.1007/978-3-030-36592-9_49
18. Kostenetskiy, P.S., Chulkevich, R.A., Kozyrev, V.I.: HPC resources of the higher school of economics. J. Phys. Conf. Ser. **1740**, 012050 (2021). https://doi.org/10.1088/1742-6596/1740/1/012050

Numerical Simulation of Quantum Dissipative Dynamics of a Superconducting Neuron

P. V. Pikunov(✉), D. S. Pashin, and M. V. Bastrakova

Lobachevsky State University of Nizhni Novgorod, Nizhny Novgorod, Russia
pavel.pikunov@internet.ru, bastrakova@phys.unn.ru

Abstract. The process of dynamic state switching in a modified scheme of a superconducting quantum parametron in a heat reservoir is studied numerically. In the Born-Markov approximation, the system evolution is reduced to an adiabatic generalized master equation for the density matrix in the instantaneous basis. The numerical solution of the Redfield equation and the possibility of using secular approximation and random phase approximation (Pauli equation) is discussed. We provide a comparison of the efficiency of two numerical simulations based on the OpenMP technology in different representations of the system Hamiltonian: coordinate and occupation number.

Keywords: Superconducting quantum neuron · Quantum dissipation · Born-Markov approximation · Generalized master equation · Redfield equation · OpenMP

1 Introduction

One of promising applications of superconducting circuits in electronics and computing is the creation of elements of artificial quantum neural networks (QNN). These systems combine the ideas of quantum and neural network computing by using the possibilities of macroscopic quantum effects in superconductors [1–3]. In the future, a new hardware implementation of the QNN with fast calculation of activation functions, suitable for working with quantum information, can radically improve efficiency of intelligent data processing systems. Such QNN can be adapted for deep machine learning algorithms [4,5], which is one of promising applications of modern quantum processors [6,7]. One way to create a quick and reliable implementation of the considered quantum-classical system is to use a well-developed technology suitable for superconducting qubits. In this case, it is necessary to take into account the influence of quantum effects on the operation of neuromorphic elements.

In this paper, we consider a physical model of a superconducting neural cell that can function in both classical [8,9] and quantum [10] modes. Note that this scheme is similar to the flux qubit used by D-Wave systems in quantum annealers [11]. In our recent paper [10], we found conditions that provide the

D. Balandin et al. (Eds.): MMST 2022, CCIS 1750, pp. 293–301, 2022.
https://doi.org/10.1007/978-3-031-24145-1_24

required sigmoidal activation function (conversion of the input magnetic flux into the average output current) for the operation of this cell in QNN as a perceptron [12]. However, an important unexplained issue in [10] remains the influence of dissipation on the dynamics of a neural cell. In the Born-Markov approximation, after averaging over the states of the environment, the system evolution is reduced to an adiabatic generalized master equation for the density matrix in the instantaneous basis. The eigenfunctions and energies for the instantaneous Hamiltonian are found numerically at each moment of time and the corresponding transition rates are calculated. The numerical algorithm is compared for different representations of instantaneous bases: coordinate and occupation number (the Fock basis). It is found out that the efficiency (program execution speed at a given accuracy) for calculating dissipative dynamics in the occupation number representation bigger than similar calculations in the coordinate representation. At the same time, OpenMP parallelization technology in C++ and intel MKL libraries were used.

2   Phisical Model

The basic element of a neural network, functioning in classical [9] and quantum [10] modes of operation, can be implemented on the basis of superconducting elements in the scheme of a parametric quantron with a SQUID instead of a Josephson junction (perceptron [8]).

The Hamiltonian of the quantum neural cell was derived in [10], and has the form

$$H(t) = \frac{p^2}{2M} + E_J \left\{ \frac{(b\varphi_{in}(t) - a\varphi)^2}{2a} + (1 - \cos\varphi) \right\}. \tag{1}$$

The system under consideration is similar to a moving particle with mass $M = \frac{\hbar^2}{2E_c}$ and momentum $p = -i\hbar\frac{\partial}{\partial\varphi}$ (where E_C charge and E_J Josephson energies) and the phase, φ, of the Josephson junction is the effective coordinate of the particle. The coefficients a, b are parameters determined experimentally. Dynamic control of the system states is carried out by a changing external magnetic flux:

$$\varphi_{in}(t) = A \left[\left(1 + e^{-2D(t-\tau_1)}\right)^{-1} + \left(1 + e^{2D(t-\tau_2)}\right)^{-1} \right] - A. \tag{2}$$

The external flux is characterized by amplitude A, rise and fall times τ_1 and $\tau_2 = 3\tau_1$, steepness parameter D.

Within the framework of the adiabatic approach, the time-independent Schrodinger equation can be solved numerically for each moment of time to find "instantaneous energy levels" $E_n(t)$ and "instantaneous eigenfunctions" $|\Psi_n(t)\rangle$ of the system:

$$H(t)|\Psi_n(t)\rangle = E_n(t)|\Psi_n(t)\rangle. \tag{3}$$

To calculate the activation function of the neuron (the dependence of output current i_{out} on the input flux $\varphi_{in}(t)$), it is necessary to calculate the evolution of the average phase of the quantum state $\langle\varphi(t)\rangle = \langle\Psi(t)|\varphi|\Psi(t)\rangle$ and the average current $\langle i(t)\rangle = b\varphi_{in}(t) - a\langle\varphi(t)\rangle$ when the input flux changes. The activation function of the neuron is determined as

$$i_{out} = \left(\varphi_{in}(t) - \frac{1+a}{b}\langle i(t)\rangle\right)\frac{2b(a-b)}{a(1+a)}. \tag{4}$$

For a correct description of the system evolution, it is necessary to take into account dissipative processes that can lead to the destruction of coherence in the system and affect the functioning of the circuit in experiments. As the simplest dissipation model, we consider the superconducting neuron to be coupled to a thermal reservoir, which is modeled as a collection of harmonic oscillators. The linear interaction between the quantum system and the reservoir can be written as

$$H_{int} = k\varphi\sum_i\left(b_i^\dagger + b_i\right), \tag{5}$$

where $b_i^\dagger$ and b_i are creation and annihilation operators of the reservoir harmonic oscillators, and k is the coupling constant. Within the framework of an adiabatic change of the external flux, we can write the system density matrix in the instantaneous basis $|\Psi_n(t)\rangle$ as

$$\rho(t) = \sum_{m,n}\rho_{mn}(t)|\Psi_m(t)\rangle\langle\Psi_n(t)|. \tag{6}$$

Under the Born-Markov approximation, dissipative dynamics is described by the adiabatic generalized master equation [13]. The density matrix in terms of the instantaneous basis in the Schrodinger picture obeys the Redfield equation:

$$\begin{aligned}
\dot{\rho}_{mn} = {}&i\frac{E_n(t) - E_m(t)}{\hbar}\rho_{mn} - \sum_{a,b}\rho_{bn}W_{bama}(t) \\
&- \sum_{c,d}\rho_{md}W_{dccn}(t) + \sum_{e,f}\left(\rho_{ef}W_{emfn}(t) + \rho_{fe}W_{enmf}(t)\right),
\end{aligned} \tag{7}$$

where we neglect the Lamb shift and the transition rates $W_{abcd}(t)$ are defined by

$$W_{abcd}(t) = \frac{\lambda}{2}\langle\Psi_a|\varphi|\Psi_b\rangle\langle\Psi_c|\varphi|\Psi_d\rangle\{\theta(\omega_{ab})[\bar{n}((\omega_{ab}) + 1] + \theta(\omega_{ba})\bar{n}((\omega_{ba})\}, \tag{8}$$

where $\lambda = \frac{2\pi gk^2}{\hbar^2}$ is the renormalized coupling constant, θ is the Heaviside step function, $\bar{n}(\omega) = \frac{1}{e^{\hbar\omega/kT}-1}$ is the Planck's distribution and g is the density of bosonic modes, which is constant in our model. Note that Eq. (7) requires the system levels are not nearly degenerate during the considered time evolution.

Note that if the energy relaxation rate of the system is much less than the frequency of transitions between levels, we can calculate the evolution within the framework of the secular approximation [14]. The generalized master equation with the secular approximation can be easily obtained from Eq. (7) by multiplying the fourth term with the Kronecker delta symbol δ_{mn} and by imposing additional conditions on the indices of summations, that is, $b = m$, $d = n$ and $e = f$. Moreover, if the system satisfies the additional condition that the phase relaxation rate is much bigger than the energy relaxation rate, then we can average over the phases (random phase approximation). In this case, keeping only the diagonal terms of the density matrix, the Pauli master equation can be obtained.

3   Numerical Implementation and Acceleration

The numerical solution of the Redfield equation for the system density matrix with the time-dependent Hamiltonian involves many challenges. Firstly, it is necessary to solve the problem on eigenvalues and eigenvectors to find transition rates. That problem is needed to be solved at each moment of time to calculate in framework of the system instantaneous basis. The correct calculation of the activation function requires taking into account a large number of the system levels $dim[\rho] = N$, hence the dimension of the system of differential equations reaches N^2. Secondly, additional computational difficulties arise due to the specifics of the coordinate representation. According to the node theorem, the wave function of the n-th excited state has n zeros, i.e. the function oscillates rapidly. Additionally, the behavior of the wave functions imposes an additional requirement for a small coordinate step $\Delta\varphi << L/N$, where L is the phase range of interest. Thus, at each moment of time it is required to find eigenvalues and eigenvectors of the matrix which dimension equals $dim[H] = L/\Delta\varphi \times L/\Delta\varphi$. All of the above shows that the problem is very demanding on computational performance.

There are several approaches to solve this problem. The first is a modification of the numerical algorithm.

A significant acceleration in the numerical calculation can be obtained by changing the representation of the Hamiltonian. Introducing the creation, $\hat{a}$, and annihilation operators, $\hat{a}^\dagger$, in Eq. (1) as

$$\varphi = \sqrt[4]{\frac{\hbar^2}{4aE_JM}}(\hat{a}^\dagger + \hat{a}), \tag{9}$$

$$p = i\sqrt[4]{\frac{\hbar^2 aE_JM}{4}}(\hat{a}^\dagger - \hat{a}), \tag{10}$$

we can get the Hamiltonian in the Fock basis. Note that in terms of the physical model and simulation results, coordinate and occupation number representations are equivalent, and only the mathematical notation changes. The advantages of the Fock basis are less the Hamiltonian dimension and high accuracy. The disadvantages include the fact that the matrix becomes hermitian instead of symmetric as in the case of the coordinate representation, which entails operations with complex numbers that cause an additional burden on a computational system.

Figure 1 shows the dependence of the program execution time, t_1, on the dimension of the density matrix $dim[\rho]$ in two different representations of the Hamiltonian. Calculations are carried out on the basis of single-threaded mode. The Hamiltonian dimension was chosen in such a way that the correct activation

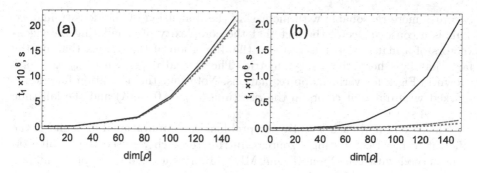

Fig. 1. Dependence of the program execution time on the dimension of the density matrix. The black line corresponds to the Redfield equation, the blue dashed line corresponds to the Redfield equation with secular approximation, the red dotted line corresponds to the Pauli equation. **(a)** The coordinate representation. **(b)** The occupation number representation. (Color figure online)

function was obtained as a result [10]. As discussed above, for the coordinate representation the required dimension of the Hamiltonian is bigger than for the occupation number representation. In this regard, the main difficulty in the coordinate representation is to find eigenvalues and eigenvectors of the Hamiltonian at each moment of time. Since this is necessary for all approaches, they differ slightly in Fig. 1a. The reverse situation occurs in the occupation number representation (Fig. 1b), where when solving the Redfield equation, the main time is spent calculating the transition rates Eq. (8). As we can see, the Fock basis

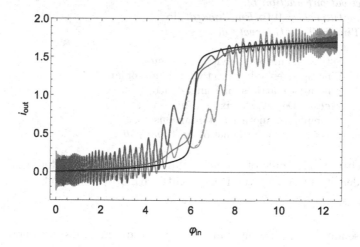

Fig. 2. The activation function was calculated in three different ways, the black-red corresponds the Pauli master equation, the solid orange-blue corresponds to the Redfield equation, dashed orange-blue corresponds to the Redfield equation with secular approximation. Numerical calculations are made for $A = 2\pi$, $\tau_1 = 500$, $D = 0.008$, $a = 0.385$, $b = 0.198$, $E_J/E_C = 1$. (Color figure online)

becomes more reasonable with increasing the dimension of the density matrix. This is a consequence of the fact that the complexity of calculating the eigenvectors of a matrix depends more on the dimension of the matrix than on the fact that it is hermitian or symmetric. The typical dependence $i_{out}(\varphi_{in})$ are shown in Fig. 2 for various approximations. Note that the activation function is marked with different colors in the rise time $(\varphi_{in} = 0 \to A)$ and the fall time $(\varphi_{in} = A \to 0)$.

The second approach to solve the problem of computational difficulty is the efficient using of computing resources. In this research we used the parallelization method with the OpenMP and MKL libraries for the C++ programming language. The OpenMP library is used to parallelize on the stages of calculating matrix elements, which can be calculated independently of each other at each moment of time. The MKL library is used to solve problems on eigenvalues and eigenvectors, based on the optimal choice according to the type of matrix (symmetric, three-diagonal, hermitian, etc.). All calculations in MKL can be done on one core or several cores on one processor. By using these libraries, it is possible to significantly optimize memory handling processes for operations with large data sets.

```
// N is a dimension of the density matrix that stores in full or
   band storage
// M is a dimension of the density matrix in full storage
// L is a number of time intervals
complex DensityMatrix[N] ;
complex k₁[M], k₂[M], k₃[M], k₄[M] = 0;
real ActivationFunction[L];
for TimeInterval = 0 to EndTime/dt do
      Time = TimeInterval * dt ;
   #pragma omp parallel
         Setting hamiltonian matrix at Time ;
         Finding eigen values/vectors of hamiltonian ;
         Finding k-matrices for Runge-Kutta ;
         Finding DensityMatrix at Time + dt ;
         Changing to another instantaneous basis;
         Finding activation function at Time + dt ;
end
```

Algorithm 1: Example of the algorithm for solving the differential equation for the density matrix by the Runge-Kutta 4th order. method

Our pseudo-code is presented in Algorithm 1 and includes the following steps.

1. Find eigenvalues/eigenvectors of hamiltonian.
   To find the eigenvalues and eigenvectors of the Hamiltonian matrix, different MKL functions are used for each of the representations. Thus, the **LAPACKE-ssbevd** function was used for the coordinate representation. It

takes a band matrix of floating-point real numbers of single precision. For the Fock representation, the function **LAPACKE-cheevd** was used, which takes a matrix of complex numbers with single-precision floating point. Both of these functions use a parallel divide and conquer algorithm.

2. Finding k-matrices for Runge-Kutta and density matrix at $Time + dt$.
   To find k matrices and density matrix at next step, we use the standard Runge-Kutta method of fourth order.

3. Changing to another instantaneous basis
   At each time step we change one instantaneous basis to another, which is given by the following expression:

$$\rho_{ij} = \sum_{n,m} \sum_{k,l} \rho'_{kl} C_{kn} C^*_{lm} C_{in} C^*_{mj},$$

where $C_{mn} = \langle \Psi'_m | \Psi_n \rangle$, $|\Psi'_m\rangle$ is the instantaneous basis in the current time step, $|\Psi_n\rangle$ is the instantaneous basis in the next time step.

It is general for finding the activation functions for the three equations. There are differences between them only in finding the k-matrices and eigenvalues/eigenvectors of the hamiltonian.

OpenMP with #**pragma omp parallel** and #**pragma omp for** directives was used to fill matrices, change instantaneous basis, find k-matrices for Runge-Kutta method, find activation function, and other operations that use matrices.

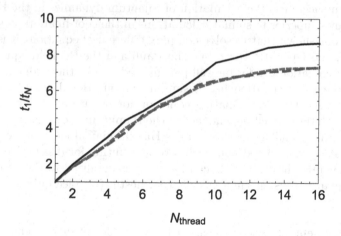

Fig. 3. Dependence of acceleration t/t_1 on the number of threads $N_{threads}$ in the system with dimension of the density matrix $dim[\rho] = 100$. Here the black line corresponds to the Redfield equation, the blue dashed line corresponds to the Redfield equation with secular approximation, the red dotted line corresponds to the Pauli equation. (Color figure online)

Figure 3 shows the dependence of acceleration, t/t_1, on the number of threads, $N_{Threads}$ involved, calculated in the occupation number representation. It can be seen that the Redfield equation has an almost linear dependence of acceleration on the number of threads due to the fact that the most complex calculation of the transition rates Eq. (8) can be performed independently by each thread. The Pauli and the Redfield equation with secular approximation have the same acceleration tendency, because these approaches require much fewer transition probabilities to be calculated, and the part related to the solution of a system of differential equations is poorly amenable to parallelization. We assume that the deviation from the linear behavior as the number of threads increases is due to the need to store eigenvectors for each thread in memory.

Numerical experiments were performed on computer with the following configuration: Intel i9-7960X (16 cores, 2.80 GHz), 32 GB RAM, OS Windows 10. We use Intel C/C++ Compiler Intel OneAPI HPC Toolkit [15].

4 Conclusion

The dissipative dynamics under adiabatic switching of a quantum neural cell is studied in this work. The numerical analysis is based on the solution of the Redfield equation and the possibility of using the secular approximation and the random phase approximation (the Pauli equation) is studied. To optimize the finding of transition rates in this system, two approaches were considered: coordinate and occupation numbers. Based on the analysis of the parallelization of the program using OpenMP technology and the MKL library, it is shown that with the same accuracy, the calculation of quantum dynamics in the Fock basis is several times superior to similar calculations in the coordinate representation.

Thus it was shown that to solve complex differential equations it is possible to use MKL and OpenMP. However, The Pauli and the Redfield equation with secular approximation cannot be well parallelized, unlike the Redfield equation. Moreover, the Pauli equation has a similar computational complexity as the Redfield with secular approximation equation for our problem. This feature is associated with the explicit dependence of the Hamiltonian on time and necessity finding eigenvalues and eigenvectors of the Hamiltonian at each moment of time.

Note that the considered approaches to accounting for dissipative processes are quite general. These methods can be easily generalized to the case of other multilevel quantum systems in a bosonic thermostat (coupled qubits, resonators, quantum dots, etc.)

Acknowledgement. The development of the main idea of superconducting neuron and the analysis of dynamics was carried out with the support of the Russian Science Foundation No. 22-72-10075. The development of the optimization algorithm is performed under support of the President of the Russian Federation No.MK-2740.2021.1.2. The M.B. acknowledges the Basis Foundation scholarship.

References

1. Schneider, M.L., et al.: Ultralow power artificial synapses using nanotextured magnetic Josephson junctions. Sci. Adv. **4**(1), e1701329 (2018)
2. Cheng, R., Goteti, U.S., Hamilton, M.C.: Superconducting neuromorphic computing using quantum phase-slip junctions. IEEE Trans. Appl. Supercond. **29**(5), 1–5 (2019)
3. Schneider, M.L., et al.: Synaptic weighting in single flux quantum neuromorphic computing. Sci. Rep. **10**(1), 1–7 (2020)
4. Müller, O., Rotter, S.: Neurotechnology: current developments and ethical issues. Front. Syst. Neurosci. **11**, 93 (2017)
5. Beer, K., et al.: Training deep quantum neural networks. Nat. Commun. **11**, 808 (2020)
6. Krantz, P., Kjaergaard, M., Yan, F., Orlando, T.P., Gustavsson, S., Oliver, W.D.: A quantum engineer's guide to superconducting qubits. Appl. Phys. Rev. **6**, 2 (2019)
7. Vozhakov, V.A., et al.: State control in superconducting quantum processors. Phys. Usp. **65**, 421–439 (2022)
8. Schegolev, A.E., Klenov, N.V., Soloviev, I.I., Tereshonok, M.V.: Adiabatic superconducting cells for ultra-low-power artificial neural networks. Beilstein J. Nanotechnol. **7**, 1397–1403 (2016)
9. Bastrakova, M., et al.: Dynamic processes in a superconducting adiabatic neuron with non-shunted Josephson contacts. Symmetry **13**(9), 1735 (2021)
10. Bastrakova, M., et al.: A superconducting adiabatic neuron in a quantum regime. Beilstein J. Nanotechnol. **13**, 653–665 (2022)
11. Boothby, K., King, A. D., Raymond, J.: Zephyr topology of D-Wave quantum processors. D-Wave Tech. Rep. Ser., 1–18 (2021)
12. da Silva, A.J., Ludermir, T.B., de Oliveira, W.R.: Quantum perceptron over a field and neural network architecture selection in a quantum computer. Neural Netw. **76**, 55–64 (2016)
13. Albash, T., Boixo, S., Lidar, D.A., Zanardi, P.: Quantum adiabatic Markovian master equations. New J. Phys. **14**(12), 123016 (2012)
14. Blum, K.: Density Matrix Theory and Applications. Plenum, New York (1981)
15. Intel oneAPI Toolkits. https://www.intel.com/content/www/us/en/developer/tools/oneapi/toolkits.html. Accessed 2022

Using Coarray Fortran for Design
of Hydrodynamics Code on Nested Grids

Igor Kulikov[1]([✉]) [iD], Igor Chernykh[1] [iD], Eduard Vorobyov[2] [iD],
and Vardan Elbakyan[3] [iD]

[1] Institute of Computational Mathematics and Mathematical Geophysics SB RAS,
Novosibirsk, Russia
kulikov@ssd.sscc.ru, chernykh@parbz.sscc.ru
[2] Department of Astrophysics, University of Vienna, Vienna, Austria
eduard.vorobiev@univie.ac.at
[3] Research Institute of Physics, Southern Federal Univeristy, Rostov-on-Don, Russia
vgelbakyan@sfedu.ru

Abstract. The paper contains details of Coarray Fortran (CAF) usage
for parallel development of the hydrodynamic code on nested grids. The
code structure is described in details. For each procedure the CAF patterns was described. Code performance analysis was carried out. The
Sedov blast wave test was used as a model problem. When using a base
mesh of 256^3, the efficiency amounts to 95%.

Keywords: HPC · Computational astrophysics · Coarray Fortran

1 Introduction

Sun, like other low-mass stars, is formed during the gravitational collapse of
dense clouds of gas and dust. During the collapse, part of the cloud material
is accreted onto the circumstellar disk and then onto the nascent protostar. It
is from this disk that planetary systems are formed. The study of circumstellar
disks is the main key to understanding the accumulation of stellar mass, the formation of planets, and the origin of life [1–5]. A key feature of collapse problems
is a strong scatter in spatial scales and density [6], which places demands on the
computational model.

In this paper, we use the nested grid approach to solve hydrodynamic equations. The main purpose of the article is a detailed description of the parallel
implementation of the nested grid approach using Coarray Fortran technology.
Although the Coarray Fortran technology appeared more than twenty years
ago, it has been included in the Fortran language standard since 2008, and in
the modern version with the use of reducing operations since 2018. Nevertheless,
the technology is actively used to solve the problems of designing materials [7],
geophysics [8], modelling by cellular automata [9], magnetic [10] and relativistic
hydrodynamics [11]. It was shown that on the problems of continuum mechanics

D. Balandin et al. (Eds.): MMST 2022, CCIS 1750, pp. 302–309, 2022.
https://doi.org/10.1007/978-3-031-24145-1_25

the performance of Coarray Fortran code is comparable with the code based on one-way MPI-3.0 communications [12]. The load balancing is also carried out rather effectively [13]. A Coarray Fortran program is a set of multiple copies of a single program code, each of which is named "image" and has its own unique serial number obtained using the "this_images" function. The total number of processes is determined using the "num_images" function. Arrays shared between multiple processes are defined using constructs like "integer a(5)[*]", where the example creates an integer array of five elements in each image. To access the element "i" of the process "p", use a construction like "a(i)[p]". The "sync all" operator is used to synchronize processes.

The second section briefly describes the computational model and the numerical method. The third section deals with details of parallel implementation and code structure. In the fourth section the efficiency of the code is demonstrated on the Sedov's blast wave problem. In the fifth section, the conclusion is formulated.

2   Numerical Model

We will consider the equations of hydrodynamics in a three-dimensional formulation:

$$\frac{\partial \rho}{\partial t} + \nabla \cdot (\rho \boldsymbol{u}) = 0,$$

$$\frac{\partial \rho \boldsymbol{u}}{\partial t} + \nabla \cdot (\rho \boldsymbol{u}\boldsymbol{u}) = -\nabla p,$$

$$\frac{\partial \rho E}{\partial t} + \nabla \cdot (\rho E \boldsymbol{u}) = -\nabla \cdot (p\boldsymbol{v}),$$

$$\rho E = \frac{1}{2}\rho \boldsymbol{u}^2 + \rho \varepsilon,$$

$$p = (\gamma - 1)\rho \varepsilon,$$

where p is the pressure, ρ is the density, $\boldsymbol{u}$ is the velocity vector, ρE is the density of total energy, ε is the density of internal energy, γ is the adiabatic index, $c = \sqrt{\frac{\gamma p}{\rho}}$ is sound speed. We introduce multilevel nested grids in the computational domain [14] (see Fig. 1). The undoubted advantage of this approach is the use

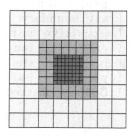

Fig. 1. Scheme of nested grids.

of well-developed numerical methods for solving hydrodynamic equations for regular grids [15,16] and the possibility of extension to more complex models [17].

3   Parallel Implementation

For the parallel implementation, we will use the geometric decomposition of the computational domain by placing each level of the nested grid on a separate Coarray Fortran image of the program (see Fig. 2). Such decomposition allows to evenly distribute the main calculations over the images of the CAF code. Here is the listing of the main computational cycle (see Fig. 3). Figure 4 shows the percentage of time spent on each procedure. As can be seen from the figure, most of the time is spent on the procedures of piecewise parabolic reconstruction of primitive variables and the solution of the Riemann problem. Next, we describe each procedure in more detail.

1. The **Projection** procedure implements the recalculation of the conservation laws from each nested grid to a grid of a higher level. The recalculation scheme is illustrated in Fig. 5. In the first part of the procedure, conservative values are averaged from four (or eight in the three-dimensional case) cells using conservation laws. In the right part of the figure, CAF code of one-sided communications is written, which transfers a part of the array to the

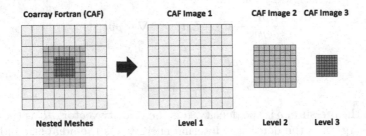

Fig. 2. Decomposition of calculations.

```
do while(timer < time)
   call Projection        ! projection
   call Boundary          ! set boundary condition
   call Primitive         ! primitive variables recovery
   call CFL               ! calculate tau
   timer = timer + tau    ! increment timer
   call Reconstruction    ! piecewise−parabolic reconstruction
   call Godunov           ! Godunov method
enddo
```

Fig. 3. The main code of the program in Fortran language.

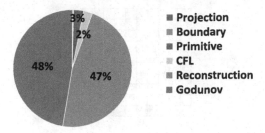

Fig. 4. Distribution of time for the main procedures.

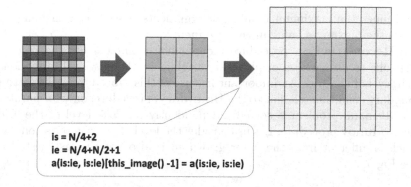

is = N/4+2
ie = N/4+N/2+1
a(is:ie, is:ie)[this_image() -1] = a(is:ie, is:ie)

Fig. 5. Scheme for recalculating the conservation laws.

previous (by number) image. Note that the recalculation of the conservation laws is carried out in sequence according to the images of the CAF program. However, taking into account the small time-fraction of the procedure, such an organization of calculations does not cause a significant increase in the total calculation time.

2. In the **Boundary** procedure, the setting of boundary conditions is implemented. For that purpose, one outer layer of cells is added at each level of the nested grid (see Fig. 6) Phantom boundary cells are shown with dark grey. The boundary conditions for the outer mesh are determined from the problem statement. The corresponding values on all nested grids are read from the grid cells from the previous CAF image. Note that the setting of boundary conditions is implemented in parallel.

3. The **Primitive** procedure implements the recalculation of physical variables from conservative variables in each computational cell. Such a procedure is implemented in parallel for each level of the nested grid without data exchanges between grids.

4. In the **CFL** procedure, the time step is calculated, which is the same for all levels of nested grids. Since each grid level is located in its own CAF image, after finding the maximum wave velocity $v = |u| \pm c$ for each level of the grid, it is necessary to determine the maximum velocity in the entire grid. To do this, the CAF reduction procedure **call co_max(v)**, is used, which

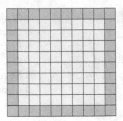

Fig. 6. Statement of boundary conditions.

determines the maximum number of arguments passed in each image, and writes the maximum value in the argument.

5. The **Reconstruction** procedure independently implements the piecewise-parabolic reconstruction of physical variables in all cells in all directions at each level of the grid. An important feature of this procedure is the setting of boundary conditions similar to the **Boundary**, procedure, only for parabolas.

6. The Riemann problem is solved independently for each level of the grid in the **Godunov** procedure. Let us consider the level of the computational grid, which is different from the outer grid, and is also not the last nested level (see Fig. 7).

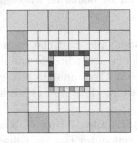

Fig. 7. Cells of different levels of nesting in the implementation of the Godunov method.

The calculation of conservation laws according to the Godunov scheme is carried out in grey cells. The solution of the Riemann problem on the boundary between two grey cells is trivial from the point of view of the calculation organization. At the boundary with a finer grid, the fluxes of conservative values are calculated on each face of the grey cells and then averaged over the boundary. At the outer boundary, two (four in the three-dimensional case) grey cells participate in the solution of the Riemann problem with the same neighbouring cell. The problem of setting external boundary conditions is removed for the external grid; the problem of setting internal boundary conditions is removed for the last nested grid.

To study to weak scalability, we consider three basic grids $M64$–64^3, $M128$–128^3, $M256$–256^3, the nesting level will vary from one to sixteen, which corresponds to the characteristic nesting levels for star formation problems. Figure 8 shows the weak scalability of the code depending on the level of nesting equal to the number of CAF images for different grids in the base. As can be seen from the figure, with a small main grid, the efficiency reaches a level of 80%, which is due to a significant number of transfers relative to the counting time, so about 10% of the cells are transferred between images (6 layers of the computational grid for the main grid 64^3). With an increase in the main grid, the number of transmitted layers is also six, which, with the main grid of size 128^3 requires the transfer of 5% of the cells and with the main grid of size 256^3 about 2% of the cells. In this case, the efficiency is about 95%.

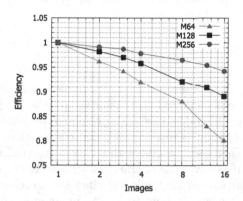

Fig. 8. Software implementation weak scalability study.

To study the parallel implementation, two nodes of the NKS-1P Siberian Supercomputer Center cluster were used, each of which is equipped with two 24-core Intel Xeon 6248R processors with Hyper-Threading support. The RAM on each node was 192 GB.

4   Sedov's Blast Wave Problem

Sedov's blast wave problem was chosen as a test. The problem statement and initial data were given in detail in [15]. The density and angular momentum profile at time $t = 0.05$ are shown in Fig. 9. As can be seen, the developed numerical method reproduces the shock wave front quite well. In general, the results of the experiments correspond to the work of [15].

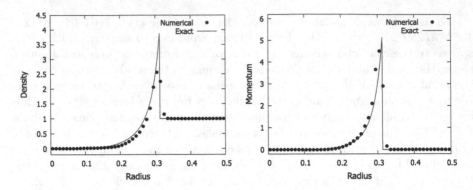

Fig. 9. Density (left) and angular momentum (right) obtained numerically by solving the Sedov's blast wave problem (circles). The exact solution is shown with the solid line.

5 Conclusions

The article describes the details of the parallel implementation of the solution of hydrodynamic equations on nested grids using the parallel programming technology Coarray Fortran. The code structure is presented and analysed; the efficiency of software implementation is investigated. Sedov's problem was considered as a model problem. When using a grid with 64^3 cells, the efficiency ampounts to 80% due to the significant volume of transfers relative to the computation time. When the grid is increased to 128^3 cells, the efficiency increases to 90%, and with the 256^3 grid, the efficiency is about 95%. The increase in efficiency is achieved thanks to the reduction in data transfer relative to the total number of cells.

Acknowledgements. The reported study was funded by RFBR and FWF according to the research project 19-51-14002 (RFBR) and I4311-N27 (FWF).

References

1. Vorobyov, E.I.: Ejection of gaseous clumps from gravitationally unstable Protostellar disks. Astron. Astrophys. **590**, 115 (2016)
2. Vorobyov, E.I., et al.: Knotty Protostellar jets as a signature of episodic Protostellar accretion? Astron. Astrophys. **613**, 18 (2018)
3. Vorobyov, E.I., Akimkin, V., Stoyanovskaya, O.P., Pavlyuchenkov, Y., Liu, H.B.: Early evolution of viscous and self-gravitating circumstellar disks with a dust component. Astron. Astrophys. **614**, 98 (2018)
4. Vorobyov, E.I., Elbakyan, V.G.: Gravitational fragmentation and formation of giant protoplanets on orbits of tens of AU. Astron. Astrophys. **618**, 7 (2018)
5. Vorobyov, E.I., Elbakyan, V.G., Omukai, K., Hosokawa, T., Matsukoba, R., Guedel, M.: Accretion bursts in low-metallicity Protostellar disks. Astron. Astrophys. **641**, 72 (2020)

6. Bate, M.R.: Collapse of a molecular cloud core to stellar densities: the formation and evolution of pre-stellar discs. Mon. Notices Royal Astron. Soc. **417**, 2036–2056 (2011)

7. Shterenlikht, A., Margetts, L., Cebamanos, L.: Modelling fracture in heterogeneous materials on HPC systems using a hybrid MPI/Fortran Coarray multi-scale CAFE framework. Adv. Eng. Softw. **125**, 155–166 (2018)

8. Reshetova, G., Cheverda, V., Khachkova, T.: Numerical experiments with digital twins of core samples for estimating effective elastic parameters. In: Voevodin, V., Sobolev, S. (eds.) RuSCDays 2019. CCIS, vol. 1129, pp. 290–301. Springer, Cham (2019). https://doi.org/10.1007/978-3-030-36592-9_24

9. Shterenlikht, A., Cebamanos, L.: MPI vs Fortran Coarrays beyond 100k cores: 3D cellular automata. Parallel Comput. **84**, 37–49 (2019)

10. Garain, S., Balsara, D., Reid, J.: Comparing Coarray Fortran (CAF) with MPI for several structured mesh PDE applications. J. Comput. Phys. **297**, 237–253 (2015)

11. Kulikov, I., et al.: A new parallel code based on a simple piecewise parabolic method for numerical modeling of colliding flows in relativistic hydrodynamics. Mathematics. **10**(11), 1865 (2022)

12. Reshetova, G., Cheverda, V., Koinov, V.: Comparative efficiency analysis of MPI blocking and non-blocking communications with Coarray Fortran. In: Voevodin, V., Sobolev, S. (eds.) RuSCDays 2021. CCIS, vol. 1510, pp. 322–336. Springer, Cham (2021). https://doi.org/10.1007/978-3-030-92864-3_25

13. Cardellini, V., Fanfarillo, A., Filippone, S.: Coarray-based load balancing on heterogeneous and many-core architectures. Parallel Comput. **68**, 45–58 (2017)

14. Matsumoto, T., Hanawa, T.: A fast algorithm for solving the Poisson equation on a nested grid. Astrophys. J. **583**, 296–307 (2003)

15. Kulikov, I., Vorobyov, E.: Using the PPML approach for constructing a low-dissipation, operator-splitting scheme for numerical simulations of hydrodynamic flows. J. Comput. Phys. **317**, 318–346 (2016)

16. Kulikov, I., Chernykh, I., Tutukov, A.: A new hydrodynamic code with explicit vectorization instructions optimizations, dedicated to the numerical simulation of astrophysical gas flow. I. Numerical method, tests and model problems. Astrophys. J. Suppl. Ser. **243**, 4 (2019)

17. Kulikov, I.: A new code for the numerical simulation of relativistic flows on supercomputers by means of a low-dissipation scheme. Comput. Phys. Commun. **257**, 107532 (2020)

Author Index

Printed in the United States
by Baker & Taylor Publisher Services